Sausages

Sausages: Nutrition, Safety, Processing and Quality Improvement

Editor

Javier Carballo

MDPI • Basel • Beijing • Wuhan • Barcelona • Belgrade • Manchester • Tokyo • Cluj • Tianjin

Editor
Javier Carballo
Food Technology
University of Vigo
Ourense
Spain

Editorial Office
MDPI
St. Alban-Anlage 66
4052 Basel, Switzerland

This is a reprint of articles from the Special Issue published online in the open access journal *Foods* (ISSN 2304-8158) (available at: www.mdpi.com/journal/foods/special_issues/sausages).

For citation purposes, cite each article independently as indicated on the article page online and as indicated below:

LastName, A.A.; LastName, B.B.; LastName, C.C. Article Title. *Journal Name* **Year**, *Volume Number*, Page Range.

ISBN 978-3-0365-1364-5 (Hbk)
ISBN 978-3-0365-1363-8 (PDF)

Contents

About the Editor

Javier Carballo

Javier Carballo is full professor of food technology at the Faculty of Sciences of the University of Vigo, Spain. He received his BS (1984) and he obtained his PhD (1989) in veterinary medicine from the University of León (Spain), both with special end-of-degree awards. Later, he completed a research stage at the Station de Recherches Laitières, Jouy-en-Josas, France. In the last 25 years, he has developed his research activity in the study and improvement of traditional foods. In this scientific field, he published more than 200 research articles in the most reputable international journals and wrote more than 25 book chapters. He was the former president (from 2008 to 2016) of the Food Microbiology Section at the Spanish Society of Microbiology, and he is an associate editor of the journal *Food Microbiology*. He has also acted assiduously as a translator of scientific books to Spanish in the field of food science and technology.

Preface to "Sausages: Nutrition, Safety, Processing and Quality Improvement"

Sausages are one of the oldest processed foods. They are strongly present in the gastronomy of each country, each region, and each geographical area, with maximum social and economic importance. Each local variety within each sausage type reflects the availability of raw materials, the climate conditions of each geographical environment, the cultural and religious conditionings, and the ancestral manufacture knowledge transmitted through generations. It is difficult to know exactly the total number of different sausages made in the world, but as an anecdotal example, Germany boasts of over 1000 different varieties. Most of the local varieties have not managed to transcend beyond their area of origin, but some others have reached a worldwide diffusion and are currently part of the gastronomic heritage of humanity. Beyond the social and cultural relevance of sausages, their economic significance is evident. According to a market report from Wise Guy Reports, the global sausage industry was valued at around USD 67.1 billion in 2016, and this figure is expected to reach about USD 76.6 billion by 2021.

The sensory excellence of the sausages that we know and enjoy is a consequence of the knowledge and good work of our ancestors, but science has also played its part. In recent years, raw materials and production processes have been improved and optimized, and this has required a programmed inquiry for reaching comprehensive knowledge of chemical, physical, and biological events that take place during the manufacturing of sausages and that are directly responsible for their nutritional features and sensory attributes.

The scientific research work carried out in recent years in this field has been extensive and fruitful, but there is still work to be done. Some sausage varieties are still unknown regarding their microbiological and biochemical features, or their production processes are insufficiently standardized. In other cases, there are safety or quality concerns that must be solved so that these sausages can be enjoyed to their fullest potential. Improvements in the sensory quality and/or the adaptation of their sensory and nutritional properties to changes in consumer preferences and requirements, all without sacrificing the personality and differential attributes of each sausage variety, is in all cases a permanent challenge. For all of these reasons, further studies and research are essential instruments to improve and continue enjoying these privileged foods. This volume aims to gather some of the latest advances and to be a useful tool for researchers and professionals in this scientific area.

Javier Carballo
Editor

foods

MDPI

Editorial

Sausages: Nutrition, Safety, Processing and Quality Improvement

Javier Carballo [1,2]

[1] Food Technology Area, Faculty of Sciences, University of Vigo, 32004 Ourense, Spain; carbatec@uvigo.es
[2] CITACA, Agri-Food Research and Transfer Cluster, Campus da Auga, University of Vigo, 32004 Ourense, Spain

Sausages are one of the oldest processed foods known to man. Several hundreds of varieties of sausages are produced worldwide with outstanding social and economic relevance. Each local variety within each sausage type (raw, scalded, cooked, or raw and fermented/ripened) reflects the availability of raw materials, the climate conditions of each geographical environment, the cultural and religious conditionings, and the ancestral manufacture knowledge transmitted through generations.

It is evident and expected that the different local varieties of sausages and the art of making them were perfected over centuries through the experience and the successes/failures in the production processes of the different generations.

However, similar to the manufacturing of other foods and to other production processes in general, the scientific knowledge of the physical, chemical and biological events that take place during the production processes and that are responsible for the nutritional value, sensory characteristics, and chemical and microbiological safety of the final products is relatively recent and, as in other cases, is based on the previous advances of basic disciplines such as chemistry and biology. Ultimately, it is the scientific knowledge of these events that allows them to be controlled and directed and to achieve sensory and nutritional excellence in the finished products.

Without disregarding specific works carried out in other countries, the dawn of scientific knowledge of sausage manufacturing was located in Germany during the central decades of the last 20th century. A good part of the scientific knowledge generated from the study of German local sausage varieties is collected in manuals [1,2] that are now timeless classics and an indisputable reference in the scientific literature of this field. These manuals also contain tips and indications for making sausages of high and constant quality, avoiding the most common defects and manufacturing accidents.

Among all the sausage types, the manufacture of dry-fermented sausages has special complexity, considering that the organoleptic characteristics of these products are the result of a series of modifications of the raw materials and ingredients promoted by the meat tissue enzymes and the microorganisms present, interacting with each other, and that these activities are modulated by the ingredients (salt, species, etc.) and the environmental conditions during the ripening process. As they do not receive heat treatment, they are also the sausages that present the greatest microbiological risks. The greatest research effort in this field has been devoted to the study of this type of sausages. After a series of pioneering studies that were duly reviewed by Lücke [3], starting in the 1980s of the last century studies of biochemical and microbiological characterization of the different local varieties of dry-fermented sausages multiplied. The results of all these studies, carried out mainly on local varieties of Italian, French, Spanish, Greek and Portuguese sausages obtained through spontaneous fermentation, coincide in pointing out the constancy, but uneven intensity, of the glycolytic, proteolytic, lipolytic and oxidation processes that take place in the different varieties of sausages during maturation and the role in these processes of the enzymes of muscle and fat and of the microorganisms.

Regarding the results of the studies on the microbiological characteristics of these local sausage varieties, the lactic acid bacteria and the microorganisms belonging to the

Citation: Carballo, J. Sausages: Nutrition, Safety, Processing and Quality Improvement. *Foods* **2021**, *10*, 890. https://doi.org/10.3390/foods10040890

Received: 28 March 2021
Accepted: 15 April 2021
Published: 19 April 2021

Staphylococcaceae and *Micrococcaceae* families were revealed as the majority microbial flora with some participation of yeast and surface moulds in concrete varieties. Lactic acid bacteria develops rapidly during the fermentation of raw-cured sausages, reaching counts of 10^8–10^9 c.f.u./g, which remain practically stable until the end of the drying/ripening stage. The participation of this microbial group is decisive for the assurance of the hygienic/sanitary quality of the sausage, as they are responsible for the production of organic acids (lactic and acetic) and the reduction of the pH values. This decrease in pH inhibits the development of spoiling bacteria, accelerates the dehydration process of the product by reducing the water holding capacity of proteins, influences color formation and stability, and contributes to the aroma and flavor of sausages, especially those of short maturation. In addition, they can produce bacteriocins, protein compounds that also contribute to its antimicrobial activity, facilitating its implantation and inhibiting the development of unwanted microorganisms. The lipolytic activity of lactic acid bacteria is considered weak, but not so its proteolytic activity, which has been studied and demonstrated especially in different *Lactobacillus* species.

Due to their greater resistance to salt and their lower oxygen demand, staphylococci largely prevail over micrococci in this type of meat product. Staphylococci reach counts of around 10^7 c.f.u./g. The growth of these microorganisms is inhibited by the reduction of the pH values, so that only in the case of sausages prepared with high nitrate and low carbohydrate concentrations these microorganisms can become the main microbiota outperforming lactic acid bacteria. The role of staphylococci in the manufacturing processes of meat products is mainly focused on three metabolic activities: (i) nitrate and nitrite reductase activities that make it possible to develop the typical red color in these products, as the nitric oxide formed reacts with the myoglobin and yields nitrosylmyoglobin, pink in color, (ii) catalase activity that degrades the peroxides accumulated in the sausages during the fermentation and that could have negative effects by oxidizing the iron and the cured color, also favoring lipid oxidation reactions, and (iii) proteolytic and lipolytic activities as they have numerous lipases and proteases that favors the triglyceride breakdown, and the formation of peptides, amino acids and other compounds, being all of them the origin of volatile compounds significantly affecting the aroma of the product.

Molds are aerobic micro-organisms, so their growth is fundamentally superficial, reaching in some sausage varieties counts of 10^5–10^7 c.f.u./cm^2. The presence of a surface microbiota in sausages carries a series of desirable effects, prevents the formation of a superficial crust favoring homogeneous dehydration of the product, avoids rancidity by protecting the sausage from the pro-oxidant effect of light, and contributes to the development of the characteristic aroma and flavor due to the lipolytic and proteolytic capacity of some mould strains. The yeast population in raw-cured sausages has been scarcely studied because its low proportion compared to the bacterial microbiota led to the undervaluation of his role by many authors. Yeasts are found in fermented sausages at levels varying from 10^3 c.f.u./g to 10^5 c.f.u./g throughout the manufacturing process. The yeasts contribute to the stabilization of the color of the sausage by displacing oxygen and degrading peroxides due to its catalase activity. Furthermore, their proteolytic and lipolytic activities contribute to the development of the characteristic flavor and aroma of each product.

In the last decades, from each of these microbial groups, isolates were taken from the most common dry-fermented sausage varieties and identified using both classical and molecular methods. With some minor differences among sausage varieties, the microbial profile of the sausages obtained by spontaneous fermentations is quite consistent regardless of the types and regions of origin. Among the lactic acid bacteria, although other genera such as *Leuconostoc, Carnobacterium, Pediococcus* and *Enterococcus* have been described, the lactobacilli and above all the homofermentative lactobacilli are the main group, being *Lactobacillus sakei, L. curvatus, L. plantarum* and *L. alimentarius* in this order the main species isolated. *Lactobacillus sakei* seems to be the better adapted species to the ecosystem represented by the dry-fermented sausages and to its special environmental conditions.

Among the isolates of staphylococci identified, *Staphylococcus xylosus* largely predominated followed by other species such as *S. carnosus*, *S. equorum*, *S. epidermidis*, *S. saprophyticus*, *S. lentus* and *S. sciuri*. Among the molds present in the meat and meat products, species of the genera *Mucor*, *Rhizopus*, *Aspergillus* and *Penicillium* have been described as dominant, being *P. nalgiovense* and, to a lesser extent, *P. chrysogenum* the main *Penicillium* species. *Debaryomyces hansenii* was reported as the main and more consistent species of yeast. It is the majority participation of one or the other of these microbial species together with the particular application of some processes such as smoking or the addition of specific spices and additives that configures the particularities that give personality to each of the sausages from the different countries and regions.

When raw-cured sausages are manufactured in the traditional way, without the addition of starter cultures, the prevailing environmental conditions in the sausage favor the selective growth of already adapted microbiota. As a way to ensure that this desired microbiota is present, a practice used for many years consisted of inoculating a portion of previously fermented meat to the fresh mix, with which products of greater consistency and stability are obtained. Jensen and Paddock [4] were the first authors to investigate the possibility of using a *Lactobacillus* strain in the production of raw-cured sausages, arousing the interest of other researchers, both Europeans as well as Americans, who began a more in-depth study of the starter cultures and their application to the meat industry.

The starter cultures are added to the mix in order to adequately control the fermentation and maturation processes of raw-cured sausages in such a way as to standardize the process and the quality of the end products. This is possible because the metabolic activities that the starter cultures develop during the processing of meat products and having an effect on various quality factors. The added microorganisms establish themselves as the predominant microbiota, directing fermentation and excluding the undesirable microbiota, thus reducing hygienic risks and accidents of manufacturing due to deficiencies of microbial origin. Furthermore, due to their fermentative, proteolytic and lipolytic activities, they improve the nutritional and sensory qualities of the product, while improving the speed and homogeneity of drying, which is a technological advantage. Nevertheless, commercial starter cultures must meet a number of safety requirements and must have technological competitiveness and economic viability so that their application generates the benefits expected. With regard to safety, microorganisms used as starters must not possess neither toxic nor pathogenic activity and the preparations must be free of any type of polluting, biological or chemical. With regard to technological functions, microorganisms inoculated should predominate over the spontaneous microbiota of the meat mass and develop their metabolic activity. Finally, in terms of economic aspects, the use of the starter cultures must be economically viable and easy to handle. In addition, the storage of frozen of lyophilized cultures should not affect the properties of the strain or cause losses from their activity.

The first starter culture with application in the meat industry that appeared on the market was a strain of *Pediococcus cerevisiae* [5], later classified as *Pediococcus acidilactici*, which was marketed by the Merck company in the U.S.A. in 1957 for the production of summer sausages and spreadable sausages. Almost at the same time, in Germany in 1961 it was commercialized a strain of *Micrococcus* (*Micrococcus* M53) [6] supplied by the Rudolf Müller company, and in 1966 a starter culture appeared for the first time, combining *Lactobacillus plantarum* with a strain of *Micrococcus* [7]. However, the widespread use of starter cultures in the meat industry did not begin to develop until the 1980s, mainly due to the fact that it is possible to obtain dry-fermented sausages of excellent quality without the addition of them [8]. The use of starter cultures, generally composed by a lactic acid bacteria (mainly a *Lactobacillus* strain and a coagulase-negative staphylococci (CNS), is a common and effective practice in the manufacture of fermented sausages in order to improve the color and flavor development, ensure safety and extend shelf-life.

The use of starter cultures and its effect on the microbiological, physicochemical and safety attributes of the different local varieties of dry-fermented sausages was one of the most recurrent research topic in the last decades. However, the use of a commercial non-autochthonous starter culture could have a negative impact on the sensory characteristics of the sausages, resulting in losses of the particularly desirable organoleptic properties that characterize each sausage type. Therefore, the development of specific starter cultures composed by strains isolated from spontaneous non-controlled elaborations of the corresponding type of sausage and adequately characterized in their metabolic performances has aroused the interest of researchers.

Currently the lines of research in relation to the use of starter cultures in the meat industry have been expanded in their objectives. Different authors have tried the addition of proteases and lipases in order to promote the development of the aroma and accelerate the processes of maturation in raw-cured sausages. Finally, another current line of research is on the use of probiotic lactic acid bacteria as starter cultures in raw-cured sausages, thus obtaining functional foods.

The safety of the sausages has also been the subject of abundant studies in recent decades. The microbiological hazards, mainly represented by foodborne pathogens such as *Listeria monocytogenes*, *Staphylococcus aureus*, *Escherichia coli*, *Salmonella* spp., *Yersinia enterocolitica*, and even less frequent pathogenic agents such as *Mycobacterium avium* subsp. *paratuberculosis*, *Aeromonas* spp., or hepatitis E virus, have been appropriately identified in sausages with different biochemical and technological features. In the same way, the chemical hazards, mainly biogenic amines, nitrosamines, mycotoxins, and polycyclic aromatic hydrocarbons in the particular case of smoked sausages, have been duly characterized. Measures of control and prevention of both microbiological and chemical hazards have been developed and essayed. The use of starter cultures appropriately selected can be a solution for most of these hazards.

In recent times, consumers are increasingly concerned about the relationship between health and diet, and they demand foods that are not harmful to their health and, ideally, even that protect and improve it. In this sense, there are concerns about the excessive salt content and the quantity and quality of fat in the diet and its proven relationship with some cardiovascular diseases and certain types of cancer. Traditional sausages are characterized by high salt and fat contents that are necessary to maintain their typical sensory attributes. In such scenario, the reduction of NaCl and/or its partial replacement with other salts, mainly KCl, $CaCl_2$, $MgCl_2$, and/or sodium ascorbate, was essayed obtaining final products having acceptable organoleptic characteristics.

Regarding the modification of the lipid fraction in sausages, the first strategies was the reduction of the fat content and/or the replacement with non-lipid substitutes. This task has not been easy because fat highly contributes to the texture, flavor and mouthfeel of these meat products. In order to reduce the fat content without compromising the organoleptic characteristics, several alternatives have been essayed consisting generally in the use of ingredients as fat substitutes and fat mimetics. In cooked meat products, some ingredients such as proteins, carbohydrates and fat-based substitutes were essayed with generally satisfactory results, especially when carbohydrates were used as fat replacers. In raw fermented sausages, the use of inulin or several other dietary fibres from fruits or cereals as fat replacers gave satisfactory results on texture and sensory properties. The use of dietary fibres as fat replacers has an added health benefit because increased proportions of fibre in the diet has been proven to reduce the risk of colon cancer, cardiovascular disease, obesity and several other disorders.

In addition to the amount of fat in the diet, the fatty acid profile of the ingested fat has a proven impact on consumer health. The formulation of the sausages has not remained unaffected by this concern. In addition to the attempt to modify the composition of animal fat used as raw material, especially through diet modifications but also by genetic selection, several studies have tried to modify the fatty acid profile of sausages by substituting animal fat by vegetable oils, mainly olive oil, or their corresponding oleogels.

The articles included in the present Special Issue show important and interesting advances and new approaches in most of the research fields mentioned.

As indicated, the study of the microbial communities of the dry-fermented sausages has been one of the most important research topics in recent years. The investigation on the microbial species diversity in fermented sausages is not only important to study the relationship between microbial species, physicochemical features and organoleptic characteristics and to gain control over quality development. It is also appropriate to better understand the link between microbial communities, specific technological features, breeding systems, geographical regions and the differential attributes that gives personality to the finished products. Comi et al. [9] studied the variability of the lactic acid bacteria in fermented sausages due to different pork breeds, breeding systems and production technology. They applied a semi-quantitative molecular method to study the alternation of the different species over time and their concentration ratios. The results of a cluster analysis in this article provide evidence of a plant-and breed-specific ecology of lactic acid bacteria. The authors have also observed that the breeding system can influence the presence of certain LAB species in the sausages.

The use of culture-independent methods presents clear advantages compared to classical methods when it comes to having a real vision of the species present and acting in a sausage throughout its production, and for this reason they have become the methods of choice in recent years. However, many of these methods still have limitations such as a taxonomic resolution not entirely satisfactory. In this context, Van Reckem et al. [10] studied the application of a previously developed high-throughput amplicon sequencing method targeting the 16S rRNA and *tuf* genes to characterize the bacterial communities of 15 fermented meats from 5 European countries. The authors also tried to associate the microbial communities of the different products with their physicochemical features and geographical origin. The obtained data broadened the view on the microbial communities associated with the products analyzed, revealing the presence of subdominant microbial species, particularly coagulase-negative *Staphylococcal* species, underreported in previous studies. On the other hand, the particular composition of the microbial communities could be linked to the special features of the particular products, mainly pH values, salt content and geographical origin.

The inclusion of a concrete microbial strain as starter culture in a fermented product requires a previous exhaustive study of its metabolic abilities in the most diverse possible environmental conditions and also in the presence of the most different substrates. All this is necessary to know what its viability and its performances are and what we can expect from that particular strain when we add it. *Lactobacillus sakei* is widely used as starter culture in fermented sausages due to its outstanding adaptation to meat environments and its ability to maintain high viability thanks to the possession of secondary pathways that are activated when hexoses are deplected. Barbieri et al. [11] studied the metabolism of *Lactobacillus sakei* Chr82, a commercial strain, in the presence of different concentrations of fermentable sugars. They inoculated this strain in a defined medium with glucose or ribose as the only carbon source, both at optimal or reduced concentrations, in order to evaluate its different metabolic and physiological responses to different growth conditions. The results evidenced different growth performances, physiological states of the cells and amino acid consumptions in relation to the carbon source and carbohydrate levels.

The probiotic bacteria have a renewed interest at the present time, given the accumulated evidence on the importance of the human intestinal microbiota on the health derived from its functionality and its interaction with the physiology of other organs, particularly skin and brain. Fermented sausages can be a good carrier for the probiotic lactic acid bacteria, and at the same time that a functional food is elaborated, the metabolic characteristics of the added strains can be exploited, making them act as starter cultures. With this aim, the use of probiotic strains as starter cultures needs a previous study on their technological characteristics to take advantage of their beneficial activities and avoid their undesirable effects on the quality and safety of sausages. Agüero et al. [12] evaluated the in vitro technological and safety characteristics in eight LAB strains having proved probiotic activity. They reported that some of them are good candidates for use as starter cultures in fermented sausages because their metabolic performances including antimicrobial activity against some of the most known pathogens. One concrete strain, *Lactobacillus rhamnosus* Lr-32, was the most promising among the tested strains but further studies are needed to check and endorse its behavior in a meat matrix.

In local varieties of fermented sausages having a strong personality and with defined and loyal markets, the use of autochthonous starter cultures is the most appropriate way to obtain manufactured products that are uniform, of high and constant quality and that retain the differentiated attributes of their homonyms artisanally obtained through spontaneous fermentations. Rodríguez-González et al. [13] studied the effect of the addition of two different autochthonous starter cultures including one strain of *Lactobacillus sakei* and one of *Staphylococcus equorum* or *Staphylococcus saprophyticus* on the biochemical changes occurring during the manufacture of Galician chorizo sausage, the most popular traditional meat product in the NW Spain. Along the manufacture of three independent batches made with each of the two starter cultures, they studied the changes in proximate composition, pH, Aw, color parameters, nitrogen fractions, free amino acids, biogenic amines, fat parameters, and free fatty acids. The two starter cultures showed good performances and seem to be suitable for increasing the quality and safety of this type of sausage.

Chemical hazards are the subject of special attention in ongoing investigations. Smoking is a process commonly applied to certain sausage types. In the past, it was applied mainly to aid in the preservation of the products due to the antimicrobial or antioxidant activity of some of the components of the smoke, and also due to a certain degree of surface dehydration that took place during the process and that contributes to the inhibition of microbial growth. At present, it is maintained as a hedonic agent, by giving the products a desirable and characteristic color, flavor and aroma. However, some of the components of smoke belonging to the family of polycyclic aromatic hydrocarbons (PAHs) are potential carcinogens and pose a chemical risk to be controlled during sausage manufacture. Usually, controlling the smoking temperature is key to limiting the deposition of PAHs during the process, but in the artisanal traditional smoking processes other strategies to achieve this end should be adopted. In the present Special Issue, Fraqueza et al. [14] make an interesting contribution in this area trying to optimize the smoking process of traditional dry-cured meat products to minimize the presence of PAHs. In this work, dry-cured sausages were submitted to four three different treatments: (i) without smoking, (ii) 20 h effective smoking, (iii) 60 h effective smoking, and (iv) effective smoking until reaching weight losses of 38–40%. The total PAHs content was generally low and did not differ significantly among the four different smoking regimes. The PAH4 (sum of four different polycyclic aromatic hydrocarbons: benzo[*a*]anthracene, chrysene, benzo[*b*]fluoranthene, and benzo[*a*]pyrene) and the benzo[*a*]pyrene contents were below the stablished legal limits in all the analyzed sausages.

The partial or total replacement of the traditional ingredients in sausages with unconventional raw materials allows the obtaining of novel and varied products capable of satisfying the needs of people who have consumption choices limited by cultural conditions, religious beliefs, personal convictions or health reasons. However, the incorporation of these new ingredients leads to the appearance of technological problems in manufacturing and also of organoleptic deficiencies in the final products that must be solved.

Within this research topic, Cullere et al. [15] studied the effect of two different fat inclusion levels, NaCl contents and LAB starter cultures on the weight loss and nutritional characteristics of an Italian-type ostrich salami. Among other results, authors observed that the lowest fat and highest NaCl level provided the greatest cumulative weight loss throughout manufacturing, but by reducing the NaCl level the weight loss was retarded without affecting the nutritional composition of the final sausage. The authors concluded that it is possible to obtain a salami from ostrich meat with lower salt and fat contents, while maintaining a satisfactory quality of the final product.

Wongkaew et al. [16] essayed the use of mango peel pectin extracted with microwave assistance as fat replacer (0%, 5%, 10% and 15% of fat replacement) in a dried Chinese sausage. Microwave-assisted extraction allows the obtaining of mango peel pectin with improved properties compared to that obtained using the conventional extraction method. Quality attributes (color, texture, and sensory characteristics) of experimental sausages were assessed and compared to those manufactured using the control formula without fat replacement. The authors concluded that the substitution of 5% of fat with pectin in the Chinese sausage could enhance color while conserve the physical qualities and sensory attributes. Therefore, mango peel pectin can be utilized in the low-fat Chinese sausage formulation as a novel fat replacer.

Vargas-Ramella et al. [17] studied the effect of partial (50%) replacement of the animal fat with healthy oils (olive, canola, and soy oil emulsions immobilized in Prosella gel) on the physicochemical attributes, texture, fatty acid profile and volatile compounds of dry-fermented deer sausage. Reformulated sausages were darker, harder, and with higher pH values than control sausages made using only animal fat. Use of soy and canola oils increased the polyunsaturated fatty acid and the w-3 fatty acid contents while decreased the w-6/w-3 ratios and the saturated fatty acid contents, thus allowing the obtaining the best nutritional properties. Regarding the effect on the volatile compounds, the animal fat replacement increased the content of the total and most of the individual volatile compounds. However, the control samples presented equal or higher contents of volatile compounds derived from the lipid oxidation processes than the reformulated sausages. Sausages reformulated with vegetable oils showed higher consumer acceptance than control sausages. In conclusion, replacement of animal fat with vegetable oils, particularly soy oil, could be an excellent strategy to elaborate healthy fermented sausages.

The use of non-meat ingredients in high percentages in sausages with the aim of obtaining healthier products may affect the microbial growth and chemical stability during storage and therefore the shelf-life of the novel sausages. Preservatives should be added in order to improve the shelf-life and safety, but consumers interested in healthier meats request at the same time products without synthetic additives. The search and the essay of natural substitutes for synthetic additives is consequently a challenge for researchers. Ranucci et al. [18] studied the effects of different concentrations of a commercial mix of extracts of pomegranate and citrus on the growth of spoilage microorganisms and on the oxidation during storage in vacuum-packaged cooked sausages made from pork meat, emmer wheat, almond and hazelnut. The authors reported that the use of the mix, particularly at a concentration of 10 g/1000 g delayed the pH decrease and the lipid oxidation processes during storage. The mix also lowered the maximum growth rate of total viable bacteria, lactic acid bacteria, and psychrotrophs. The sensory analysis carried out by sausage consumers showed an increase in the shelf-life of 6 and 16 days for the sausages made by the addition of 5‰ and 10‰ of the extract mix, respectively.

This Special Issue ends with an interesting review made by Martuscelli et al. [19]. Some religious convictions and beliefs and the restrictions imposed on the food to the practitioners entail limitations on the raw materials to be used in the manufacture of meat products (animal species and the ways in which they are slaughtered), and in processing conditions. Various scientific studies were conducted until now on ritual slaughtering practices and manufacturing of meat products for Jewish and Muslim religious communities. On the other hand, the authenticity and traceability of meat is one of the priorities of halāl food certification systems in order to prevent fraudulent practices motivated by both economic and technological reasons. In their article, Martuscelli et al. [19] reviewed studies conducted on the safety, quality and analytical authentication of halāl meat products, with particular attention on salami. The authors discussed the halāl meat products and regulations in Europe, halāl salami processing (halāl raw material, preservatives, use of spices and/or plant extracts, casings, sensory profile and the hazard represented by the biogenic amines), food safety in halāl assurance, and analytical authentication of halāl meat, salami and other meat products.

Study of the sausages is an exciting task that still offers many possibilities and motivations for researchers. Given the economic importance of sausages, the complexity and variety of the physical, chemical and microbiological processes that take place during their production that make them surprising and captivating for scientists, and the fact that many of them are deeply rooted in culture and in the heritage of various countries and geographical areas, it is certain that these products will continue to be a source of attention for scientific researchers in the future.

Some sausage varieties are still hardly known in terms of their microbiological and biochemical features, or their production processes are insufficiently standardized. In other cases, there are safety or quality concerns that must be solved so that these sausages can be enjoyed to the fullest of their potential. The improvement of the sensory quality and/or the adaptation of their sensory and nutritional properties to the changes in consumer preferences and requirements, all without sacrificing the personality and differential attributes of each sausage variety, is in all cases a permanent challenge. For all these reasons, further studies and research are essential to improve and continue enjoying these privileged foods.

Funding: The author acknowledges the Xunta de Galicia, Spain (The Regional Government) the financial support through the CITACA Strategic Partnership (grant number ED431E 2018/07).

Conflicts of Interest: The author declares no conflict of interest.

References

1. Coretti, K. *Rohwurstreifung und Fehlerzeugnisse bei der Rohwurstherstellung*; Verlag der Rheinhessischen Druckwerkstätte: Alzey, Germany, 1971; p. 183.
2. Frey, W. *Die Sichere Fleischwarenherstellung: Leitfaden für den Praktiker*; Hans Holzmann Verlag GMBH & Co. KG.: Bad Wörishofen, Germany, 1983; p. 168.
3. Lücke, F.K. Fermented sausages. In *Microbiology of Fermented Foods*; Wood, B.J.B., Ed.; Elsevier Applied Science: London, UK, 1985; Volume 2, pp. 41–83.
4. Jensen, L.B.; Paddock, L. Sausage Treatment. U.S. Patent 2,225,783, 24 December 1940.
5. Niven, C.F.; Deibel, R.H.; Wilson, G.D. Production of Fermented Sausage. U.S. Patent 2,907,661, 6 October 1959.
6. Niinivaara, F.P.; Pohja, M.; Komulainen, S.E. Some aspects about using bacterial pure cultures in the manufacture of fermented sausages. *Food Technol.* **1964**, *18*, 147–153.
7. Nurmi, E. Effect of Bacterial Inoculation on Characteristics and Microbial Flora of Dry Sausage. Ph.D. Thesis, University of Helsinki, Helsinki, Finland, 1966.
8. Jessen, B. Starter cultures for meat fermentation. In *Fermented Meats*; Campbell-Platt, G., Cook, P.E., Eds.; Blackie Academic & Professional: Glasgow, UK, 1995; pp. 130–159.
9. Comi, G.; Muzzin, A.; Corazzin, M.; Iacumin, L. Lactic acid bacteria: Variability due to different pork breeds, breeding systems and fermented sausage production technology. *Foods* **2020**, *9*, 338. [CrossRef] [PubMed]
10. Van Reckem, E.; Charmpi, C.; Van der Veken, D.; Borremans, W.; De Vuyst, L.; Weckx, S.; Leroy, F. Application of a high-throughput amplicon sequencing method to chart the bacterial communities that are associated with European fermented meats from different origins. *Foods* **2020**, *9*, 1247. [CrossRef] [PubMed]
11. Barbieri, F.; Laghi, L.; Gardini, F.; Montanari, C.; Tabanelli, G. Metabolism of *Lactobacillus sakei* Chr82 in the presence of different amounts of fermented sugars. *Foods* **2020**, *9*, 720. [CrossRef] [PubMed]
12. Agüero, N.L.; Frizzo, L.S.; Ouwehand, A.C.; Aleu, G.; Rosmini, M.R. Technological characterisation of probiotic lactic acid bacteria as starter cultures for dry-fermented sausages. *Foods* **2020**, *9*, 596. [CrossRef] [PubMed]
13. Rodríguez-González, M.; Fonseca, S.; Centeno, J.A.; Carballo, J. Biochemical changes during the manufacture of Galician chorizo sausage as affected by the addition of autochthonous starter cultures. *Foods* **2020**, *9*, 1813. [CrossRef] [PubMed]
14. Fraqueza, M.J.; Laranjo, M.; Alves, S.; Fernandes, M.H.; Agulheiro-Santos, A.C.; Fernandes, M.J.; Potes, M.E.; Elias, M. Dry-cured meat products according to the smoking regime: Process optimization to control polycyclic aromatic hydrocarbons. *Foods* **2020**, *9*, 91. [CrossRef] [PubMed]
15. Cullere, M.; Novelli, E.; Dalle Zotte, A. Fat inclusion level, NaCl content and LAB starter cultures in the manufacturing of Italian-type ostrich salami: Weight loss and nutritional traits. *Foods* **2020**, *9*, 476. [CrossRef] [PubMed]
16. Wongkaew, M.; Sommano, S.R.; Tangpao, T.; Rachtanapun, P.; Jantanasakulwong, K. Mango peel pectin by microwave-assisted extraction and its use as fat replacement in dried Chinese sausage. *Foods* **2020**, *9*, 450. [CrossRef] [PubMed]
17. Vargas-Ramella, M.; Munekata, P.E.S.; Gagaoua, M.; Franco, D.; Campagnol, P.C.B.; Pateiro, M.; da Silva Barretto, A.C.; Domínguez, R.; Lorenzo, J.M. Inclusion of healthy oils for improving the nutritional characteristics of dry-fermented deer sausage. *Foods* **2020**, *9*, 1487. [CrossRef] [PubMed]

18. Ranucci, D.; Roila, R.; Andoni, E.; Braconi, P.; Branciari, R. *Punica granatum* and *Citrus* spp. extract mix affects spoilage microorganisms growth rate in vacuum-packaged cooked sausages made from pork meat, emmer wheat (*Triticum dicoccum* Schübler), almond (*Prunus dulcis* Mill.) and hazelnut (*Corylus avellana* L.). *Foods* **2019**, *8*, 664. [CrossRef] [PubMed]
19. Martuscelli, M.; Serio, A.; Capezio, O.; Mastrocola, D. Safety, quality and analytical authentication of halāl meat products, with particular emphasis on salami: A review. *Foods* **2020**, *9*, 1111. [CrossRef] [PubMed]

foods

MDPI

Article

Lactic Acid Bacteria: Variability Due to Different Pork Breeds, Breeding Systems and Fermented Sausage Production Technology

Giuseppe Comi, Alessia Muzzin, Mirco Corazzin and Lucilla Iacumin *

Department of Agriculture, Food, Environmental and Animal Science, University of Udine, via Sondrio 2/A, 33100 Udine, Italy

* Correspondence: lucilla.iacumin@uniud.it; Tel.: +39-0432-558126

Received: 3 February 2020; Accepted: 10 March 2020; Published: 13 March 2020

check for updates

Abstract: Changes in the ecology of the various lactic acid bacteria (LAB) species, which are involved in traditional fermented sausages, were investigated in the light of the use of different breeds of pork, each of which was raised in two different environments and processed using two different technologies. The semi-quantitative molecular method was applied in order to understand how the different species alternate over time, as well as their concentration ratios. A significant increase in LAB over the first days of fermentation characterized the trials where the starter culture wasn't added (T), reaching values of 10^7–10^8 cfu g^{-1}. On the other hand, in the trials in which sausages were produced with starter addition, LAB counts had a less significant incremental jump from about 10^6 cfu g^{-1} (concentration of the inoculum) to 10^8 cfu g^{-1}. *Lactobacillus sakei* and *Lb. curvatus* were detected as the prevalent population in all the observed fermentations. *Pediococcus pentosaceus, Lb. casei, Leuconostoc mesenteroides, Lactococcus garviae*, and *Lb. graminis* also appeared, but their concentration ratios varied depending on the diverse experimental settings. The results of cluster analysis showed that a plant- and breed-specific LAB ecology exists. In addition, it was also observed that the breeding system can influence the presence of certain LAB species.

Keywords: lactic acid bacteria; fermented sausages; ecology; breed; breeding system

1. Introduction

Traditional fermented sausages are well-known and popular meat products in Italy. In north-eastern Italy, Friuli Venezia Giulia region, traditional fermented sausages are made without the use of microbial starters to preserve the typical characteristics of these products, that are delicate sourness gentle acidity, and elastic consistency. These foodstuffs are made from about 60% of minced fresh pork meat and 40% of minced lard mixed with sugar, NaCl, spices and additives (i.e., nitrate, nitrite). Starter cultures are used mainly for large-scale productions [1–3].

The study of microbial ecology during the fermentation of cured meat products began in the 70s [4]. Since then, fifty years of studies have shown that the two groups responsible for the characteristic transformations of these products, regardless of the specific technologies and recipes applied in the different regional cultures, are lactic acid bacteria (LAB) and coagulase-negative catalase-positive cocci (CNCPC) [5–9]. Among them, the species that develop are closely related to the process conditions applied. Specifically, the first parameter of selection is the temperature, which selectively favours the development of some LAB species at the expense of others during the first days of ripening. Sausages subjected to a short ripening using high ripening temperature allow the development of lactobacilli right from the early stages of fermentation and, at the end of ripening, an acidic flavour with slight aroma predominates in salami. Indeed, the pH decreases due to the transformation of sugars caused by the high concentration of lactobacilli and consequently inhibits the growth of CNCPC [9–12].

This phenomenon is directly proportional to the increase in temperature, in fact, the higher the temperature the more the selection pushes towards more acidifying species, with an increasingly more rapid inhibition, not only of the pathogenic and spoiling species, but also of the flavouring ones. On the other hand, sausages with longer maturation and ripening conducted at a lower temperature are characterized by low acidity and higher numbers of CNCPC resulting in a more aromatic product [12]. CNCPC participate in desirable reactions, such as lipolysis and proteolysis, which influence the aroma, besides being makers of the production and stabilization of colour, reducing nitrates in nitrite [13–15].

LAB, therefore, play a primary role in fermented sausages production because they synthesize lactic acid, which causes the lowering of the pH value and the consequent inhibition of pathogenic bacteria and spoilage microorganisms [9,15,16]. Moreover, the acidity accelerates the reduction of the nitrite, and then the process of developing and fixing colour. Finally, LAB release substances, which affect the flavour profile, such as acetic acid, ethanol, acetoin, butanediol, and diacetyl, or having antimicrobial effect, such as bacteriocins [17–19].

Over the last 20 years, LAB ecology has been studied by several authors, using both traditional and molecular microbiological techniques. Among them, polymerase chain reaction (PCR) and Denaturing Gradient Gel Electrophoresis (PCR-DGGE) have been extensively used both as a culture-dependent and a culture-independent method [5,20–24]. High-throughput sequencing (HTS) techniques represent a relatively new approach in the way microbiologists address ecology and diversity in different environments, and consequently also in foods. As explained by Franciosa et al. [25], in HTS, mixed nucleic acid molecules from a complex ecosystem can be sequenced, and therefore can lead to a detailed profile of the microbial populations (identified as operational taxonomic units, OTU) present. In the mid-2000s, HTS technologies became ubiquitous in microbial ecology studies, but limitations remain: a) DNA is a chemically stable molecule, which can be found for a long time after the death of a cell; b) quantification of the genera and/or species is only relatively possible. For these reasons this approach cannot be generally applied, but a reasoned choice of method is needed, in particular when ecological studies are made during a relatively short time, and also important is the detection and identification of species whose counts are $<10^3$ cfu/g.

The previous studies have identified species belonging to *Lactobacillus* genus as the main ones involved in natural sausage fermentation. The species mainly spotted in this kind of products are *Lactobacillus sakei*, *Lb. curvatus* and *Lb. plantarum* [26].

Species and strain dynamics and counts in fermented meat products, as demonstrated by many authors, vary according to function of process conditions, particularly NaCl concentration and temperature, which have a direct impact on strain selection, pH, and water activity (A_W) [27–31]. Other authors illustrated the roles of plant, meat, and ingredients in the selection of a specific microbiota [32–35].

The role of pork breed and breeding system in influencing LAB ecology during fermented sausages' ripening has been not yet explored. This is a gap that deserves to be investigated, because they could have direct and undirect effects on the microbial ecology of meat products. In fact, several studies showed the relevant presence of species belonging to the genus *Lactobacillus* in both gut and reproductive tract microbiota of pigs [36–38] and demonstrated that breed affects colonic microbiota and immune status since postnatal period (14–49 days) [36,39]. A possible link between the gut microbiota and feed efficiency in pigs was demonstrated by Tan et al. [38], who discovered that, in Landrace pigs, the high-feed conversion rate pigs had a greater abundance of *Lactobacillus* and *Streptococcus* than the low-feed conversion rate pigs, having an evident repercussion on pork meat quality [40]. In parallel, it is now well known that the diet also has a very important influence on the modulation of the intestinal flora, both for humans and for animals, including pigs [41,42]; and the breeding system plays a fundamental role on the type of feeding provided [43]. In the case of the outdoor breeding system, compared to the indoor breeding system, there is a greater variability of nutrients supplied as well as a greater movement of the animals with possible repercussions both in the composition of the intestinal microbiota and on the skin of the animal itself.

Moreover, the microbiological quality of the meat is strictly dependent on the environmental conditions. The meat of healthy and non-fatigued animals is usually sterile, only the lymphatic ganglia and some organs such as liver and spleen can contain microorganisms. However, microorganisms, especially pathogens and/or opportunists, can invade the muscles when the animal is still alive, before slaughter. The causes for this are various and concern, in particular, weakening or stress before slaughter or primary contamination, which can escape the ante-mortem visit: undiagnosed bacteremia and septicemia can be a vehicle for microorganisms deep in the tissues, and psychological stresses can promote the migration of the same from the intestine. The secondary contaminations during all processing phases, remain those to which the cause of the main contamination is due. In particular, during slaughtering, microorganisms present on/in the animal are transferred to the meat, defining its own microbiota [44–46].

Thus, the core of our study was to investigate LAB species ecology and dynamics under different experimental conditions. In particular, fermented sausages produced in both artisanal and industrial plants with meat from pigs belonging to three different breeds, and breeding in either indoor or outdoor systems, were investigated.

2. Materials and Methods

2.1. Fermented Sausages and Sampling Procedures

The experimental design was described in Iacumin et al. [7]. In particular, all pigs (male) were raised under the same conditions on a farm in Friuli Venezia Giulia Region. Animals were at the same age and kept on the same diet and breeding conditions, specifically outdoor or indoor. All the trials were conducted at the same time. Pigs were slaughtered in the same abattoir, on the same day. The carcasses were divided into half-carcasses and evenly distributed between the two processing plants, so that each one had half of each animal available. Two local meat factories in northeaster Italy were selected for the study: plant D (industrial) and plant T (artisanal). The annual sausage production in the industrial plant was approximately 450–500 tons per year. In contrast, the annual sausage production in the artisanal plant was approximately 35–45 tons per year. Starter culture (*Pediococcus pentosaceus*) was used only in the industrial plant. Traditional techniques were used for the production of fermented sausages. The final product was ready for sale after 45–50 (plant D) or 90 (plant T) days. Common ingredients for sausage production included 60 kg pork meat, 40 kg lard, 2.5 kg sodium chloride, 70 g black pepper, and 200 ppm nitrite and nitrate. Sugar was not added for fermentation in plant T, whereas in plant D, 1.5 kg sugar was added. The fermentation mixtures were stuffed into natural casings, and the resulting fresh sausages were 25 cm long and 5 cm in diameter. All of the sausages made in plant D were identical in shape and mass (1 kg each), whereas the sausages made in plant T were not identical in shape, and their masses ranged from 0.8 kg to more than 1 kg each. Ripening parameters were also different in the two plants. The first stage of ripening consisted of 2 days of drying with relative humidity (RH) of 85% at 18–20 °C (plant T) or 22 °C (plant D). The temperature was then decreased to 14 °C (plant T) or 12 °C (plant D) at a rate of 2 °C per day with RH between 60% and 90%. Further ripening was then performed for the rest of the period at those temperatures (14 °C plant T, 12 °C plant D) in storerooms with 65%–85% RH. Plant D had a more effective system to control the environmental temperature and RH than plant T.

In total, 6 batches of products, in triplicate, were monitored in plant D (using industrial production technology) as follows: (i) Cinta Senese meat, indoor breeding system (D-CS-I); (ii) Cinta Senese meat, outdoor breeding system (D-CS-O); (iii) Goland meat, indoor breeding system (D-G-I); (iv) Goland meat, outdoor breeding system (D-G-O); (v) Mora Romagnola meat, indoor breeding system (D-MR-I); and (vi) Mora Romagnola meat, outdoor breeding system (D-MR-O). In plant T (using artisanal production technology), only 4 batches, in triplicate, were monitored: (i) Cinta Senese meat, indoor breeding system (T-CS-I); (ii) Cinta Senese meat, outdoor breeding system (T-CS-O); (iii) Goland meat, indoor breeding system (T-G-I); and (iv) Goland meat, outdoor breeding system (T-G-O).

Every batch of products from each plant was exposed to microbiological analysis. Samplings were performed at 0, 2, 9, 23, 30, 60, and 90 days in plant T and 0, 9, 30, 60, and 90 days in plant D, and three fermented sausages per batches at each sampling point were used for microbiological analysis. A major number of sampling points was applied in plant T, because the fermentation process was slower than in D and this required an extension of the monitoring, in particular at the beginning of the fermentation process.

2.2. *Microbial Enumeration and Bulk Cell Collection*

Serial dilutions of each sample in 0.25X Ringer's solution (Oxoid, Milan, Italy) were used to inoculate deMan, Rogosa, Sharpe (MRS agar, Oxoid, Milan, Italy) plates, which are widely used to cultivate LAB. Two series of agar plates were inoculated and incubated at 30 °C for 48 h in Jair with AnaeroGen^{TM} 3.5 L kit (Oxoid, Milan, Italy). Portions (0.1 mL) of appropriate dilutions were spread on plates in triplicate. Colonies were counted, and results were calculated as the means of three determinations. After counting, all plates were used for bulk formation as previously described [23]. In brief, a bulk formation was performed using all plates from the serial dilutions (–2 to the last). For each dilution, all colonies grown on the plates' surface were suspended in 2 mL of quarter strength Ringer's solution, harvested with a sterile pipette and frozen at −20 °C. To minimise the effects of different concentrations, all suspensions were standardised at 1 unit of optical density (600 nm). Then, 1 mL of the bulk suspension was used for DNA extraction as described below and subjected to molecular analysis.

2.3. *DNA Extraction from Bulk Cultures*

One milliliter of each bulk suspension was centrifuged at 14,000× g for 10 min at 4 °C to pellet the cells, and the pellet was subjected to DNA extraction according to Andrighetto et al. [26]. Briefly: the pellet was added to 30 μL of Lysozyme (0.1 g/mL Lysozyme, 25% Sucrose, Sigma-Aldrich, Milan, Italy) and incubated 30 min at 30 °C for lysis in a Thermomixer (Thermomixer Confort, Eppendorf, Eppendorf AG, Hamburg, Germany).

2.4. *PCR*

Primers P1V1 and P2V1 [24] spanning the V3 region of the 16S rDNA were used in this study. A GC-clamp (5′-CGC CCG CCG CGC CCC GCG CCC GTC CCG CCG CCC CCG CCC G-3) was attached to the 5′ end of primer P1V1 to obtain amplicons to be subjected to DGGE analysis. Amplifications were carried out in a final volume of 25 μL, containing 1 μL (100 ng total) template DNA, 10 mM Tris HCl (pH 8.3), 50 mM KCl, 1.5 mM $MgCl_2$, 0.2 mM deoxynucleoside triphosphates (dNTPs), 1.25 U *Taq* polymerase (Invitrogen, Milan, Italy), and 0.2 μM each primer, using the thermal cycler DNA Engine DYAD^{TM} SYSTEM. The amplification cycle included an initial denaturation step at 95 °C for 5 min, followed by 35 series composed by denaturation, performed at 95 °C for 1 min, annealing at 45 °C for 1 min, and extension performed at 72 °C for 1 min. Finally, an extension cycle, performed at 72 °C for 7 min, was added. The PCR products were resolved by agarose (Sigma-Aldrich, Milan, Italy) (2% *w/v*) gel electrophoresis at 100 V for 2 h. In each electrophoresis gel, 3 μL of PCR Marker (100 bp) were loaded in the first and in the last well. After running, amplicons were visualized under UV light using the Syngene G: Box Chemi-XX9 (Syngene, Cambridge, United Kindom) and digitally captured by using the software GeneSys version 1.5.7.0 (Syngene, Cambridge, United Kindom).

2.5. *DGGE Analysis*

The DCode universal mutation detection system (Bio-Rad Laboratories S.r.l., Milan, Italy) was used for DGGE analysis. For PCR products obtained with the primers P1V1-GC and P2V1, electrophoresis was performed in a 0.8-mm-thick polyacrylamide gel (8% (wt/vol) acrylamide-bisacrylamide (37.5:1)), with a denaturing gradient from 30% to 50% (100% corresponded to 7 M urea and 40% (wt/vol) formamide) increasing in the direction of the electrophoretic run. Gels were subjected to a constant voltage of 130 V for 3 h and 30 min at 60 °C. After the electrophoresis, gels were stained for 30 min in 1.25X Tris-acetate-EDTA containing 1X SYBR Green (final concentration; Molecular Probes, Milan, Italy). Pictures of the gels were visualized under UV light using the Syngene G: Box Chemi-XX9 (Syngene, Cambridge, United Kindom) and digitally captured by using the software GeneSys version 1.5.7.0 (Syngene, Cambridge, United Kindom).

A reference pattern was established consisting of amplicons from 4 different bacterial species: *Lactobacillus brevis* (DSMZ 20054), *Lb. casei* (DSMZ 20011), *Lb. curvatus* (DSMZ 6179) and *Lb. sakei* (DSMZ 6333). By including this standard reference pattern three times on each DGGE gel, resulting DGGE fingerprint band profiles from the different sausages were digitally normalised using Gel Compare 4.1 software Version 4.1 (Applied Maths, Kortrijk, Belgium). Additionally, this reference pattern was used to obtain a preliminary identification of the species. When possible, almost three bands migrating the same position in every single gel were excised and subjected to sequencing and sequence analysis to confirm the preliminary identification and to identify amplicons, which did not correspond to those of the reference strains used. DGGE analyses were performed at least twice.

2.6. Sequence Analysis of DGGE Bands

Blocks of polyacrylamide gels containing selected DGGE bands were excised with gel cutting tips. Blocks were then transferred to 100 µL sterile water, and the DNA in the bands was left to diffuse overnight at 4 °C. Two microliters of the eluted DNA were used for re-amplification, and PCR products generated with a GC-clamped primer were verified by DGGE. Only products migrating as a single band and in the same position of the control were amplified, as described above, with the primer without the GC clamp. Products were then cloned into the pGEM-T Easy vector (Promega, Milan, Italy) following the manufacturer's instructions. Clones were checked as described above (co-migration with control), and the inserts in appropriate clones were sequenced at a commercial facility (Eurofins, Edersberg, Germany). Sequence comparisons were performed in GenBank using the Blast program version 2.2.18 (https://blast.ncbi.nlm.nih.gov/Blast.cgi) [47].

2.7. Statistical Analysis

The statistical tests were performed using R software, vers. 3.4.0 (R core team, 2017, Vienna, Austria). Normality of data distribution and homoscedasticity were tested using Shapiro–Wilk and Levene test, respectively. Data were assessed within the plant. The LAB count in the plant D was analysed with a two-way ANOVA model, with experimental group (batches) and ripening time as factors. Also, the interaction experimental group × ripening time was considered. For multiple comparisons, the P-values were adjusted using the Holm method, and, when appropriate, the White-corrected P-values [48] were considered. The same model, but with the Scheirer–Ray–Hare test for non-parametric data [49] was used for LAB count in the plant T. Multiple comparisons were performed using Mann–Whitney U test, and the P-values were adjusted using the Holm method.

As a result of the "bulks", the molecular analysis of different fingerprints was performed. Analysis of the patterns and their corresponding dilutions give information about the dominant species occurring and also allows to ascertain the concentration of every single species found in the DGGE profile of the original sample. However, these concentration values are not properly quantitative, they are ordinal value and in statistical analysis they have to be treated as qualitative parameters. For this reason, and to avoid a subjective conversion of the data (from bands in numeric values), a direct cluster analysis from the obtained fingerprints was chosen [7].

The different profiles of each dilution from the same sample were previously joined to get a total sample profile. The profiles of the different times are not treated in a bulk for each sample, but one by one. Pictures of the gels were analysed by using the pattern analysis software package Gel Compare Version 4.1 (Applied Maths, Kortrijk, Belgium). Calculations of similarities in band profiles were based on Pearson (correlation coefficient at 53%) and Dice (correlation coefficient 60%) product-moment correlation coefficients. Dendrograms were obtained via the unweighted pair group method using an arithmetic average (UPGMA) clustering algorithm [50]. Two different analyses, using different correlation coefficients (Pearson and Dice), were applied to verify the strength of the results.

3. Results and Discussion

3.1. LAB Counts

LAB counts in the different trials and at the different sampling points corresponding to the different fermented sausages production steps were shown in Table 1. Fermentations in plant T were characterised by lower initial LAB counts compared to that in plant D, where *P. pentosaceus* was added as starter culture ($P < 0.01$).

In plant T, the statistical analyses highlighted a significant effect of ripening time ($P < 0.01$), but not of experimental group ($P > 0.05$) on LAB count. Also, the interaction *ripening time × experimental group* did not reach a level of significance ($P > 0.05$). LAB count increased from t0 to t23 ($P < 0.05$), after which counts remained stable till T30 ($P > 0.05$) and then decreased at t90 ($P < 0.05$), where LAB count was similar to that observed at t9 ($P > 0.05$). LAB counts slightly decresed at t90 ($P < 0.05$), where LAB count was similar to that observed at t9 ($P > 0.05$). At the same time, this difference was not significant compared to LAB count at t60. From a microbiological point of view, this can be explained by the fact that at t90 the physic-chemical characteristics of the products, such as the combined effect of pH (5.32 ± 0.13, data not shown), activity water (Aw, 0.925 ± 0.024, data not shown) and the absence of sugars led to a progressive slow inactivation of LAB.

In plant D, LAB count was affected both by experimental group ($P < 0.01$) and by ripening time ($P < 0.01$). Moreover, also the interaction *experimental group × ripening time* was significant ($P < 0.01$). It means that the main effect of a factor can be interpreted only at each level of the other factor. Therefore, in Table 1, the interaction from the perspective of the experimental group is represented in a row, and the interaction from the perspective of ripening time is represented in a column. The first of these shows that LAB count increased from t9 to t30 ($P < 0.05$) and, then, remained stable from t30 to t90 ($P > 0.05$) for the sausages produced with Goland breed, irrespectively of the husbandry system adopted. Furthermore, in the D-MR-I group, LAB count remained stable from t30 to t90 ($P > 0.05$), but, in this group, LAB count started to increase already from t0 to t9 ($P < 0.05$). The D-CS-I and D-MR-O groups showed LAB count that increased from t0 to t9 ($P < 0.05$), but decreased from t60 to t90 ($P < 0.05$). Conversely, LAB count in D-CS-O decreased from t0 to t9 ($P < 0.05$) and the value recorded at t90 was similar to that shown at t9 ($P > 0.05$). Considering the results about the interaction *experimental group × ripening time* from the perspective of the ripening time, at t0, the lowest LAB count values were observed in sausages produced with the Goland and Cinta Senese breed indoor-housed ($P < 0.05$), conversely, the higher values were recorded in the sausages produced with the same breeds, but with outdoor-housed animals ($P < 0.05$). Lower differences between experimental groups were found in the other ripening times with LAB count that was lower in D-G-I than D-MR-O ($P < 0.05$) at t30, and in D-CS-O than in D-CS-I ($P < 0.05$) and D-MR-O ($P < 0.05$) at t60. Moreover, sausages produced with Cinta Senese breed had lower LAB count that those observed in D-G-O group at t90 ($P < 0.05$).

Table 1. Mean ± standard deviation of Lactic acid bacteria (LAB) colony counts, expressed as Log CFU/g, of sausages at different ripening days (t) and produced in two plants, industrial (D) and traditional (T) whose meats derived from pigs of three breeds, Goland (G), Cinta Senese (CS), and Mora Romagnola (MR) that were indoor (I) or outdoor (O) housed.

	Processing Time						
	t0 *	t2	t9	t23	t30	t60	t90
T-G-I	3.61 ± 0.21	4.84 ± 0.03	7.52 ± 0.13	8.11 ± 0.09	8.72 ± 0.05	7.73 ± 0.14	7.75 ± 0.01
T-G-O	3.99 ± 0.06	5.52 ± 0.08	8.52 ± 0.24	8.95 ± 0.01	8.12 ± 0.10	7.96 ± 0.10	7.95 ± 0.16
T-CS-I	3.53 ± 0.07	6.69 ± 0.07	8.00 ± 0.02	8.56 ± 0.02	8.54 ± 0.07	8.87 ± 0.03	7.78 ± 0.03
T-CS-O	4.50 ± 0.05	5.07 ± 0.03	7.93 ± 0.03	8.72 ± 0.03	8.42 ± 0.06	8.20 ± 0.15	8.29 ± 0.07
Mean	3.91 ± 0.41 $^{\alpha}$	5.53 ± 0.74 $^{\beta}$	7.99 ± 0.39 $^{\gamma}$	8.62 ± 0.33 $^{\delta}$	8.45 ± 0.24 $^{\delta}$	8.19 ± 0.46 $^{\gamma\delta}$	7.94 ± 0.23 $^{\gamma}$
D-G-I	6.17 ± 0.18 $^{A\alpha}$	n.d.	8.14 ± 0.24 $^{\alpha}$	8.47 ± 0.12 $^{\beta}$	8.65 ± 0.14 $^{A\beta}$	8.45 ± 0.05 $^{AB\beta}$	8.54 ± 0.13 $^{AB\beta}$
D-G-O	6.88 ± 0.10 $^{C\alpha}$	n.d.	8.42 ± 0.30 $^{\alpha}$	n.d.	8.55 ± 0.14 $^{AB\beta}$	8.62 ± 0.10 $^{AB\beta}$	8.58 ± 0.05 $^{B\beta}$
D-CS-I	6.14 ± 0.09 $^{A\alpha}$	n.d.	8.53 ± 0.04 $^{\gamma}$	n.d.	8.63 ± 0.06 $^{AB\gamma}$	8.58 ± 0.02 $^{B\gamma}$	8.25 ± 0.02 $^{B\beta}$
D-CS-O	8.76 ± 0.04 $^{D\alpha}$	n.d.	8.17 ± 0.08 $^{\beta}$	n.d.	8.61 ± 0.05 $^{AB\gamma}$	8.27 ± 0.12 $^{A\beta}$	8.12 ± 0.07 $^{A\beta}$
D-MR-I	6.33 ± 0.20 $^{AB\alpha}$	n.d.	8.54 ± 0.08 $^{\beta}$	n.d.	8.62 ± 0.13 $^{AB\beta}$	8.52 ± 0.11 $^{AB\beta}$	8.50 ± 0.11 $^{AB\beta}$
D-MR-O	6.50 ± 0.05 $^{B\alpha}$	n.d.	8.58 ± 0.13 $^{\beta\gamma}$	n.d.	8.11 ± 0.16 $^{B\beta}$	8.78 ± 0.10 $^{B\gamma}$	8.32 ± 0.12 $^{AB\beta}$
Mean	6.80 ± 0.94 $^{\alpha}$		8.39 ± 0.23 $^{\beta}$	8.47 ± 0.12 $^{\beta}$	8.53 ± 0.22 $^{\beta}$	8.53 ± 0.18 $^{\beta}$	8.38 ± 0.19 $^{\beta}$

Legend: $^{A, B, C, D}$: $P < 0.05$ within plant and column; $^{\alpha, \beta, \gamma, \delta}$: $P < 0.05$ within plant and row; n.d.: not done; *t: days.

The comparison along the time is not the effect under study, but it allowed to verify if there were deviations in the correct development of fermentation. If this had happened, this could have been attributed to the different composition (e.g., intramuscular fat, type of fat.) of the meat due to the breed. In Figure 1 it can be observed that, apart from t0, the coefficients of variation among the trials per each time is higher in plant T than in plant D. This can be attributed to the different procedures and technologies applied in the two plants. The industrial plant (D), using exactly the same procedures at each production phase, as well as an optimized protocol of drying and ripening in chambers with a more accurate system to control humidity and temperature, resulted in a major standardization of the microbial fermentation and, consequently, of the final product. On the other hand, although in the artisanal plant (T) there is a higher coefficient of variability, it is possible to note that fermentation has progressed flawlessly. In this case, no microbial starters have been added, but nevertheless the production technology has allowed the correct development of the indigenous LAB, which in nine days have reached the same values as those inoculated in plant D. Finally, looking at day 0, it seems that the counts of the trials performed using pork meat from outdoor breeding system were significantly higher than those of the indoor breeding system. Probably the outdoor breeding system has brought the pigs to smear/soil more than the indoor breeding system, and therefore to increase the microbial load of the skin, which during the slaughter has favoured a greater contamination of the carcass.

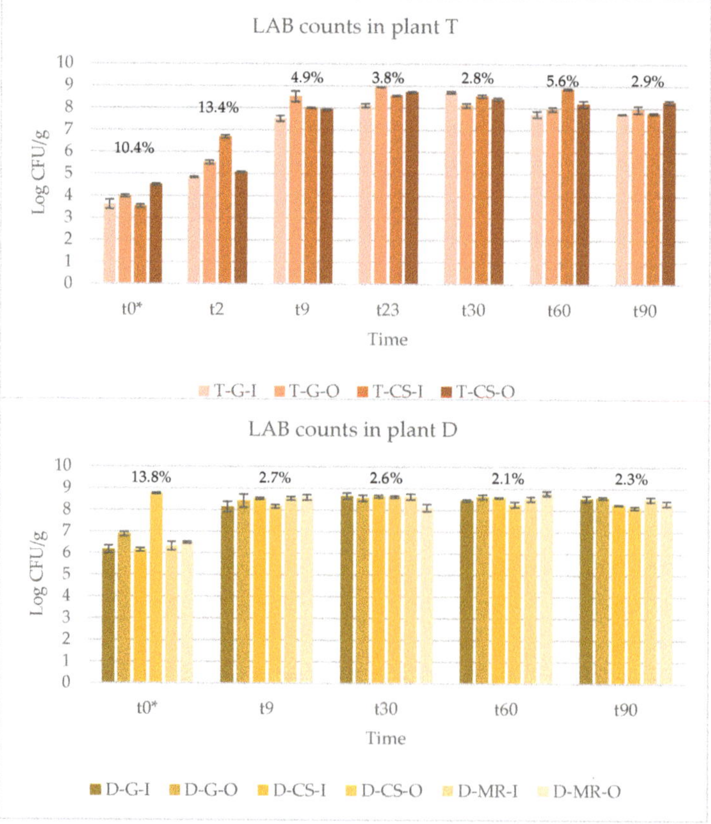

Figure 1. LAB development during ripening in the different trials. Industrial plant (D) and traditional plant (T) whose meats derived from pigs of three breeds, Goland (G), Cinta Senese (CS), and Mora Romagnola (MR) that were indoor (I) or outdoor (O) housed. The coefficient of variation among trials per time is shown above histograms.

3.2. Dynamics of the Detected Species

In order to obtain information regarding the species involved in the fermentation and their relative concentration, DGGE fingerprints were obtained from cells collected in bulk from the count plates for each dilution, after colony enumeration. From the DGGE fingerprints (Figure 2), the following species, as responsible for the fermentation of sausages, were detected: *Lb. sakei*, *Lb. casei*, *Lb. cuvatus*, *Lb. graminis*, *Lactococcus garviae*, *Leuconostoc mesenteoides*, and *Pediococcus pentosaceus*.

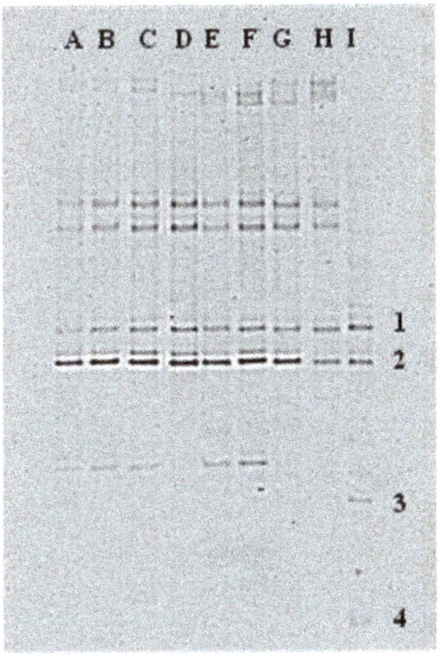

Figure 2. Examples of PCR-DGGE fingerprints obtained from the bulk suspensions. Lines A, B, C, D, E, F, G correspond to the samples coming from samples T-CS-I. Line I correspond to the standard used as a control: 1—*Lb. curvatus*, DSM 20019; 2—*Lb. sakei*, DSM 6333; 3—*Lb. brevis*, DSM 20054; 4—*Lb. casei*, DSM 20011.

In particular, *Lb. sakei* constituted the predominant specie over the monitored period. In this study, it has been therefore confirmed that *Lb. sakei* remains the overriding species in fermented meat products, as demonstrated by several authors, who reported percentages of isolation starting from 42% in Italian-Greek suasages, increasing to a percentage of 76%–89% in Spanish sausages and reaching the 100% in French sausages [21,27,51–54].

In samples from the industrial plant (plant D, Table 2) a limited number of species, ranging from two to four, were detected in the different trials. *Lb. sakei* and *P. pentosaceus* were present in all monitored fermentations and showed the highest counts during the entire fermentation process. Considering that in these trials *P. pentosaceus* was inoculated as starter culture at a level of about 10^6 CFU/g, the results highlighted the capability of *Lb. sakei* to dominate in this peculiar ecological niche being the more adaptable species at the specific environmental conditions of fermented meats. In fact, although from the beginning of fermentation high counts of *P. pentosaceus* were present, *Lb. sakei* was able to compete and achieve comparable concentrations, remaining almost stable throughout the fermentation process. The development of *Lb. sakei* was favoured by the temperature of the first stage of ripening (22 °C), which is more adapted to *Lb. sakei* than *P. pentosaceus*. Indeed, the choice to use a first ripening temperature lower than the optimal temperature for *P. pentosaceous* growth was determined by the aim of reducing the rate of acidification by this species, in order to obtain a more flavourful product.

Table 2. Plate dilutions at which each identified species were detected per sampling point from the industrial plant (D).

Identified Species (NCBI Accession Number)	t0	t9	t30	t60	t90
	Plate Dilution at Which the Identified Species were Detected				
D-G-I					
Lactobacillus sakei (NR_113821.1)	10^{-2}	n.d.	n.d.	n.d.	10^{-6}
Lactobacillus curvatus (NR_113334.1)	n.d.	n.d.	n.d.	n.d.	10^{-7}
Lactobacillus casei (NR_041893.1)	n.d.	n.d.	10^{7}	n.d.	10^{-7}
Pediocossus pentosaceus (NR_042058.1)	10^{-7}	10^{-7}	10^{-7}	10^{-7}	10^{-7}
D-G-O					
Lactobacillus sakei (NR_113821.1)	10^{-3}	10^{-6}	10^{-7}	10^{-7}	10^{-7}
Lactobacillus curvatus (NR_113334.1)	n.d.	n.d.	10^{-6}	10^{-6}	10^{-6}
Pediocossus pentosaceus (NR_042058.1)	10^{-6}	10^{-6}	10^{-7}	10^{-7}	10^{-7}
D-CS-I					
Lactobacillus sakei (NR_113821.1)	10^{-2}	10^{-6}	10^{-7}	10^{-7}	10^{-6}
Pediocossus pentosaceus (NR_042058.1)	10^{-5}	10^{-7}	10^{-7}	10^{-7}	10^{-7}
D-CS-O					
Lactobacillus sakei (NR_113821.1)	n.d.	n.d.	n.d.	10^{-5}	10^{-4}
Lactobacillus curvatus (NR_113334.1)	n.d.	n.d.	n.d.	10^{-5}	10^{-5}
Pediocossus pentosaceus (NR_042058.1)	10^{-6}	10^{-7}	10^{-7}	10^{-6}	10^{-7}
D-MR-I					
Lactobacillus sakei (NR_113821.1)	n.d.	n.d.	n.d.	10^{-6}	10^{-4}
Pediocossus pentosaceus (NR_042058.1)	10^{-5}	10^{-7}	10^{-7}	10^{-7}	10^{-6}
D-MR-O					
Lactobacillus sakei (NR_113821.1)	n.d.	10^{-7}	10^{-7}	10^{-7}	10^{-7}
Lactobacillus curvatus (NR_113334.1)	n.d.	10^{-7}	10^{-7}	10^{-7}	10^{-7}
Pediocossus pentosaceus (NR_042058.1)	10^{-6}	10^{-7}	10^{-7}	10^{-7}	10^{-7}

n.d.: not detected; D-CS-I: Industrial plant—Cinta Senese meat—indoor breeding system; D-CS-O: Industrial plant—Cinta Senese meat—outdoor breeding system; D-G-I: Industrial plant—Goland meat—indoor breeding system; D-G-O: Industrial plant—Goland meat—outdoor breeding system.

Other than these two species, also *Lb. curvatus* was found, confirming to be one of the most recurrently species, suggesting that, in association with *Lb. sakei*, it predominantly leads meat fermentations [33,55–58]. In fact, *Lb. curvatus* was present during the ripening of industrial products, except for those obtained from Cinta Senese and Mora Romagnola indoor breeding system productions (D-CS-I, D-MR-I).

In the artisanal products (plant T, Table 3), this species had a more constant trade, from the stuffing to the end of ripening, where it amounted to the highest dilutions (10^{-6}–10^{-7}). Instead, in the industrial productions (D), it was constantly founded in Goland outdoor, Cinta Senese indoor and Mora Romagnola outdoor breeding system products (D-G-O, D-CS-I, D-MR-O), whereas in the other cases *Lb. sakei* was only detected at the end of the ripening.

Table 3. Plate dilutions at which each identified species were detected per sampling point from the artisanal plant (T).

Identified Species (NCBI Accession Number)	t0	t2	t9	t23	t30	t60	t90
	Plate Dilution at Which the Identified Species were Detected						
			T-G-I				
Lactobacillus sakei (NR_113821.1)	10^{-2}	10^{-5}	10^{-5}	10^{-5}	10^{-6}	10^{-6}	n.d.
Lactobacillus curvatus (NR_113334.1)	10^{-2}	10^{-4}	10^{-5}	10^{-5}	10^{-6}	10^{-6}	10^{-6}
Lactobacillus casei (NR_041893.1)	10^{-2}	10^{-4}	10^{-6}	n.d.	n.d.	n.d.	n.d.
Lactobacillus graminis (NR_042438.1)	n.d.	n.d.	n.d.	n.d.	n.d.	n.d.	10^{-6}
Lactococcus garviae (KU898985.1)	n.d.	10^{-3}	10^{-6}	10^{-7}	n.d.	n.d.	n.d.
Leuconostoc mesenteroides (DQ297412.1)	n.d.	n.d.	n.d.	10^{-7}	10^{-6}	10^{-6}	n.d.
Pediocossus pentosaceus (NR_042058.1)	n.d.	n.d.	n.d.	10^{-5}	10^{-5}	n.d.	n.d.
			T-G-O				
Lactobacillus sakei (NR_113821.1)	10^{-2}	10^{-4}	10^{-7}	10^{-7}	10^{-6}	10^{-6}	10^{-6}
Lactobacillus curvatus (NR_113334.1)	10^{-2}	10^{-4}	10^{-7}	10^{-7}	10^{-6}	n.d.	n.d.
Leuconostoc mesenteroides (DQ297412.1)	10^{2}	n.d.	n.d.	n.d.	n.d.	n.d.	n.d.
			T-CS-I				
Lactobacillus sakei (NR_113821.1)	10^{-2}	10^{-6}	10^{-6}	10^{-7}	10^{-7}	10^{-7}	10^{-7}
Lactobacillus curvatus (NR_113334.1)	n.d.	10^{-6}	10^{-6}	10^{-7}	10^{-7}	10^{-7}	10^{-7}
Lactococcus garviae (KU898985.1)	10^{-3}	10^{-4}	n.d.	n.d.	n.d.	n.d.	n.d.
Leuconostoc mesenteroides (DQ297412.1)	10^{-2}	10^{-6}	n.d.	10^{-5}	10^{-7}	n.d.	n.d.
			T-CS-O				
Lactobacillus sakei (NR_113821.1)	n.d.	10^{-3}	10^{-7}	10^{-7}	10^{-7}	10^{-6}	10^{-7}
Lactobacillus curvatus (NR_113334.1)	n.d.	10^{-3}	10^{-7}	10^{-7}	10^{-7}	10^{-6}	10^{-7}
Lactococcus garviae (KU898985.1)	n.d.	10^{-3}	n.d.	n.d.	n.d.	n.d.	n.d.
Leuconostoc mesenteroides (DQ297412.1)	10^{-3}	n.d.	n.d.	n.d.	n.d.	n.d.	n.d.

n.d.: not detected; D-CS-I: Industrial plant—Cinta Senese meat—indoor breeding system; D-CS-O: Industrial plant—Cinta Senese meat—outdoor breeding system; D-G-I: Industrial plant—Goland meat—indoor breeding system; D-G-O: Industrial plant—Goland meat—outdoor breeding system; D-MR-I: Industrial plant—Mora Romagnola meat—indoor breeding system; D-MR-O: Industrial plant—Mora Romagnola meat, outdoor breeding system.

Considering the artisanal plant, *Lb. sakei* and *Lb. curvatus* were detected in all the monitored products, especially after the ninth day of ripening, always at the higher dilutions (10^{-5}–10^{-7}), once more confirming their dominant role also in spontaneous fermentations. *Lb. casei* was the least common species identified in the sausages from both the plants, but it appeared at high concentration at 90 days for the industrial D-G-I trial, and only from zero to nine days in the artisanal spontaneous fermentation T-G-I. Considering the relevant aspects related to the probiotic characteristics of this specie, their presence as spontaneous microflora in this products supports gives clear evidence that this species could be used as starter culture to provide the market with a new nutritional alternative that increases the variety of existing probiotic products, as well as increasing the added value of sausages [59,60].

Other species were detected during the process, but they were involved only in some of the considered productions. The equilibrium among the different species as well as the presence or absence of some of them are responsible for the production of different aroma characteristics in the final product; characteristics that are even more evident in the comparison of artisanal rather than industrial products.

Leuconostoc mesenteroides was identified in each of the artisanal productions, till the 60th day of ripening. *Lactococcus garviae* was detected in each monitored sausage, except for the Goland outdoor breeding system products (T-G-O). *Lb. casei*, *Lb. graminis*, and *Pediococcus pentosaceus* were only found in the Goland indoor breeding system products (T-G-I).

Lb. curvatus was constantly participating in artisanal productions (T), especially in sausages obtained from Cinta Senese breed, reaching the highest dilutions (10^{-7}) at the end of ripening. Products from Goland breed showed some differences depending on the breeding system: in T-G-I, *Lb. curvatus* was always detected during ripening, whereas in T-G-O, this species was founded only in the first step of the process. Fermented sausages from the industrial plan (D) showed an irregular presence of *Lb. curvatus* during ripenings. In some cases, it was absent, in other ones, it was detected only at the last steps of ripening, and in other trials, this species was found all long the production process. *Pediococcus pentosaceus* had a specific role in industrial sausage production (D) due to its use as a microbial starter; indeed, in this case, this species was detected, from the stuffing to the end of the ripening, at the greater dilutions. In the artisanal productions (T), *P. pentosaceus* was founded only in sausages obtained by Goland pork breed at the sampling point named t30. *Leuconostoc mesenteroides* was observed only in the artisanal products (T), with an irregular presence. In samples from the outdoor breeding system, this species was detected at the moment of stuffing, in low dilutions (10^{-2}–10^{-3}); in samples from Cinta senese indoor breeding system (T-CS-I), *L. mesenteroides* was present till t30 sampling time, in increasing dilutions (from 10^{-2} to 10^{-7}); in sausages from Goland indoor breeding system (T-G-I), this species appeared from t23 to t60 sampling points, in high dilutions (10^{-6}–10^{-7}). *Lb. casei* was detected in both plants (T, D), but only in products obtained from Goland indoor breeding system, during the first step of ripening. *Lactococcus garviae* was detected in artisanal products (T) deriving from indoor breeding, at the first sampling times and in the highest dilutions (10^{-7}). *Lactococcus garvieae* was rarely associated to fermented sausages [15,51] and it has been classified as an emerging pathogen found in animals (cows, buffalos, farmed fish), as well as in human clinical isolates. Since then, many papers stated that this bacterium is a human gastrointestinal commensal or transient bacterium, which can cause a variety of infections [61]. *Lb. graminis* was present only in artisanal fermented sausages (T), from Goland indoor breeding system, during the middle period of ripening (t2–t23), in high dilutions (10^{-6}–10^{-7}). This species was at first isolated at the end fermentation of grass silage. Noteworthy that, this heterofermentative species, which is not frequently isolated from meat products, has a low DNA-DNA homology value with the type strains of *L. sakei* and *L. curvatus*, but they were similar with respect to D(-)- and L(+)- lactate formation, Rf-values of the D- and L-LDH, G+C content of the DNA and the L-Lys-D-Asp murein type [62].

3.3. Fingerprint Analyses

Statistical analysis was also applied to the DNA fingerprints obtained by PCR-DGGE. The results of the cluster analysis using the Pearson product-moment correlation coefficient were shown in Figure 3. A similarity coefficient of 53% was arbitrarily selected. As shown (Figure 3), a total of 35 clusters were discerned. Except for clusters XV, XVII, XXVII and XXIX, all the others grouped samples produced in the same plant, clearly suggesting that they were plant-specific clusters. Half of the groups included only two, three, or four samples, the others are composed of six or more samples. The most relevant clusters were the number XIII (17 fingerprints, plant T), XX (16 fingerprints, plant D), XXV (13 fingerprints, plant D) and XXVII (18 fingerprints, plant T). Furthermore, 22 clusters were found to be also breed-specific. Only nine samples did not belong to any cluster. Subsequently, a different cluster analysis based on the Dice correlation coefficient (60% coefficient of similarity) was performed, and the results are presented in Figure 4. The second analysis confirmed the previous results: 20 clusters of 33 resulted to be plant-specific, 15 of them included two, three, or four samples and the others had five or more samples. The most relevant clusters are the number XI (15 fingerprints, plant D), XXIV (14 fingerprints, plant D) and XXVIII (14 fingerprints, plant D). This analysis confirmed also the presence of breed-specific clusters (22 of 33) and 12 samples did not belong to any cluster.

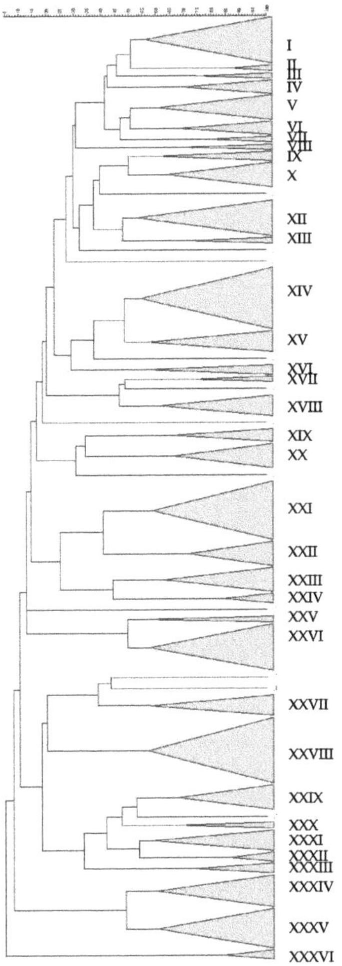

cluster	number of samples per cluster	cluster composition				
		T		D		
		G	CS	G	CS	MR
I	13	-	-	-	1	12
II	2	2	-	-	-	-
III	2	2	-	-	-	-
IV	4	4	-	-	-	-
V	7	1	6	-	-	-
VI	4	4	-	-	-	-
VII	2	2	-	-	-	-
VIII	2	-	-	-	-	2
IX	3	2	1	-	-	-
X	7	-	-	-	-	7
XI	10	-	-	-	-	10
XII	2	-	-	-	-	2
XIII	17	12	5	-	-	-
XIV	6	2	4	-	-	-
XV	3	-	1	2	-	-
XVI	2	2	-	-	-	-
XVII	6	-	5	1	-	-
XVIII	4	-	-	2	2	-
XIX	7	-	-	5	2	-
XX	16	-	-	-	13	3
XXI	7	-	-	-	7	-
XXII	7	-	-	-	7	-
XXIII	3	-	3	-	-	-
XXIV	2	-	-	-	2	-
XXV	13	-	-	-	13	-
XXVI	6	-	4	1	1	-
XXVII	18	4	14	-	-	-
XXVIII	7	6	-	1	-	-
XXIX	2	-	2	-	-	-
XXX	6	6	-	-	-	-
XXXI	3	-	3	-	-	-
XXXII	3	-	3	-	-	-
XXXIII	9	-	-	9	-	-
XXXIV	11	-	-	8	3	-
XXXV	3	-	-	-	-	3

Legend: T, artisanal plant; D, industrial plant; G, Goland breed; CS, Cinta senese breed; MR, Mora Romagnola breed.

Figure 3. Cluster analysis using Pearson product–moment correlation coefficients and unweighted pair group method using an arithmetic average (UPGMA) of the profiles obtained from the fingerprinting analysis of the different fermentations. A similarity coefficient of 53% was arbitrarily chosen. Identified clusters are indicated with roman numerals. Cluster composition is defined in the table within the figure.

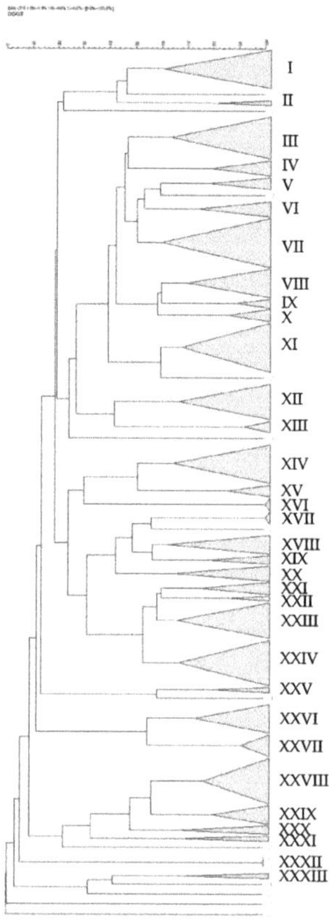

cluster	number of isolates per clusters	cluster composition				
		T		D		
		G	CS	G	CS	MR
I	12	7	5	-	-	-
II	2	2	-	-	-	-
III	13	10	1	2	-	-
IV	5	-	5	-	-	-
V	4	-	4	-	-	-
VI	5	-	5	-	-	-
VII	15	-	5	-	10	-
VIII	12	3	9	-	-	-
IX	3	-	3	-	-	-
X	4	-	4	-	-	-
XI	15	-	-	11	4	-
XII	11	10	1	-	-	-
XIII	4	-	4	-	-	-
XIV	12	-	-	-	-	12
XV	4	-	-	-	-	4
XVI	4	-	-	-	4	-
XVII	4	-	4	-	-	-
XVIII	6	2	4	-	-	-
XIX	3	-	3	-	2	-
XX	5	4	-	-	-	1
XXI	4	-	-	4	-	-
XXII	2	-	-	-	2	-
XXIII	11	-	-	-	11	-
XXIV	14	-	-	13	1	-
XXV	2	-	-	-	13	-
XXVI	9	-	-	-	-	9
XXVII	7	-	-	-	-	7
XXVIII	14	-	-	-	10	4
XXIX	6	-	-	-	6	-
XXX	3	-	-	-	3	-
XXXI	2	-	-	-	-	2
XXXII	2	2	-	-	-	-
XXXIII	2	1	1	-	-	-

Legend: T, artisanal plant; D, industrial plant; G, Goland breed; CS, Cinta senese breed; MR, Mora Romagnola breed.

Figure 4. Cluster analysis using Dice product–moment correlation coefficients and unweighted pair group method using an arithmetic average (UPGMA) of the profiles obtained from the fingerprinting analysis of the different fermentations. A similarity coefficient of 60% was arbitrarily chosen. Identified clusters are indicated with roman numerals. Cluster composition is defined in the table within the figure.

The congruence of the data obtained with the use of both Pearson and Dice product-moment-correlation coefficients proves the significance of the result, demonstrating the existence of plant- and breed-specificity among the monitored fermentations. If the plant effect could be simply explained as an environmental selection during the time of the microflora, which forms a biofilm on structures and equipment, the breed effect could result from the different chemical-physical characteristics of the raw meats, more precisely the type and quantity of intramuscular fat, which are strictly correlated to pork breed. Actually, during grinding and/or mixing, the rise in temperature due to friction can cause the melting of low melting point fats, with consequent difficulties in the dehydration process during drying and ripening, leading to irregular changes in water activity (Aw) that influence the microbial species equilibrium. Finally, most of the breed-specific clusters, obtained by both the correlation coefficients, showed homogeneous samples also concerning the different breeding systems. Therefore, it's possible to say that the microbial ecology is influenced also by the breeding system, as speculated in the introduction. However, it is possible that other factors may intervene and change the microflora characterizing the animal and, consequently, the meat. In our experiment, for example, seasonality was not taken into account, to avoid adding too many variables. It should be considered, however, that, among others, humidity and environmental temperature can influence both the environmental microflora and the state of health of the animal, with consequent modification of the intestinal microflora, leading to variations of the isolated species. Furthermore, it is well known that small variations can always be present, considering the differences between the autochthonous and allochthonous microflora, but the former will always have a character of stability, due to the fact that being more adapted to that particular environment, it will also be more incisive than the allochthonous ones. These results confirm the data from previous studies, which stated that the production technologies applied in a specific production plant and the pork breed select a characteristic and specific microbial ecology [6,7,34,63].

Author Contributions: Conceptualization, L.I.; methodology, L.I.; validation, L.I. and G.C.; formal analysis, M.C. and L.I.; investigation, A.M. and D.B.; resources, L.I. and G.C.; data curation, L.I. and M.C.; writing—original draft preparation, A.M. and L.I.; writing—review and editing, L.I. and G.C.; visualization, L.I.; supervision, L.I.; project administration, L.I.; funding acquisition, L.I. and G.C. All authors have read and agreed to the published version of the manuscript.

Funding: This research was funded by ERSA-FVG.

Conflicts of Interest: The authors declare no conflict of interest.

References

1. Cocolin, L.; Rantsiou, K.; Iacumin, L.; Urso, R.; Cantoni, C.; Comi, G. Study of the ecology of fresh sausages and characterization of populations of lactic acid bacteria by molecular methods. *Appl. Environ. Microbiol.* **2004**, *70*, 1883–1894. [CrossRef] [PubMed]
2. Iacumin, L.; Manzano, M.; Comi, G. Catalase positive cocci in fermented sausages: Variability due to different pork breeds, breeding system and sausage production technology. *Food Microbiol.* **2012**, *29*, 178–186. [CrossRef] [PubMed]
3. Cocolin, L.; Manzano, M.; Cantoni, C.; Comi, G. Denaturing gradient gel electrophoresis analysis of the 16S rRNA gene V1 region to monitor dynamic changes in the bacterial population during fermentation of Italia sausages. *Appl. Environ. Microbiol.* **2001**, *67*, 5113–6282. [CrossRef]
4. Lücke, F.K. Fermented sausages. In *Microbiology of Fermented Food*; Wood, B.J.B., Ed.; Elsevier Applied Science: New York, NY, USA, 1985.
5. Iacumin, L.; Osualdini, M.; Bovolenta, S.; Boscolo, D.; Chiesa, L.; Panseri, S.; Comi, G. Microbial, chemico-physical and volatile aromatic compounds characterization of Pitina PGI, a peculiar sausage-like product of North East Italy. *Meat Sci.* **2020**, *162*, 108081. [CrossRef]
6. Coppola, R.; Marconi, E.; Rossi, F.; Dellaglio, F. Artisanal production of Naples-types salami: Chamical and microbiological aspects. *Ital. J. Food Sci.* **1995**, *7*, 57–62.
7. Coppola, R.; Iorizzo, M.; Saotta, R.; Sorrentino, E.; Grazia, L. Characterization of micrococci and staphylococci isolated from sopressata molisana, a Southern Italy fermented sausage. *Food Microbiol.* **1997**, *14*, 47–53. [CrossRef]
8. Iacumin, L.; Comi, G.; Cantoni, C.; Cocolin, L. Ecology and dynamics of coagulase-negative cocci isolated from naturally fermented Italian sausages. *Syst. Appl. Microbiol.* **2006**, *29*, 480–486. [CrossRef]
9. Lücke, F.K.; Hechelmann, H. Starter cultures for dry sausages and raw ham. *Fleischwirt* **1987**, *66*, 1505–1508.
10. Demeyer, D.I.; Verplaetse, A.; Gistelink, M. Fermentation of meat: An integrated process. *Belg. J. Food Chem. Biotechnol.* **1986**, *41*, 131–140.
11. Baka, A.M.; Papavergou, E.J.; Pragalaki, T.; Bloukas, J.G.; Kotzekidou, P. Effect of selected autochthonous starter cultures on processing and quality characteristics of Greek fermented sausages. *LWT Food Sci. Technol.* **2011**, *44*, 54–61. [CrossRef]
12. Geisen, R.; Lüke, F.K.; Krockel, L. Starter and protective cultures for meat and meat products. *Fleischwirt* **1992**, *72*, 894–898.
13. Selgas, M.D.; Sanz, B.; Ordonez, J.A. Selected characteristics of micrococci isolated from Spanish dry fermented sausages. *Food Microbiol.* **1988**, *5*, 185–193. [CrossRef]
14. Varnam, A.H.; Sutherland, J.P. *Meat and Meat Products*; Chapman & Hall: London, UK, 1995.

15. Jones, R.J.; Hussein, H.M.; Zaigorec, M.; Brightnell, G.; Tagg, J.R. Isolation of lactic acid bacteria with inhibitory activity against pathogens and spoilage organisms associated with fresh meat. *Food Microbiol.* **2008**, *25*, 228–234. [CrossRef] [PubMed]
16. Rantsiou, K.; Urso, R.; Iacumin, L.; Cantoni, C.; Cattaneo, P.; Comi, G.; Cocolin, L. Culture-Dependent and -Independent Methods to Investigate the Microbial Ecology of Italian Fermented Sausages. *Appl. Environ. Microbiol.* **2005**, *4*, 1977–1986. [CrossRef]
17. Comi, G.; Iacumin, L. The use of bioprotective cultures. In *Strategies to Obtaining Healthier Foods*; Nova Science Publishers, Inc. Hauppauge: New York, NY, USA, 2017.
18. Holzapfel, R.W.; Geisen, R.; Schillinger, U. Biological preservation of foods with reference to protective cultures, becteriocins and food-grade enzymes. *Int. J. Food Microbiol.* **1995**, *24*, 343–362. [CrossRef]
19. Stiles, M.E. Biopreservation by lactic acid bacteria. *Antonie Von Leeuwenhoek* **1996**, *70*, 331–345. [CrossRef] [PubMed]
20. Ben Omar, N.; Ampe, F. Microbial community dynamics during production of the Mexican fermented maize dough pozol. *Appl. Eviron. Microbiol.* **2000**, *66*, 3664–3673. [CrossRef] [PubMed]
21. Coppola, S.; Mauriello, G.; Aponte, M.; Moschetti, G.; Villani, F. Microbial succession during ripening of Naples-type salami, a southern Italian fermented sausage. *Meat Sci.* **2000**, *56*, 321–329. [CrossRef]
22. Fontana, C.; Vignolo, G.; Cocconcelli, P.S. PCR–DGGE analysis for the identification of microbial populations from Argentinean dry fermented sausages. *J. Microbiol. Methods* **2005**, *63*, 254–263. [CrossRef]
23. Iacumin, L.; Cecchini, F.; Manzano, M.; Osualdini, M.; Boscolo, D.; Orlic, S.; Comi, G. Description of the microflora of sourdoughs by culture-dependent and culture-independent methods. *Food Microbiol.* **2009**, *26*, 128–135. [CrossRef]
24. Klijn, N.; Weerkamp, A.H.; de Vos, W.M. Identification of mesophilic lactic acid bacteria by using polymerase chain reaction-amplified variable regions of 16S rRNA and specific DNA probes. *Appl. Environ. Microbiol.* **1991**, *57*, 3390–3393. [CrossRef] [PubMed]
25. Franciosa, I.; Alessandria, V.; Dolci, P.; Rantsiou, K.; Cocolin, L. Sausage fermentation and starter cultures in the era of molecular biologymethods. *Int. J. Food Microbiol.* **2018**, *279*, 26–32. [CrossRef] [PubMed]
26. Andrighetto, C.; Zampese, L.; Lombardi, A. RAPD-PCR characterization of lactobacilli isolated from artisanal meat plants and traditional fermented sausages of Veneto region. *Lett. Appl. Microbiol.* **2001**, *33*, 26–30. [CrossRef] [PubMed]
27. Papamanoli, E.; Tzanetakis, N.; Litopoulou-Tzanetaki, E.; Kotzekidou, P. Characterization of lactic acid bacteria isolated from a Greek dry-fermented sausage in respect of their technological and probiotic properties. *Meat Sci.* **2003**, *65*, 859–867. [CrossRef]
28. Leroy, F.; de Vuyst, L. Temperature and pH Conditions That Prevail during Fermentation of Sausages Are Optimal for Production of the Antilisterial Bacteriocin Sakacin K. *Appl. Environ. Microbiol.* **1999**, *65*, 974–981. [CrossRef] [PubMed]
29. Leroy, F.; de Vuyst, L. The Presence of Salt and a Curing Agent Reduces Bacteriocin Production byLactobacillus sakei CTC 494, a Potential Starter Culture for Sausage Fermentation. *Appl. Environ. Microbiol.* **1999**, *65*, 5350–5356. [CrossRef]
30. Martin, B.; Garriga, M.; Hugas, M.; Bover-Cid, S.; Veciana-Nogués, M.T.; Aymerich, T. Molecular, technological and safety characterization of Gram-positive catalase-positive cocci from slightly fermented sausages. *Int. J. Food Microbiol.* **2006**, *107*, 148–158. [CrossRef]
31. Ravyts, F.; Steen, L.; Goemaere, O.; Paelinck, H.; De Vuyst, L.; Leroy, F. The application of staphylococci with flavour-generating potential is affected by acidification in fermented dry sausages. *Food Microbiol.* **2010**, *27*, 945–954. [CrossRef]
32. Fista, G.A.; Bloukas, J.G.; Siomos, A.S. Effect of leek and onion processing and quality characteristics of Greek traditional sausages. *Meat Sci.* **2004**, *68*, 163–172. [CrossRef]
33. Iacumin, L.; Comi, G.; Cantoni, C.; Cocolin, L. Molecular and technological characterization of *Staphylococcus xylosus* isolated from naturally fermented Italian sausages by RAPD, Rep-PCR and Sau-PCR analysis. *Meat Sci.* **2006**, *74*, 281–288. [CrossRef]
34. Morot-Bizot, S.; Talon, R.; Leroy-Setrin, S. Development of specific PCR primers for a rapid and accurate identification of *Staphylococcus xylosus*, a species used in food fermentation. *J. Microbiol. Methods* **2003**, *55*, 279–286. [CrossRef]

35. Verluyten, J.; Leroy, F.; Vuyst, L. Effects of Different Spices Used in Production of Fermented Sausages on Growth of and Curvacin A Production by *Lactobacillus curvatus* LTH 1174. *Appl. Environ. Microbiol.* **2004**, *70*, 4807–4813. [CrossRef] [PubMed]
36. Mu, C.; Bian, G.; Su, Y.; Zhu, W. Differential Effects of Breed and Nursing on Early-Life Colonic Microbiota and Immune Status as Revealed in a Cross- Fostering Piglet. *Appl. Environ. Microbiol.* **2019**, *85*, e02510–e02518. [CrossRef] [PubMed]
37. Yang, H.; Huang, X.; Fang, S.; Xin, W.; Huang, L.; Chen, C. Uncovering the composition of microbial community structure and metagenomics among three gut locations in pigs with distinct fatness. *Sci. Rep.* **2016**, *6*, 27427. [CrossRef] [PubMed]
38. Tan, Z.; Yang, T.; Wang, Y.; Xing, K.; Zhang, F.; Zhao, X.; Ao, H.; Chen, S.; Liu, J.; Wanet, C. Metagenomic analysis of cecal microbiome identified microbiota and functional capacities associated with feed efficiency in landrace finishing pigs. *Front. Microbiol.* **2017**, *8*, 1546. [CrossRef] [PubMed]
39. Bian, G.R.; Ma, S.Q.; Zhu, Z.G.; Su, Y.; Zoetendal, E.G.; Mackie, R.; Liu, J.H.; Mu, C.L.; Huang, R.H.; Smidt, H.; et al. Age, introduction of solid feed and weaning are more important determinants of gut bacterial succession in piglets than breed and nursing mother as revealed by a reciprocal cross-fostering model. *Environ. Microbiol.* **2016**, *18*, 1566–1577. [CrossRef] [PubMed]
40. Monin, G.; Mejenes-Quijano, A.; Talmant, A.; Sellier, P. Influence of breed and muscle metabolic type on muscle glycolytic potential and meat pH in pigs. *Meat Sci.* **1987**, *20*, 149–158. [CrossRef]
41. Zentek, J.; Ferrara, F.; Pieper, R.; Tedin, L.; Meyer, W.; Vahjen, W. Effects of dietary combinations of organic acids and medium chain fatty acids on the gastrointestinal microbial ecology and bacterial metabolites in the digestive tract of weaning piglets. *J. Anim. Sci.* **2013**, *91*, 3200–3210. [CrossRef]
42. Nguyen, D.H.; Lee, K.Y.; Tran, H.N.; Kim, I.H. Effect of a protected blend of organic acids and medium-chain fatty acids on growth performance, nutrient digestibility, blood profiles, meat quality, faecal microflora, and faecal gas emission in finishing pigs. *Can. J. Anim. Sci.* **2019**, *99*, 448–455. [CrossRef]
43. Albrecht, A.; Hebel, M.; Mittler, M.; Hurck, C.; Kustwan, K.; Heitkonig, B.; Bitschinski, D.; Kreyenschmidt, J. Influence of Different Production Systems on the Quality and Shelf Life of Poultry Meat: A Case Study in the German Sector. *J. Food Qual.* **2019**, 3718057. [CrossRef]
44. Gounadaki, A.S.; Skandamis, P.N. Microbial ecology of food contact surfaces and products of small-scale facilities producing traditional sausages. *Food Microbiol.* **2008**, *25*, 313–323. [CrossRef] [PubMed]
45. Duffy, E.A.; Belk, K.E.; Sofos, J.N.; Bellinger, G.R.; Pape, A.; Smith, G.C. Extent of Microbial Contamination in United States Pork Retail Products. *J. Food Prot.* **2001**, *64*, 172–178. [CrossRef]
46. Saide-Albornoz, J.J.; Knipe, C.L.; Murano, E.A.; Beran, G.W. Contamination of Pork Carcasses during Slaughter, Fabrication, and Chilled Storage. *J. Food Prot.* **1995**, *58*, 993–997. [CrossRef] [PubMed]
47. Altschul, S.F.; Madden, T.L.; Shaffer, A.A.; Zhang, J.; Zhang, Z.; Miller, W.; Lipman, D.J. Gapped BLAST and PSI-BLAST: A new generation of protein database search programs. *Nucleic Acids Res.* **1997**, *25*, 3389–3402. [CrossRef] [PubMed]
48. White, H. *Asymptotic Theory for Econometricians*; Academic Press: Orlando, FL, USA, 1980.
49. Dytham, C. *Choosing and Using Statistics: A Biologist's Guide*, 2nd ed.; Blackwell: Malden, MA, USA, 2003.
50. Vauterin, L.; Vauterin, P. Computer-aided objective comparison of electrophoretic patterns for grouping and identification of microorganisms. *Eur. Microbiol.* **1992**, *1*, 37–41.
51. Ammor, S.; Rachman, C.; Chaillou, S.; PreÅLvost, H.; Dousset, X.; Zagorec, M.; Dufour, E.; Chevallier, I. Phenotypic and genotypic identification of lactic acid bacteria isolated from a small-scale facility producing traditional dry sausages. *Food Microbiol.* **2005**, *22*, 373–382. [CrossRef]
52. Comi, G.; Urso, R.; Iacumin, L.; Rantsiou, K.; Cattaneo, P.; Cantoni, C.; Cocolin, L. Characterization of naturally fermented sausages produced in North East of Italy. *Meat Sci.* **2005**, *69*, 381–392. [CrossRef]
53. Greco, M.; Mazette, R.; De Santis, E.P.L.; Corona, A.; Cosseddu, A.M. Evolution and identification of lactic acid bacteria isolated during the ripening of Sardinian sausages. *Meat Sci.* **2005**, *69*, 733–739. [CrossRef]
54. Talon, R.; Leroy, S.; Lebert, I. Microbial ecosystems of traditional fermented meat products: The importance of indigenous starters. *Meat Sci.* **2007**, *77*, 55–62. [CrossRef]
55. Casaburi, A.; Aristoy, M.C.; Cavella, S.; Di Monaco, S.; Di Monaco, R.; Ercolini, D.; Toldra, F.; Villani, F. Biochemical and sensory characteristics of traditional fermented sausage of Vallo di Diano (Southern Italy) as affected by the use of starter cultures. *Meat Sci.* **2006**, *76*, 295–307. [CrossRef]

56. Kostinek, M.; Specht, I.; Edward, V.A.; Schillinger, U.; Hertel, C.; Holzapfel, W.H.; Franz, C.M. Diversity and technological proprities of predominant lactic acid bacteria from fermented cassava used for the preparation of Gari, a traditional African food. *Syst. Appl. Microbiol.* **2005**, *6*, 527–540. [CrossRef] [PubMed]
57. Torriani, S.; Di Bucchianico, R.; Pattarini, F.; Zabeo, G.; Dell'Aglio, F. Presence and biotechnological charaterization of lactic acid bacteria and micrococcaceaea strains in Abruzzo traditional raw dry sausages. *Ind. Delle Conserve* **1994**, *69*, 3–9.
58. Villani, F.; Casaburi, A.; Pennacchia, C. Microbial ecology of the soppressata of Vallo di Diano, a traditional dry fermented sausage from southern Italy, and in vitro and in situ selection of autochthonous starter cultures. *Appl. Environ. Microbiol.* **2007**, *73*, 5453–5463. [CrossRef] [PubMed]
59. Mosquera, C.; Donoso, E.; Valenzuela, Y.M.; Arizaga, R.; Salous, A.E. Effect of probiotic and prebiotic in the formulation and elaboration of sausage as an alternative of consumption. *Int. J. Res. Pharm. Sci.* **2019**, *10*, 2781–2785. [CrossRef]
60. Sidira, M.; Mitropoulou, G.; Galanis, A.; Kanellaki, M.; Kourkoutas, Y. Effect of sugar content on quality characteristics and shelf-life of probiotic dry-fermented sausages produced by free or immobilized *Lactobacillus casei* ATCC 393. *Foods* **2019**, *8*, 219. [CrossRef]
61. Mehmeti, I.; Muji, S.; Diep, D.B.; Nesa, I.F. High frequency of the potential pathogen *Lactococcus garvieae* in raw milk from Kosovo. *Food Control.* **2015**, *53*, 189–194. [CrossRef]
62. Beck, R.; Weiss, N.; Winter, J. *Lactobacillus graminis* sp. nov., a New Species of Facultatively Heterofermentative Lactobacilli Surviving at Low pH in Grass Silage. *Syst. Appl. Microbiol.* **1988**, *10*, 279–283. [CrossRef]
63. Lebert, I.; Leroy, S.; Giammarinaro, P.; Lebert, A.; Charcornac, J.P.; Bover-Cid, S.; Vidal-Carou, M.C.; Talon, R. Diversity of microorganisms in the environment and dry fermented sausages of small traditional French processing units. *Meat Sci.* **2007**, *76*, 112–122. [CrossRef]

foods

MDPI

Article

Application of a High-Throughput Amplicon Sequencing Method to Chart the Bacterial Communities that Are Associated with European Fermented Meats from Different Origins

Emiel Van Reckem, Christina Charmpi, David Van der Veken, Wim Borremans, Luc De Vuyst, Stefan Weckx and Frédéric Leroy *

Research Group of Industrial Microbiology and Food Biotechnology (IMDO), Faculty of Sciences and Bioengineering Sciences, Vrije Universiteit Brussel, Pleinlaan 2, B-1050 Brussels, Belgium; Emiel.Van.Reckem@vub.be (E.V.R.); Christina.Charmpi@vub.be (C.C.); David.Van.Der.Veken@vub.be (D.V.d.V.); Wim.Borremans@vub.be (W.B.); Luc.De.Vuyst@vub.be (L.D.V.); Stefan.Weckx@vub.be (S.W.)

* Correspondence: frederic.leroy@vub.be; Tel.: +32-629-3612

Received: 18 August 2020; Accepted: 3 September 2020; Published: 7 September 2020

check for updates

Abstract: Insight into the microbial species diversity of fermented meats is not only paramount to gain control over quality development, but also to better understand the link with processing technology and geographical origin. To study the composition of the microbial communities, the use of culture-independent methods is increasingly popular but often still suffers from drawbacks, such as a limited taxonomic resolution. This study aimed to apply a previously developed high-throughput amplicon sequencing (HTS) method targeting the 16S rRNA and *tuf* genes to characterize the bacterial communities in European fermented meats in greater detail. The data obtained broadened the view on the microbial communities that were associated with the various products examined, revealing the presence of previously underreported subdominant species. Moreover, the composition of these communities could be linked to the specificities of individual products, in particular pH, salt content, and geographical origin. In contrast, no clear links were found between the volatile organic compound profiles of the different products and the country of origin, distinct processing conditions, or microbial communities. Future application of the HTS method offers the potential to further unravel complex microbial communities in fermented meats, as well as to assess the impact of different processing conditions on microbial consortia.

Keywords: staphylococci; lactic acid bacteria; fermented meats; high-throughput sequencing; microbiota

1. Introduction

The geographical origin of fermented foods plays an important role in conferring cultural and gastronomic meanings as well as in influencing certain product characteristics [1–3]. Naturally fermented dairy products, for instance, have been part of the culinary heritage of many communities throughout the world, showcasing a rich microbial species diversity that is linked to their origin [4]. Similarly, European fermented meat products often provide explicit references to geography [5], which parallels differences in fermentative microbiota [6]. The fermentative microorganisms consist primarily of lactic acid bacteria (LAB) and coagulase-negative staphylococci (CNS). Although many distinctive regional practices exist, key differences in processing can often be reduced to discrepancies in processing between Northern and Southern Europe [7,8]. North-European fermented meat products are usually subjected to a faster fermentation at higher temperatures, leading to a more acid (pH of

5.0 or lower) product than is the case for South-European variants [9]. They also habitually rely on a smoking step that prevents molding, whereas fermented meats in Southern Europe are commonly typified by desirable molding and extensive maturation [2,8,10,11].

Such distinct differences in processing are known to influence the growth and composition of the predominant LAB and CNS communities [6]. For instance, CNS communities in South-European fermented meats are dominated by *Staphylococcus xylosus* and *Staphylococcus equorum*, whereas in Northern Europe there is typically a much higher prevalence of *Staphylococcus carnosus* [6,8,12–15]. Furthermore, as most fermented sausages are currently produced on an industrial scale, LAB and CNS communities are also heavily influenced by potential starter culture application, which in the case of LAB mostly consists of *Latilactobacillus sakei* and *Pediococcus pentosaceus* and in the case of CNS is limited to *S. xylosus* and/or *S. carnosus* [16–18]. In contrast, in traditional fermented sausages, fermentation occurs spontaneously by microorganisms originating from the meat and production environment [19,20]. Depending on the applied processing conditions, traditional fermented sausages can support a much broader CNS species diversity, with *S. equorum*, *Staphylococcus saprophyticus*, and *S. xylosus* being the most prevalent species [17,21].

Culture-dependent as well as culture-independent identification methods have been used to study the bacterial communities of fermented meats [10,22,23]. The former has several disadvantages, as they are often biased towards certain microorganisms because of the selective media and growth conditions used and their inability to recover microorganisms that are less numerically abundant or in a viable but non-cultivable state [23,24]. The latter generally offers a more complete view of the microbiota present, although methods such as denaturing gradient gel electrophoresis of targeted PCR amplicons (PCR-DGGE) can suffer from several drawbacks, such as being labor-intensive, time-consuming, and offering low resolution [23,25].

Recently, culture-independent high-throughput sequencing (HTS) technologies have been applied to analyze microbial communities in fermented foods, including fermented meats, in greater detail [26–28]. For this purpose, several studies have relied on the sequencing of one or more hypervariable region(s) of the 16S rRNA gene [25,29]. However, the 16S rRNA gene cannot be used to distinguish between different species within the *Staphylococcus* genus. It is, therefore, necessary to apply other phylogenetic marker genes, such as the *rpoB* or *tuf* gene, when analyzing diverse staphylococcal communities [30–32].

The objective of this study was to assess if a previously developed amplicon-based HTS method targeting the 16S rRNA and *tuf* genes [33] can be successfully applied to chart the bacterial communities in European fermented meats. A second goal was to assess if this also allows one to reveal a relationship between those communities and the country of origin and the applied processing parameters of each product (i.e., pH and salt concentration). Additionally, potential links with the volatile organic compound (VOC) profiles of European fermented meat products were investigated.

2. Materials and Methods

2.1. Sampling and Experimental Set-Up

A total of 15 fermented pork products was obtained from supermarkets in and around Brussels, Belgium. The commercial products originated from Belgium, Germany, France, Italy, and Spain. Three fermented meat products were selected per country of origin for dedicated HTS analysis. Products originating from Belgium, France, and Germany came from at least two different producers per country and products originating from Spain and Italy came from at least three different producers per country. For each product, the label was analyzed to monitor relevant information, i.e., salt concentration and country of origin, and triplicate measurements of pH and the bacterial loads of presumable LAB, determined using de Man–Rogosa–Sharpe (MRS) agar, and CNS, determined using mannitol–salt–agar (MSA), were obtained as described previously [6].

Most fermented meat products were typical representatives of their geographical region of production with respect to their bacterial loads, acidity, and salt content, and were selected based on the data from an earlier study [6]. For all fermented meat products, except one, albeit on different purchases representing different production batches, the bacterial composition has been established previously using a culture-dependent $(GTG)_5$-fingerprinting method [6]. These data were used for comparison with the HTS method applied in the present study.

2.2. Extraction of Bacterial Genomic DNA

From each fermented meat product, two samples were taken, representing biological duplicates, to analyze the bacterial composition. Bacterial DNA was extracted by first centrifuging a homogenized mixture of sample and recovery diluent at 900× *g* for 10 min to remove coarse impurities. This recovery diluent consisted of a sterile solution of 0.85% (*m*/*v*) NaCl (VWR International, Darmstadt, Germany) and 0.1% (*m*/*v*) bacteriological peptone (Oxoid, Basingstoke, Hampshire, UK). Thereafter, the resulting suspensions were filtered through a 20-µm average pore-size 50-mL Steriflip filter (Merck, Darmstadt, Germany) to remove remaining contaminants. Cell pellets were then obtained by centrifuging at 4000× *g* at 4 °C for 20 min.

Thereafter, DNA extraction was carried out, using a combination of enzymatic, chemical, and mechanical cell lysis. This was followed by phenol/chloroform/isoamyl alcohol extraction and column purification of the DNA, as described in [34], except that the enzymatic steps using lyticase and Zymolyase were excluded from the protocol as molds or yeasts were not targeted in this study.

2.3. PCR Assay and Sequencing

To target the overall bacterial communities, the V4 region of the 16S rRNA gene was amplified using the primer set F515/R806 [35], further referred to as V4, that was extended with an Illumina platform-specific 5′ tag, as described previously [36]. In short, PCR assay conditions encompassed an initial denaturation step at 94 °C for 5 min, succeeded by 35 cycles of denaturation at 94 °C for 45 s, annealing at 50 °C for 60 s, and extension at 72 °C for 90 s, and a final extension step at 72 °C for 10 min. To unravel the staphylococcal communities in more detail, a 379-bp region of the *tuf* gene with sufficient discriminatory power to distinguish different CNS species was amplified using primer set Tuf387/765 (Table 1). This primer set has been designed previously, based on a custom database containing 2556 *tuf* gene sequences, representing 48 staphylococcal species selected from the European Nucleotide Archive of the European Bioinformatics Institute (ENA/EBI), and allows distinguishing staphylococcal communities to species level [33]. The PCR assay conditions comprised an initial denaturation step at 94 °C for 2 min, 25 cycles of denaturation at 94 °C for 30 s, annealing at 55 °C for 60 s, and extension at 72 °C for 3 min, and a final extension at 72 °C for 7 min. Next, the amplicons were purified (Wizard Plus SV gel and PCR clean-up system; Promega, Madison, WI, USA), subjected to size selection (Agencourt AMPure XP PCR purification magnetic beads; Beckman Coulter, Brea, CA, USA), and sequenced (Illumina MiSeq sequencing platform; VUB-ULB BRIGHTcore sequencing facility, Jette, Belgium). A mock community (RM3), containing known ratios of *S. carnosus*, *Staphylococcus epidermidis*, *S. equorum*, *Staphylococcus haemolyticus*, *Staphylococcus hominis*, *Staphylococcus lugdunensis*, *S. saprophyticus*, *Staphylococcus sciuri*, *Staphylococcus succinus*, and *S. xylosus*, was used to check the performance of the PCR assays and subsequent sequencing.

Table 1. Sequence and amplicon size of the Tuf387/765 primer set.

Primer	Primer Sequence	Amplicon Size (bp)
Tuf387	5′-YCCAATGCCWCAAACKCGTGA-3′	379
Tuf765	5′-RAYTTGHCCACGTTCAACAC-3′	

2.4. Bioinformatic Analysis

Processing of the V4 amplicons was similar to the method described in Zhang et al. [37], whereby amplicon sequence variants (ASVs) were determined using the DADA2 package (version 1.10.1) [38] and the SILVA database (version 138) [39]. For the amplicons generated with primer set Tuf387/765, taxonomy was assigned using the custom database containing 2556 *tuf* gene sequences as described above. In the case of the *tuf* gene, ASVs were only classified to species level if a 100% match with a sequence in the custom database was found. If taxonomy could not be assigned using the custom database, ASVs with more than 1000 reads were aligned to the non-redundant nucleotide database nt of the National Center for Biotechnological Information (NCBI, Bethesda, MA, USA), using blastn as an alignment tool [40]. ASVs were assumed to be present in a sample if they amounted to more than 0.01% of the total reads.

2.5. Volatile Organic Compound Profiling

Non-targeted VOC profiling was conducted in triplicate by headspace/solid-phase microextraction coupled to gas chromatography and time-of-flight mass spectrometry (HS/SPME-GC-TOF-MS), using a Trace 1300 gas chromatograph (Thermo Fisher, Waltham, MA, USA) equipped with a Stabilwax-MS column (30 m by 0.25 mm, film thickness of 0.25 μm; Restek, Bellefonte, PA, USA) and coupled to a BenchTOF-HD mass spectrometer operating with electron impact ionization at 70 eV (Markes International, Llantrisant, Wales, UK). For analysis, approximately 10 g of the fermented meat product sample was frozen using liquid nitrogen (Airliquide, Paris, France) and grounded into a fine powder with a coffee grinder (DeLongi KG49, Treviso, Italy). Subsequently, 1.5 g of grounded powder was incubated in a 10-mL screw-top headspace vial at 60 °C for 10 min, after which the sample was exposed to an SPME fiber (DVB/CAR/PDMS; Sigma-Aldrich, St. Louis, MO, USA) at 60 °C for 20 min. To each sample, 10 μL of a 10 ppm toluene-D8 solution (Sigma-Aldrich) was added as an internal standard. The VOCs adsorbed to the SPME fiber were thermally desorbed at 250 °C with a split rate of 50 mL/min. The GC oven temperature program consisted of an initial step at 40 °C for 1.5 min, followed by a continuous increase to 225 °C at 10 °C/min, and finally, the temperature was kept constant at 225 °C for 15 min. Helium gas (Praxair, Danbury, CT, USA) was used as the carrier gas at a flow rate of 1 mL/min. The TOF-MS was scanned in the m/z range of 35 to 400 after 2 min. The raw data were deconvoluted by TOF-DS software (Markes). Afterward, peaks were identified based on the NIST library (National Institute of Standard and Technology, Gaitherburg, MD, USA) and the use of commercially available standards [37,41]. For each identified compound, the peak area was normalized to the peak area of the internal standard.

2.6. Statistics

RStudio software (version 3.5.2, Rstudio, Boston, USA) was used for all statistical analyses and tests [42].

Assessment of inter-sample diversity (beta-diversity) was achieved by performing a permutational multivariate analysis of variance (PERMANOVA) based on Bray–Curtis dissimilarity scores to ascertain differences in bacterial composition between the fermented meat products investigated. Subsequently, a series of pairwise PERMANOVA comparisons and a similarity percentage analysis (SIMPER) were performed to assess the differences in bacterial communities between fermented meat products produced in the countries mentioned above. The vegan (version 2.5-4) [43] and RVAideMemoire packages (version 0.9-73) [44] were applied for this purpose.

Heatmaps of the non-targeted VOC compound profiling were calculated and clustered using the packages ComplexHeatmap (version 2.0.0) [45] and circlize (version 0.4.7) [46]. Hierarchical clustering was performed according to the average distance between the points in the clusters, based on a correlation matrix of the entire data set. PERMANOVA was used to assess differences in VOC profiles according to the country of origin and/or the prevailing microbial communities, based on Bray–Curtis dissimilarity scores.

3. Results

3.1. Bacterial Communities in Fermented Meat Products from Different Origins

The characteristics with respect to the pH, salt content, and bacterial loads of each sample of the commercial fermented meat products examined are provided in Table 2. Most products were typical representatives of their geographical region of production. Only the Belgian fermented meat product encoded BE3, which claimed to be produced "à l'ancienne" (i.e., according to "tradition"), was somewhat atypical, as it had a considerably higher pH (5.7) than the other Belgian products with a pH of 5.1 (BE1) and 5.0 (BE2) (Table 2).

In addition, the bacterial communities present in the various fermented meat products were assessed through amplicon-based HTS, based on 83,642 and 71,971 raw V4 and Tuf387/765 sequence reads on average per sample, respectively. The V4 sequence reads provided a first overview of the bacterial genera present in the different fermented meat products (Figure 1). It unveiled that *Lactobacillus* was present in all samples, no matter the geographical origin of the product. In contrast, *Pediococcus* was only found in samples from Belgian and German fermented meat products and *Pseudomonas* was exclusive to one sample of a Belgian product. Samples from Spanish fermented meat products showcased the greatest microbial diversity, with genera such as *Bacillus*, *Brochothrix*, *Carnobacterium*, *Corynebacterium*, *Leuconostoc*, and *Tetragenococcus* being present. *Staphylococcus* was only found sporadically via this approach in Belgian, German, and Spanish fermented meat products, whereas it was present in all French and Italian ones.

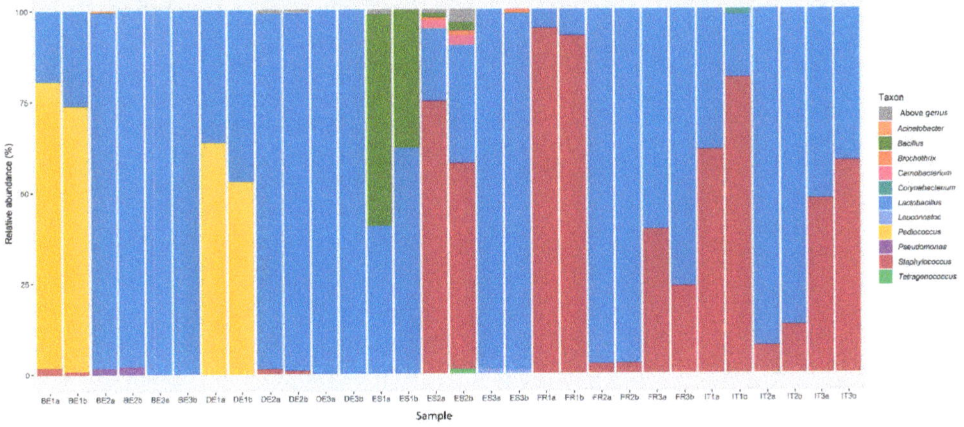

Figure 1. Distribution of amplicon sequencing variants (ASVs) of the V4 region of the 16S rRNA gene in selected commercial fermented meat products originating from BE (Belgium), DE (Germany), ES (Spain), FR (France), and IT (Italy), with "a" and "b" annotations in the sample names representing biological duplicate samples from the same product.

Table 2. Individual and average bacterial counts, pH, and salt content with standard deviation (SD) of selected commercial fermented meat products originating from BE (Belgium), DE (Germany), ES (Spain), FR (France), and IT (Italy). MSA, mannitol-salt-agar (representing presumptive staphylococcal species); MRS, de Man–Rogosa–Sharpe (representing presumptive lactic acid bacterial species).

Sample	MSA Counts [log(CFU/g)]			MRS Agar Counts [log(CFU/g)]			pH			Salt Content [g/100g]		
	Individual	Average	SD	Individual	Average	SD	Individual	Average	SD	Individual	Average	SD
BE1	5.40			6.75			5.12			4.10		
BE2	5.13	5.18	0.20	8.27	7.76	0.87	5.00	5.27	0.37	4.60	4.30	0.26
BE3	5.02			8.25			5.70			4.20		
DE1	5.57			7.14			4.78			3.80		
DE2	5.00	5.17	0.35	7.91	7.64	0.43	4.57	4.66	0.11	3.25	3.43	0.32
DE3	4.94			7.88			4.63			3.25		
ES1	5.72			6.53			5.82			3.00		
ES2	5.80	5.83	0.12	5.60	6.83	1.41	6.04	5.85	0.18	3.00	3.27	0.46
ES3	5.95			8.37			5.69			3.80		
FR1	8.16			7.70			5.70			5.70		
FR2	6.52	7.31	0.82	8.26	8.16	0.42	5.51	5.59	0.10	4.60	4.83	0.78
FR3	7.26			8.52			5.56			4.20		
IT1	8.63			8.44			5.66			3.40		
IT2	6.68	7.34	1.12	8.56	8.34	0.30	5.26	5.57	0.28	3.30	3.67	0.55
IT3	6.70			8.00			5.80			4.30		

The Tuf387/765 sequence reads obtained by partial *tuf* gene sequencing gave a more detailed view of the CNS communities present in the different fermented meat products (Figure 2). All species present in the mock community (RM3) were correctly identified. Sequences corresponding with several non-*Staphylococcus* genera were also encountered, namely *Bacillus*, *Brochothrix*, *Carnobacterium*, and *Enterococcus*. *Brochothrix* was encountered in Belgian, German, Italian, and Spanish fermented meat products. *Bacillus*, *Carnobacterium*, and *Enterococcus* were only found in Spanish and German fermented meat products, although in the latter only a small number of reads (<0.3%) could be ascribed to these genera.

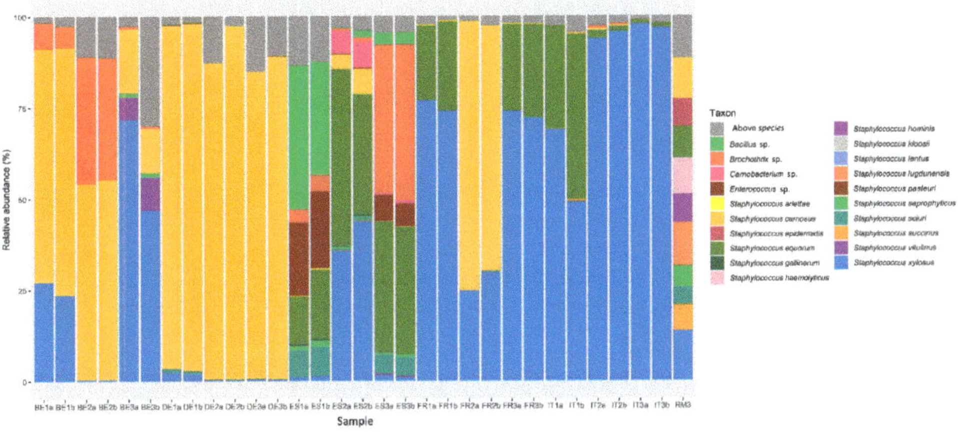

Figure 2. Distribution of amplicon sequencing variants (ASVs) of a partial region of the *tuf* gene in selected commercial fermented meat products originating from BE (Belgium), DE (Germany), ES (Spain), FR (France), and IT (Italy), with "a" and "b" annotations in the sample names representing biological duplicate samples from the same product. RM3 represents the mock community used to check the performance of the PCR assay and subsequent sequencing.

Taking into account the only sequence reads attributed to *Staphylococcus* species using partial *tuf* gene sequencing, differences were found between the CNS communities of the fermented meat products from different countries (Figure 3). In Belgian fermented meat products, most sequence reads were assigned to *S. carnosus*, followed by *S. xylosus*. Additionally, *Staphylococcus pasteuri*, *S. sciuri*, *S. saprophyticus*, and *Staphylococcus vitulinus* were found, albeit mostly in minor relative abundances. German fermented meat products were characterized by a large relative abundance of reads originating from *S. carnosus*, with *S. equorum*, *Staphylococcus gallinarum*, *S. sciuri*, and *S. xylosus* occurring as minor fractions. CNS communities in Spanish fermented meat products exhibited the greatest species diversity, including *Staphylococcus arlettae*, *S. carnosus*, *S. equorum*, *S. gallinarum*, *Staphylococcus lentus*, *S. saprophyticus*, *S. sciuri*, *S. vitulinus*, and *S. xylosus*. In French and Italian fermented meat products, the majority of sequence reads was related to *S. xylosus*, followed by *S. equorum*, *S. carnosus*, *S. saprophyticus*, *S. vitulinus*, *S. succinus*, *S. gallinarum*, *Staphylococcus kloosii*, and *S. sciuri*.

To highlight the greater resolution achieved when studying CNS communities by applying the aforementioned amplicon-based HTS method, the CNS species found in the present study were compared to those previously found culture-dependently in the same products (Table 3) [6].

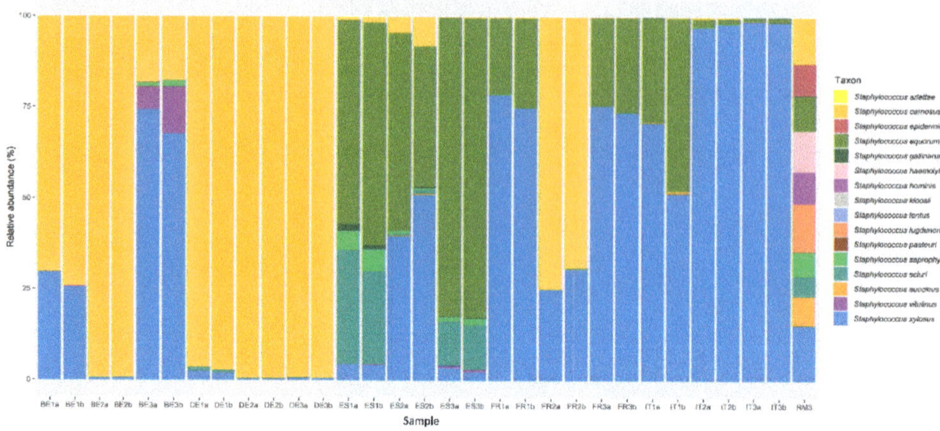

Figure 3. Distribution of amplicon sequencing variants (ASVs) of a partial region of the *tuf* gene assigned to staphylococcal species in selected commercial fermented meat products originating from BE (Belgium), DE (Germany), ES (Spain), FR (France), and IT (Italy), with "a" and "b" annotations in the sample names representing biological duplicate samples from the same product. RM3 represents the mock community used to check the performance of the PCR assay and subsequent sequencing.

A PERMANOVA indicated that the country of origin had a significant ($p < 0.05$) impact on the composition of the microbial communities. Pairwise PERMANOVA tests showed that microbial communities from Spanish fermented meat products were significantly different from those found in fermented meat products from all other countries. German fermented meat products had different ($p < 0.05$) microbial profiles than French, Italian, and Spanish variants. The microbial communities of Belgian fermented meat products were not significantly different from their German or French counterparts, whereas French fermented meat products had comparable microbial profiles to the Italian ones. SIMPER highlighted that differences in microbial communities were primarily due to the presence of *S. equorum* in Spanish fermented meat products, *S. carnosus* in Belgian and German ones, and *S. xylosus* in French and Italian fermented meat products (Tables S1–S10).

3.2. Volatile Organic Compound Profiling

A total of 186 VOCs were detected across the different European fermented meat products examined, of which 25% were terpenes (e.g., limonene, careen, and pinene), 16% were aromatic hydrocarbons (e.g., 3-methylphenol and eugenol), 10% were esters (e.g., methyl butanoate and methyl octanoate), 9% were alcohols (e.g., 2,3-butanediol and 1-octen-3-ol), and 9% were ketones (e.g., acetoin). Minor VOCs consisted of aldehydes (e.g., 3-methylbutanal, hexanal, and nonanal), alkanes (e.g., octane and decane), carboxylic acids (e.g., acetic acid and 2-methylbutanoic acid), furans (e.g., furfural), sulphuric compounds (e.g., diallyl sulphide), and nitrogen compounds (e.g., 2,6-dimethylpyrazine) (Figure S1). Despite individual products displaying differences in their VOC profiles, no VOC profiles or specific VOCs could be linked to either the country of origin or the prevailing microbial communities of the products or to distinct processing conditions using PERMANOVA.

Table 3. Comparison of the *Staphylococcus* (*S.*) species diversity in selected commercial fermented meat products obtained using culture-dependent methods (data obtained from Van Reckem et al., 2019) *versus* the use of the amplicon-based high-throughput sequencing (HTS) method targeting the *tuf* gene (present study). Fermented meat products originated from BE (Belgium), DE (Germany), ES (Spain), FR (France), and IT (Italy).

Sample	Culture-Dependent Method (Total Number of Isolates)	Amplicon-Based HTS Method
BE1	*S. carnosus* (18)	*S. carnosus, S. vitulinus, S. xylosus*
BE2	*S. carnosus* (29)	*S. carnosus, S. saprophyticus, S. xylosus*
BE3	n.a. [1]	*S. carnosus, S. pasteuri, S. saprophyticus, S. sciuri, S. xylosus*
DE1	*S. carnosus, S. xylosus* (13)	*S. carnosus, S. gallinarum, S. sciuri, S. xylosus*
DE2	*S. carnosus* (20)	*S. carnosus, S. equorum, S. xylosus*
DE3	*S. carnosus* (15)	*S. carnosus, S. xylosus*
ES1	*S. equorum, S. saprophyticus* (21)	*S. arlettae, S. carnosus, S. equorum, S. gallinarum, S. lentus, S. saprophyticus, S. sciuri, S. xylosus*
ES2	*S. carnosus, S. equorum, S. xylosus* (15)	*S. carnosus, S. equorum, S. gallinarum, S. saprophyticus, S. sciuri, S. succinus, S. xylosus*
ES3	*S. equorum* (14)	*S. carnosus, S. equorum, S. gallinarum, S. saprophyticus, S. sciuri, S. vitulinus, S. xylosus*
FR1	*S. equorum, S. xylosus* (23)	*S. carnosus, S. equorum, S. saprophyticus, S. xylosus*
FR2	*S. equorum, S. xylosus* (13)	*S. carnosus, S. vitulinus, S. xylosus*
FR3	*S. equorum, S. xylosus* (10)	*S. carnosus, S. equorum, S. saprophyticus, S. xylosus*
IT1	*S. equorum, S. xylosus* (12)	*S. equorum, S. succinus, S. xylosus*
IT2	*S. xylosus* (13)	*S. carnosus, S. equorum, S. gallinarum, S. kloosii, S. saprophyticus, S. sciuri, S. xylosus*
IT3	*S. xylosus* (20)	*S. equorum, S. saprophyticus, S. vitulinus, S. xylosus*

[1] data not available.

4. Discussion

Due to variations in ingredients and processing technologies, the geographical origin of fermented foods can have a profound impact on their technological properties and microbial diversity [3,4]. This is also true for fermented meat products, where clear-cut differences have for instance been found in the characteristics of Northern and Southern European variants [5,6].

The differences in technological properties, namely salt content and pH, of fermented meat products in the present study are in line with the findings of an earlier study, in which the same fermented meat products were investigated, albeit at another moment in time, thus representing other batches [6]. Generally, the salt content of European cured meat products differs within and across different countries [6,47]. However, these differences are not expected to largely influence the technologically important microbiota, consisting mainly of LAB and CNS, as these microorganisms are commonly well adapted to high-salt levels [21]. The usual contrast in acidification extent between Northern and Southern European fermented meat products was illustrated by the differences in pH between German and Belgian commercial products, on the one hand, and French, Italian, and Spanish ones, on the other hand [6,9,11].

With respect to the microbial diversity, application of amplicon-based HTS methods allowed studying the microbial communities with greater resolution compared to previous culture-dependent studies [6,25]. Sequencing of the V4 hypervariable region of the bacterial 16S rRNA gene showed that the genus *Lactobacillus* was omnipresent in fermented meat products across all producing countries examined. Although the genus *Lactobacillus* has recently been reclassified into 25 genera, of which *Latilactobacillus* and *Lactiplantibacillus* contain the most relevant species for meat fermentation [48], the SILVA database used in the present study was not yet updated at the moment the bioinformatics analysis was performed, hence still *Lactobacillus* is reported here. The presence of *Lactobacillaceae* is not surprising as they are frequently encountered as one of the most important genera in fermented sausages, with multiple species such as *Latilactobacillus sakei* (formerly known as *Lactobacillus sakei*), showing great adaptation to the meat matrix [21,25,49]. The presence of *Pediococcus* in some Belgian and German fermented meat products might be a consequence of its addition as a starter culture, as pediococci, such as *P. pentosaceus*, are sometimes added in Northern Europe and Northern America to hasten acidification under higher fermentation temperatures [6,50]. Spanish fermented meat products were remarkable because of the presence of several uncommon genera, such as *Bacillus*, *Brochothrix*, *Carnobacterium*, *Corynebacterium*, and *Tetragenococcus*. These species have nonetheless all been encountered previously to some degree in the natural microbiota of fermented meat products [51–55].

The fact that the genus *Staphylococcus* was not encountered across all Belgian and German fermented meat product samples when targeting the V4 region, did not necessarily infer its absence, as differences in biomass and 16S rRNA copy numbers may cause less abundant genera to be under the detection limit [56,57]. Indeed, whenever a substantial number of reads were ascribed to *Staphylococcus*, the MSA and MRS counts were comparable, with less than 1.0 log (cfu/g) of difference. Furthermore, high-throughput amplicon sequencing of a part of the *tuf* gene revealed diversified CNS communities in all fermented meat products examined. The CNS communities in French and Italian fermented meat products demonstrated a high relative abundance of *S. xylosus*, which was almost always accompanied by *S. equorum*. Other CNS species, such as *S. carnosus* and *S. saprophyticus*, were encountered intermittently and mostly in low relative abundances. This was in line with previous studies, where *S. xylosus* and *S. equorum* were frequently found to characterize Southern-European fermented meat products because they preferred relatively high pH and low fermentation temperatures [6,10,14,54,58,59]. Spanish fermented meat products displayed a high relative abundance of *S. equorum*, but exhibited a greater CNS species diversity. The latter was largely due to the presence of *S. xylosus*, *S. saprophyticus*, *S. sciuri*, *S. gallinarum*, *S. carnosus*, and *S. vitulinus*, which have also previously been found in spontaneously fermented Southern-European fermented meat products [14,60–63].

When compared to French and Italian fermented meat products, the occurrence of less common genera and species and the overall lower bacterial loads in Spanish products suggest differences on the level of starter culture application, either because the process is based on spontaneous fermentation or because of poor adaptation of the starter culture bacteria involved to the processing conditions [6,16,64]. In contrast, the relatively high abundance of *S. carnosus* in Belgian and German fermented meat products does indicate starter culture use, as this species is not common in the natural microbiota of spontaneously fermented meat products but routinely used as a starter culture in Northern Europe [8,54,65,66]. Noteworthy in that respect was the case of the Belgian product BE3, which was characterized by a somewhat atypical higher pH of 5.7 for its region of production. Although *S. carnosus* was still found, the relative abundance of *S. xylosus* increased and *S. saprophyticus* and *S. vitulinus* also emerged, all of which have been encountered in previous studies in spontaneously fermented meat products of similar acidity [14,67,68]. As fermented meat products, BE1 and BE3 were manufactured by the same producer, it is likely that the same starter culture was used but that different communities were obtained, due to differences in the processing conditions, although details about the latter are not known. This pronounced difference in the composition of the CNS communities underlines the influence of the pH as a major processing factor, as well as the influence of the processing conditions on the structure of the CNS communities in meat fermentation, whether or not starter cultures are used [7,15,64].

Compared to the culture-dependent methodology applied previously on the same commercial fermented meat products [6], amplicon-based HTS allowed for the charting of the CNS communities with far greater resolution (Table 3). Whereas the dominant CNS species were comparable, amplicon-based HTS allowed for the uncovering of several subdominant CNS species. Failure of detection of the latter by culture-dependent methods was due to limitations imposed by the cultivation step and the subsequent work needed to obtain a sufficient number of isolates, fingerprints, and sequencing data. This showcases the capability of amplicon-based HTS methods to rapidly unravel complex microbial communities in fermented foods in a satisfactory degree of detail [25].

Several VOCs were detected in the fermented meat products examined, which could either be related to microbial metabolism (e.g., acetoin, acetic acid, and 3-methylbutanal), the addition of certain spices and herbs (e.g., terpenes and sulfur compounds), or the processing applied (e.g., the presence of phenolic compounds due to smoking) [41,69–73]. However, no clear trends relating VOC compositions to the presence of certain microbial groups or unique processing practices could be distinguished. For this purpose, a more quantitative approach targeting specific VOCs may be needed.

5. Conclusions

The present study demonstrated that an amplicon-based HTS method targeting regions of the 16S rRNA and *tuf* genes allows for an improved exploration of the species diversity of the microbial communities in fermented meats, in particular the staphylococcal communities. Amplicon-based HTS broadened the view on the microbial communities to also encompass several subdominant CNS species that previously may have been underreported. Furthermore, it permitted to emphasize the influence of the processing conditions on the bacterial diversity, indicative of pH and starter culture impact. In the present study, only bacterial communities were targeted using partial 16S rRNA and *tuf* gene sequencing. In the future it might be valuable to target other marker genes as well, to obtain a higher resolution image of the microbial communities, including an overview of yeast species diversity. Future applications of the aforementioned amplicon-based HTS methods offer great potential to further unravel complex microbial communities in fermented meat products and other fermented foods as well as to assess the impact of different processing conditions on the entirety of the microbial consortia present.

Supplementary Materials: The following are available online at http://www.mdpi.com/2304-8158/9/9/1247/s1, Table S1: Similarity percentage analysis (SIMPER) for staphylococcal species identified in Belgian and German fermented meat products, their contribution (in%) to within-group similarity per country, and the cumulative total (in%) of the contributions, Table S2: Similarity percentage analysis (SIMPER) for staphylococcal species identified in Belgian and Spanish fermented meat products, their contribution (in%) to within-group similarity per country, and the cumulative total (in%) of the contributions, Table S3: Similarity percentage analysis (SIMPER) for staphylococcal species identified in Belgian and French fermented meat products, their contribution (in%) to within-group similarity per country, and the cumulative total (in%) of the contributions, Table S4: Similarity percentage analysis (SIMPER) for staphylococcal species identified in Belgian and Italian fermented meat products, their contribution (in%) to within-group similarity per country, and the cumulative total (in%) of the contributions, Table S5: Similarity percentage analysis (SIMPER) for staphylococcal species identified in German and Spanish fermented meat products, their contribution (in%) to within-group similarity per country, and the cumulative total (in%) of the contributions, Table S6: Similarity percentage analysis (SIMPER) for staphylococcal species identified in German and French fermented meat products, their contribution (in%) to within-group similarity per country, and the cumulative total (in%) of the contributions, Table S7: Similarity percentage analysis (SIMPER) for staphylococcal species identified in German and Italian fermented meat products, their contribution (in%) to within-group similarity per country, and the cumulative total (in%) of the contributions, Table S8: Similarity percentage analysis (SIMPER) for staphylococcal species identified in Spanish and French fermented meat products, their contribution (in%) to within-group similarity per country, and the cumulative total (in%) of the contributions, Table S9: Similarity percentage analysis (SIMPER) for staphylococcal species identified in Spanish and Italian fermented meat products, their contribution (in%) to within-group similarity per country, and the cumulative total (in%) of the contributions, Table S10: Similarity percentage analysis (SIMPER) for staphylococcal species identified in French and Italian fermented meat products, their contribution (in%) to within-group similarity per country, and the cumulative total (in%) of the contributions, Figure S1: Hierarchical clustering analysis and heatmap visualization of semi-quantitative volatile organic compound profiles in fermented meat products originating from BE (Belgium), DE (Germany), ES (Spain), FR (France), and IT (Italy), determined using HS/SPME-GC-TOF-MS.

Author Contributions: Conceptualization, E.V.R. and F.L.; methodology, E.V.R. and W.B.; software, E.V.R.; validation, E.V.R.; formal analysis, E.V.R.; investigation, C.C., D.V.d.V., and E.V.R.; resources, L.D.V., F.L., and S.W.; data curation, E.V.R.; writing—original draft preparation, E.V.R.; writing—review and editing, L.D.V., F.L., and S.W.; visualization, C.C., D.V.d.V., and E.V.R.; supervision, F.L. and S.W.; project administration, L.D.V., F.L., and S.W.; funding acquisition, E.V.R., L.D.V., F.L., and S.W. All authors have read and agreed to the published version of the manuscript.

Funding: This work was supported by the Research Council of the Vrije Universiteit Brussel (OZR, SRP, IRP, HOA, and IOF projects), the Hercules Foundation (projects UABR 09/004 and UAB13002), and the Research Foundation-Flanders (G021518N). E.V.R. is the recipient of an SB Fellowship of the FWO (1S06717N).

Conflicts of Interest: The authors declare no conflict of interest.

Data Availability: The amplicon sequencing data sets were submitted to the European Nucleotide Archive of the European Bioinformatics Institute (ENA/EBI) and are accessible under the study accession number PRJEB39544 (http://www.ebi.ac.uk/ena/data/view/ PRJEB39544).

References

1. Guerrero, L.; Guàrdia, M.D.; Xicola, J.; Verbeke, W.; Vanhonacker, F.; Zakowska-Biemans, S.; Sajdakowska, M.; Sulmont-Rossé, C.; Issanchou, S.; Contel, M.; et al. Consumer-driven definition of traditional food products and innovation in traditional foods. A qualitative cross-cultural study. *Appetite* **2009**, *52*, 345–354.
2. Leroy, F.; Geyzen, A.; Janssens, M.; De Vuyst, L.; Scholliers, P. Meat fermentation at the crossroads of innovation and tradition: A historical outlook. *Trends Food Sci. Technol.* **2013**, *31*, 130–137. [CrossRef]
3. Fenger, M.H.; Aschemann-Witzel, J.; Hansen, F.; Grunert, K.G. Delicious words—Assessing the impact of short storytelling messages on consumer preferences for variations of a new processed meat product. *Food Qual. Pref.* **2015**, *41*, 237–244. [CrossRef]
4. Zhong, Z.; Hou, Q.; Kwok, L.; Yu, Z.; Zheng, Y.; Sun, Z.; Menghe, B.; Zhang, H. Bacterial microbiota compositions of naturally fermented milk are shaped by both geographic origin and sample type. *J. Dairy Sci.* **2016**, *99*, 7832–7841. [CrossRef]
5. Leroy, F.; Scholliers, P.; Amilien, V. Elements of innovation and tradition in meat fermentation: Conflicts and synergies. *Int. J. Food Microbiol.* **2015**, *212*, 2–8. [CrossRef]
6. Van Reckem, E.; Geeraerts, W.; Charmpi, C.; Van der Veken, D.; De Vuyst, L.; Leroy, F. Exploring the link between the geographical origin of European fermented foods and the diversity of their bacterial communities: The case of fermented meats. *Front. Microbiol.* **2019**, *10*, 2302. [CrossRef]
7. Leroy, F.; Verluyten, J.; De Vuyst, L. Functional meat starter cultures for improved sausage fermentation. *Int. J. Food Microbiol.* **2006**, *106*, 270–285. [CrossRef]

8. Aquilanti, L.; Garofalo, C.; Osimani, A.; Clementi, F. Ecology of lactic acid bacteria and coagulase-negative cocci in fermented dry sausages manufactured in Italy and other Mediterranean countries: An overview. *Int. Food Res. J.* **2016**, *23*, 429–445.
9. Holck, A.; Heir, E.; Johannessen, T.C.; Axelsson, L. Northern European products. In *Handbook of Fermented Meat and Poultry*, 2nd ed.; Toldrà, F., Hui, Y.H., Astiasarán, I., Sebranek, J.G., Talon, R., Eds.; John Wiley and Sons: Chichester, UK, 2015; pp. 313–320.
10. Janssens, M.; Myter, N.; De Vuyst, L.; Leroy, F. Community dynamics of coagulase-negative staphylococci during spontaneous artisan-type meat fermentations differ between smoking and moulding treatments. *Int. J. Food Microbiol.* **2013**, *166*, 168–175. [CrossRef]
11. Hierro, E.; Fernández, M.; de la Hoz, L.; Ordóñez, J.A. Mediterranean products. In *Handbook of Fermented Meat and Poultry*, 2nd ed.; Toldrà, F., Hui, Y.H., Astiasarán, I., Sebranek, J.G., Talon, R., Eds.; John Wiley and Sons: Chichester, UK, 2015; pp. 301–308.
12. Janssens, M.; Myter, N.; De Vuyst, L.; Leroy, F. Species diversity and metabolic impact of the microbiota are low in spontaneously acidified Belgian sausages with an added starter culture of *Staphylococcus carnosus*. *Food Microbiol.* **2012**, *29*, 167–177. [CrossRef]
13. Fonseca, S.; Cachaldora, A.; Gómez, M.; Franco, I.; Carballo, J. Monitoring the bacterial population dynamics during the ripening of Galician chorizo, a traditional dry fermented Spanish sausage. *Food Microbiol.* **2013**, *33*, 77–84. [CrossRef] [PubMed]
14. Greppi, A.; Ferrocino, I.; La Storia, A.; Rantsiou, K.; Ercolini, D.; Cocolin, L. Monitoring of the microbiota of fermented sausages by culture independent rRNA based approaches. *Int. J. Food Microbiol.* **2015**, *212*, 67–75. [CrossRef] [PubMed]
15. Stavropoulou, D.A.; De Maere, H.; Berardo, A.; Janssens, B.; Filippou, P.; De Vuyst, L.; De Smet, S.; Leroy, F. Species pervasiveness within the group of coagulase-negative staphylococci associated with meat fermentation is modulated by pH. *Front. Microbiol.* **2018**, *9*, 2232. [CrossRef] [PubMed]
16. Ojha, K.S.; Kerry, J.P.; Duffy, G.; Beresford, T.; Tiwari, B.K. Technological advances for enhancing quality and safety of fermented meat products. *Trends Food Sci. Tech.* **2015**, *44*, 105–116. [CrossRef]
17. Stavropoulou, D.A.; De Vuyst, L.; Leroy, F. Nonconventional starter cultures of coagulase-negative staphylococci to produce animal-derived fermented foods, a SWOT analysis. *J. Appl. Microbiol.* **2018**, *125*, 1570–1586. [CrossRef]
18. Laranjo, M.; Potes, M.E.; Elias, M. Role of starter cultures on the safety of fermented meat products. *Front. Microbiol.* **2019**, *10*, 853. [CrossRef]
19. Belleggia, L.; Ferrocino, I.; Reale, A.; Boscaino, F.; Di Renzo, T.; Corvaglia, M.R.; Cocolin, L.; Milanović, V.; Cardinali, F.; Garofalo, C.; et al. Portuguese cacholeira blood sausage: A first taste of its microbiota and volatile organic compounds. *Food Res. Int.* **2020**, *136*, 109567. [CrossRef]
20. Settanni, L.; Barbaccia, P.; Bonanno, A.; Ponte, M.; Di Gerlando, R.; Franciosi, E.; Di Grigoli, A.; Gaglio, R. Evolution of indigenous starter microorganisms and physicochemical parameters in spontaneously fermented beef, horse, wild boar and pork salamis produced under controlled conditions. *Food Microbiol.* **2020**, *87*, 103385. [CrossRef]
21. Charmpi, C.; Van der Veken, D.; Van Reckem, E.; De Vuyst, L.; Leroy, F. Raw meat quality and salt levels affect the bacterial species diversity and community dynamics during the fermentation of pork mince. *Food Microbiol.* **2020**, *89*, 103434. [CrossRef]
22. Rantsiou, K.; Urso, R.; Iacumin, L.; Cantoni, C.; Cattaneo, P.; Comi, G.; Cocolin, L. Culture-dependent and -independent methods to investigate the microbial ecology of Italian fermented sausages. *Appl. Environ. Microbiol.* **2005**, *71*, 1977–1986. [CrossRef]
23. Cocolin, L.; Alessandria, V.; Dolci, P.; Gorra, R.; Rantsiou, K. Culture independent methods to assess the diversity and dynamics of microbiota during food fermentation. *Int. J. Food Microbiol.* **2013**, *167*, 29–43. [CrossRef] [PubMed]
24. Giraffa, G.; Carminati, D. Molecular techniques in food fermentation: Principles and applications. In *Molecular Techniques in the Microbial Ecology of Fermented Foods*; Cocolin, L., Ercolini, D., Eds.; Springer: New York, NY, USA, 2008; pp. 1–30.
25. Połka, J.; Rebecchi, A.; Pisacane, V.; Morelli, L.; Puglisi, E. Bacterial diversity in typical Italian salami at different ripening stages as revealed by high-throughput sequencing of 16S rRNA amplicons. *Food Microbiol.* **2015**, *46*, 342–356. [CrossRef] [PubMed]

26. Ercolini, D.; De Filippis, F.; La Storia, A.; Iacono, M. "Remake" by high-throughput sequencing of the microbiota involved in the production of water buffalo mozzarella cheese. *Appl. Environ. Microbiol.* **2012**, *78*, 8142–8145. [CrossRef] [PubMed]
27. De Filippis, F.; Parente, E.; Ercolini, D. Recent past, present, and future of the food microbiome. *Annu. Rev. Food Sci. Technol.* **2018**, *9*, 589–608. [CrossRef]
28. Maidana, S.D.; Ficoseco, C.A.; Bassi, D.; Cocconcelli, P.S.; Puglisi, E.; Savoy, G.; Vignolo, G.; Fontana, C. Biodiversity and technological-functional potential of lactic acid bacteria isolated from spontaneously fermented chia sourdough. *Int. J. Food Microbiol.* **2020**, *316*, 108425. [CrossRef]
29. dos Santos Cruxen, C.E.; Funck, G.D.; Haubert, L.; da Silva Dannenberg, G.; de Lima Marques, J.; Chaves, F.C.; Padilha da Silva, W.; Fiorentini, Â.M. Selection of native bacterial starter culture in the production of fermented meat sausages: Application potential, safety aspects, and emerging technologies. *Food Res. Int.* **2019**, *122*, 371–382. [CrossRef]
30. Heikens, E.; Fleer, A.; Paauw, A.; Florijn, A.; Fluit, A.C. Comparison of genotypic and phenotypic methods for species-level identification of clinical isolates of coagulase-negative staphylococci. *J. Clin. Microbiol.* **2005**, *43*, 2286–2290. [CrossRef]
31. Ghebremedhin, B.; Layer, F.; König, W.; König, B. Genetic classification and distinguishing of *Staphylococcus* species based on different partial *gap*, 16S rRNA, *hsp60*, *rpoB*, *sodA*, and *tuf* gene sequences. *J. Clin. Microbiol.* **2008**, *46*, 1019–1025. [CrossRef]
32. McMurray, C.L.; Hardy, K.J.; Calus, S.T.; Loman, N.J.; Hawkey, P.M. Staphylococcal species heterogeneity in the nasal microbiome following antibiotic prophylaxis revealed by *tuf* gene deep sequencing. *Microbiome* **2016**, *4*, 63. [CrossRef]
33. Van Reckem, E.; De Vuyst, L.; Leroy, F.; Weckx, S. Amplicon-based high-throughput sequencing method capable of species-level identification of coagulase-negative staphylococci in diverse communities. *Microorganisms* **2020**, *8*, 897. [CrossRef]
34. Vermote, L.; Verce, M.; De Vuyst, L.; Weckx, S. Amplicon and shotgun metagenomic sequencing indicates that microbial ecosystems present in cheese brines reflect environmental inoculation during the cheese production process. *Int. Dairy J.* **2018**, *87*, 44–53. [CrossRef]
35. Caporaso, J.G.; Lauber, C.L.; Walters, W.A.; Berg-Lyons, D.; Lozupone, C.A.; Turnbaugh, P.J.; Fierer, N.; Knight, R. Global patterns of 16S rRNA diversity at a depth of millions of sequences per sample. *Proc. Natl. Acad. Sci. USA* **2011**, *108*, 4516–4522. [CrossRef] [PubMed]
36. De Bruyn, F.; Zhang, S.J.; Pothakos, V.; Torres, J.; Lambot, C.; Moroni, A.V.; Callanan, M.; Sybesma, W.; Weckx, S.; De Vuyst, L. Exploring the impacts of postharvest processing on the microbiota and metabolite profiles during green coffee bean production. *Appl. Environ. Microbiol.* **2017**, *83*, e02398-16. [CrossRef]
37. Zhang, S.J.; De Bruyn, F.; Pothakos, V.; Torres, J.; Falconi, C.; Moccand, C.; Weckx, S.; De Vuyst, L. Following coffee production from cherries to cup: Microbiological and metabolomic analysis of wet processing of Coffea arabica. *Appl. Environ. Microbiol.* **2019**, *85*, e02635-18. [CrossRef]
38. Callahan, B.J.; McMurdie, P.J.; Holmes, S.P. Exact sequence variants should replace operational taxonomic units in marker-gene data analysis. *ISME J.* **2017**, *11*, 2639–2643. [CrossRef] [PubMed]
39. Quast, C.; Pruesse, E.; Yilmaz, P.; Gerken, J.; Schweer, T.; Yarza, P.; Peplies, J.; Glöckner, F.O. The SILVA ribosomal RNA gene database project: Improved data processing and web-based tools. *Nucleic Acids Res.* **2012**, *41*, D590–D596. [CrossRef] [PubMed]
40. Altschul, S.F.; Gish, W.; Miller, W.; Myers, E.W.; Lipman, D.J. Basic local alignment search tool. *J. Mol. Biol.* **1990**, *215*, 403–410. [CrossRef]
41. Geeraerts, W.; De Vuyst, L.; Leroy, F.; Van Kerrebroeck, S. Monitoring of volatile production in cooked poultry products using selected ion flow tube-mass spectrometry. *Food Res. Int.* **2019**, *119*, 196–206. [CrossRef]
42. R Core Team. *R: A Language and Environment for Statistical Computing*; R Foundation for Statistical Computing: Vienna, Austria, 2018.
43. Oksanen, J.; Blanchet, F.G.; Friendly, M.; Kindt, R.; Legendre, P.; McGlinn, D.; Minchin, P.R.; O'Hara, R.B.; Simpson, G.L.; Solymos, P.; et al. Vegan: Community Ecology Package. R package version 2.5-4. 2019. Available online: https://CRAN.R-project.org/package=vegan (accessed on 10 June 2020).
44. Hervé, M. RVAideMemoire: Testing and Plotting Procedures for Biostatistics. R package version 0.9-73. 2019. Available online: https://CRAN.R-project.org/package=RVAideMemoire (accessed on 10 June 2020).

45. Gu, Z.; Eils, R.; Schlesner, M. Complex heatmaps reveal patterns and correlations in multidimensional genomic data. *Bioinformatics* **2016**, *32*, 2847–2849. [CrossRef]
46. Gu, Z.; Gu, L.; Eils, R.; Schlesner, M.; Brors, B. Circlize implements and enhances circular visualization in R. *Bioinformatics* **2014**, *30*, 2811–2812. [CrossRef]
47. Delgado-Pando, G.; Fischer, E.; Allen, P.; Kerry, J.P.; O'Sullivan, M.; Hamill, R.M. Salt content and minimum acceptable levels in whole-muscle cured meat products. *Meat Sci.* **2018**, *139*, 179–186. [CrossRef] [PubMed]
48. Zheng, J.; Wittouck, S.; Salvetti, E.; Franz, C.M.; Harris, H.M.; Mattarelli, P.; O'Toole, W.; Pot, B.; Vandamme, P.; Walter, J.; et al. A taxonomic note on the genus *Lactobacillus*: Description of 23 novel genera, emended description of the genus Lactobacillus Beijerinck 1901, and union of Lactobacillaceae and Leuconostocaceae. *Int. J. Syst. Evol. Microbiol.* **2020**, *70*, 2782–2858. [CrossRef] [PubMed]
49. Chaillou, S.; Champomier-Vergès, M.C.; Cornet, M.; Crutz-Le Coq, A.M.; Dudez, A.M.; Martin, V.; Beaufils, S.; Darbon-Rongère, E.; Bossy, R.; Loux, V.; et al. The complete genome sequence of the meat-borne lactic acid bacterium *Lactobacillus sakei* 23K. *Nat. Biotechnol.* **2005**, *23*, 1527–1533. [CrossRef] [PubMed]
50. Kumar, P.; Chatli, M.K.; Verma, A.K.; Mehta, N.; Malav, O.P.; Kumar, D.; Sharma, N. Quality, functionality, and shelf life of fermented meat and meat products: A review. *Crit. Rev. Food Sci. Nutr.* **2017**, *57*, 2844–2856. [CrossRef]
51. Masson, F.; Johansson, G.; Montel, M.C. Tyramine production by a strain of *Carnobacterium divergens* inoculated in meat–fat mixture. *Meat Sci.* **1999**, *52*, 65–69. [CrossRef]
52. Amadoro, C.; Rossi, F.; Piccirilli, M.; Colavita, G. *Tetragenococcus koreensis* is part of the microbiota in a traditional Italian raw fermented sausage. *Food Microbiol.* **2015**, *50*, 78–82. [CrossRef]
53. Pisacane, V.; Callegari, M.L.; Puglisi, E.; Dallolio, G.; Rebecchi, A. Microbial analyses of traditional Italian salami reveal microorganisms transfer from the natural casing to the meat matrix. *Int. J. Food Microbiol.* **2015**, *207*, 57–65. [CrossRef]
54. Stavropoulou, D.A.; Filippou, P.; De Smet, S.; De Vuyst, L.; Leroy, F. Effect of temperature and pH on the community dynamics of coagulase-negative staphylococci during spontaneous meat fermentation in a model system. *Food Microbiol.* **2018**, *76*, 180–188. [CrossRef]
55. Geeraerts, W.; Stavropoulou, D.A.; De Vuyst, L.; Leroy, F. Meat and Meat Products. In *How Fermented Foods Feed a Healthy Gut Microbiota*; Azcarate-Peril, M.A., Arnold, R.R., Bruno-Bárcena, J.M., Eds.; Springer Nature: Cham, Switzerland, 2019; pp. 57–90.
56. Amend, A.S.; Seifert, K.A.; Bruns, T.D. Quantifying microbial communities with 454 pyrosequencing: Does read abundance count? *Mol. Ecol.* **2010**, *19*, 5555–5565. [CrossRef]
57. Elbrecht, V.; Leese, F. Can DNA-based ecosystem assessments quantify species abundance? Testing primer bias and biomass—Sequence relationships with an innovative metabarcoding protocol. *PLoS ONE* **2015**, *10*, e0130324. [CrossRef]
58. Baruzzi, F.; Matarante, A.; Caputo, L.; Morea, M. Molecular and physiological characterization of natural microbial communities isolated from a traditional Southern Italian processed sausage. *Meat Sci.* **2006**, *72*, 261–269. [CrossRef] [PubMed]
59. Iacumin, L.; Comi, G.; Cantoni, C.; Cocolin, L. Ecology and dynamics of coagulase-negative cocci isolated from naturally fermented Italian sausages. *Syst. Appl. Microbiol.* **2006**, *29*, 480–486. [CrossRef] [PubMed]
60. Aymerich, T.; Martin, B.; Garriga, M.; Hugas, M. Microbial quality and direct PCR identification of lactic acid bacteria and nonpathogenic staphylococci from artisanal low-acid sausages. *Appl. Environ. Microbiol.* **2003**, *69*, 4583–4594. [CrossRef] [PubMed]
61. Drosinos, E.H.; Mataragas, M.; Xiraphi, N.; Moschonas, G.; Gaitis, F.; Metaxopoulos, J. Characterization of the microbial flora from a traditional Greek fermented sausage. *Meat Sci.* **2005**, *69*, 307–317. [CrossRef]
62. Drosinos, E.H.; Paramithiotis, S.; Kolovos, G.; Tsikouras, I.; Metaxopoulos, I. Phenotypic and technological diversity of lactic acid bacteria and staphylococci isolated from traditionally fermented sausages in Southern Greece. *Food Microbiol.* **2007**, *24*, 260–270. [CrossRef]
63. Janssens, M.; Van der Mijnsbrugge, A.; Mainar, M.S.; Balzarini, T.; De Vuyst, L.; Leroy, F. The use of nucleosides and arginine as alternative energy sources by coagulase-negative staphylococci in view of meat fermentation. *Food Microbiol.* **2014**, *39*, 53–60. [CrossRef]
64. Sawitzki, M.C.; Fiorentini, Â.M.; Bertol, T.M.; Sant'Anna, E.S. *Lactobacillus plantarum* strains isolated from naturally fermented sausages and their technological properties for application as starter cultures. *Food Sci. Technol.* **2009**, *29*, 340–345. [CrossRef]

65. Talon, R.; Leroy, S.; Lebert, I. Microbial ecosystems of traditional fermented meat products: The importance of indigenous starters. *Meat Sci.* **2007**, *77*, 55–62. [CrossRef]
66. Ravyts, F.; De Vuyst, L.; Leroy, F. Bacterial diversity and functionalities in food fermentations. *Eng. Life Sci.* **2012**, *12*, 356–367. [CrossRef]
67. Leroy, S.; Giammarinaro, P.; Chacornac, J.P.; Lebert, I.; Talon, R. Biodiversity of indigenous staphylococci of naturally fermented dry sausages and manufacturing environments of small-scale processing units. *Food Microbiol.* **2010**, *27*, 294–301. [CrossRef]
68. Marty, E.; Buchs, J.; Eugster-Meier, E.; Lacroix, C.; Meile, L. Identification of staphylococci and dominant lactic acid bacteria in spontaneously fermented Swiss meat products using PCR–RFLP. *Food Microbiol.* **2012**, *29*, 157–166. [CrossRef] [PubMed]
69. Djinović, J.; Popović, A.; Jira, W. Polycyclic aromatic hydrocarbons (PAHs) in different types of smoked meat products from Serbia. *Meat Sci.* **2008**, *80*, 449–456. [CrossRef] [PubMed]
70. Casaburi, A.; Piombino, P.; Nychas, G.-J.; Villani, F.; Ercolini, D. Bacterial populations and the volatilome associated to meat spoilage. *Food Microbiol.* **2015**, *45*, 83–102. [CrossRef] [PubMed]
71. Flores, M.; Olivares, A. Flavor. In *Handbook of Fermented Meat and Poultry*, 2nd ed.; Toldrà, F., Hui, Y.H., Astiasarán, I., Sebranek, J.G., Talon, R., Eds.; John Wiley and Sons: Chichester, UK, 2015; pp. 217–225.
72. Montanari, C.; Bargossi, E.; Gardini, A.; Lanciotti, R.; Magnani, R.; Gardini, F.; Tabanelli, G. Correlation between volatile profiles of Italian fermented sausages and their size and starter culture. *Food Chem.* **2016**, *192*, 736–744. [CrossRef]
73. Belleggia, L.; Milanović, V.; Ferrocino, I.; Cocolin, L.; Haouet, M.N.; Scuota, S.; Maolonia, A.; Garofalo, C.; Cardinalia, F.; Aquilantia, L.; et al. Is there any still undisclosed biodiversity in Ciauscolo salami? A new glance into the microbiota of an artisan production as revealed by high-throughput sequencing. *Meat Sci.* **2020**. [CrossRef]

foods

MDPI

Article

Metabolism of *Lactobacillus sakei* Chr82 in the Presence of Different Amounts of Fermentable Sugars

Federica Barbieri [1], Luca Laghi [1,2], Fausto Gardini [1,2], Chiara Montanari [2,*] and Giulia Tabanelli [2,3]

1 Department of Agricultural and Food Sciences, University of Bologna, 47521 Cesena, Italy; federica.barbieri16@unibo.it (F.B.); l.laghi@unibo.it (L.L.); fausto.gardini@unibo.it (F.G.)
2 Interdepartmental Center for Industrial Agri-Food Research, University of Bologna, 47521 Cesena, Italy; giulia.tabanelli2@unibo.it
3 Department of Agricultural and Food Sciences, University of Bologna, 40127 Bologna, Italy
* Correspondence: chiara.montanari8@unibo.it; Tel.: +39-34-0649-7745

Received: 26 March 2020; Accepted: 27 May 2020; Published: 2 June 2020

check for updates

Abstract: *Lactobacillus sakei* is widely used as a starter culture in fermented sausages since it is well adapted to meat environments and able to maintain high viability thanks to secondary pathways activated when hexoses are depleted (i.e., metabolism of pentoses and amino acids). In this study, a commercial strain of *L. sakei* was inoculated in a defined medium with ribose or glucose as the carbon source, at optimal or reduced concentrations, to evaluate its different physiological and metabolic responses in relation to different growth conditions. The results obtained with different approaches (HPLC, ^{1}H-NMR, flow cytometry) evidenced different growth performances, amino acid consumptions and physiological states of cells in relation to the carbon source as an active response to harsh conditions. As expected, higher concentrations of sugars induced higher growth performances and the accumulation of organic acids. The low sugars amount induced the presence of dead cells, while injured cells increased with ribose. Arginine was the main amino acid depleted, especially in the presence of higher ribose, and resulted in the production of ornithine. Moreover, the ^{1}H-NMR analysis evidenced a higher consumption of serine at the optimal sugars concentration (pyruvate production). This information can be helpful to optimize the use of these species in the industrial production of fermented sausages.

Keywords: *Lactobacillus sakei*; sugar metabolism; amino acid metabolism; ^{1}H-NMR; flow cytometry

1. Introduction

Lactobacillus sakei is a species with a high level of adaptation for meat environments in which it can rapidly grow and efficiently compete with other species present as components of the microbial communities of this raw material. Due to this aptitude, selected strains of this species are widely used as starter cultures in meat fermentation for dry sausages production [1]. The technological and safety advantage of the use of this species as a starter culture consists in its ability to inhibit pathogenic as well as spoilage microorganisms, to grow at a low temperature and to colonize the habitat during all of the ripening period, competing with undesired species [2]. This latter aspect, essential for guaranteeing the quality of fermented sausages throughout all the steps of production and commercialization, depends on its ability to efficiently produce metabolic energy even when the hexoses, which are fermented through the homofermentative pathway, are completely depleted. In fact, this species can also ferment pentoses contained in the nucleosides via the phosphoketolase pathway, as demonstrated by McLeod et al. [3] and Rimaux et al. [4].

Moreover, the arginine deiminase (ADI) pathway is active in *L. sakei*, even if with different efficiency. This pathway is an important additional energy source giving a competitive advantage in matrices with a low fermentable sugar concentration but high arginine content, such as meat [5,6].

The mean genotype size of *L. sakei* is relatively small (approximately 2020 kb) and reflects this specialization, even if a great variation in genome dimension is observed within the species (about 25%) [7]. The adaptation to grow in proteinaceous matrices (meat and fish) explains the absence of genes responsible for amino acid anabolism, in particular, transaminases: *L. sakei* strains are auxotrophic for all the amino acids, except for aspartate and glutamate [8]. Nevertheless, the metabolisms of some of these compounds are crucial for explaining the success of this species in the colonization of fermented meat.

In addition to the use of arginine for energy supply, other amino acids may be useful for the energetic strategies of this bacterium. Serine may be deaminated by L-serine dehydratase, yielding a surplus of pyruvate, and relevant uptakes of this molecule in defined media by *L. sakei* have been observed [5,6]. Threonine [5] and cysteine [6] were depleted in remarkable amounts (higher than those required by the generation of the intracellular amino acid pool) by this species under defined conditions. The presence of a gene coding for L-threonine dehydrogenase active in some *L. sakei* strains has been described: this protein catalyzes the conversion to glycine via 2-amino-3-ketobutyrate with a concomitant NAD^+ reduction to NADH [5].

Survival and growth in environments poor in fermentable sugars have been also explained by an efficient pyruvate metabolism carried out for generating further ATP and gaining reducing power (regeneration of NAD^+). The pyruvate formate lyase (PFL) pathway leads to the possible accumulation of by-products such as formate, acetate and ethanol in anaerobic or reducing conditions, while in an aerobic condition, CO_2 and acetate may be produced through the pyruvate oxidase (POX) pathway and the pyruvate dehydrogenase complex (PDC). Enzymes involved in these pathways were found in *L. sakei* and their transcription was enhanced in the presence of pentoses as the fermentable substrate [9,10].

The same authors also demonstrated that glucose availability can affect different parameters such as growth rate, fermentative pathway (i.e., shift from homolactic to mixed acid fermentation), amino acid consumption and gene expression, but no effect on cell viability (in terms of percentage of alive cells) was observed. They hypothesized that this condition of low glucose availability is for *L. sakei* analogous to the so-called "complete caloric restriction", that in eukaryotes, from single-celled yeast to humans, is a conserved mechanism that results in an expanded healthy life span in response to a reduction in energy intake [5].

A previous work was focused on the evaluation of the metabolic response of resting cells of six *L. sakei* strains in relation to the sugar presence (glucose or ribose) [6]. Cells in the stationary phase were inoculated at high concentrations (about 9 log CFU/mL) in a defined medium (DM) and incubated for 24 h to assess the consumption of sugars and amino acids and the resulting accumulation of organic acids and other metabolites.

Based on the obtained results, the strain *L. sakei* Chr82 (used as a commercial starter culture in fermented sausage production) was chosen for the present work in order to study its growth, survival and metabolic response when inoculated at about 7 log CFU/mL in different DMs containing glucose or ribose at two initial concentrations (25 mM or 2.5 mM), to simulate an optimal or limited growth condition.

With the aim to better investigate the physiological response of this strain, different analytical approaches were used. In particular, cell cultivability was tested by plate count while flow cytometry was used to assess cell membrane permeability and depolarization as well as cell viability. Moreover, organic acid accumulation and amino acid variation were quantified by HPLC and the results were compared and discussed with metabolome analyses, performed by ^{1}H-NMR. This latter aspect, that is the complete set of small metabolites consumed or produced [11], was done on the assumption that the metabolome would be the best representation of the microorganism phenotype, being downstream of the genome, transcriptome, and proteome. The analytical platform selected for this purpose was ^{1}H-NMR, whose high reproducibility was expected to counterbalance the low sensitivity [12]. Moreover, this technique does not require derivatization or molecular separation and allows the untargeted, simultaneous detection of molecules pertaining to a broad range of chemical classes.

2. Materials and Methods

2.1. Microorganism Used

The commercial strain *Lactobacillus sakei* Chr82, supplied by the company Chr. Hansen (Parma, Italy), used as a starter culture in the production of fermented cured meats, was used.

2.2. Growth Media

L. sakei Chr82 was pre-grown in MRS broth prepared according to Oxoid formulation (peptone 10 g/L, lab-lemco powder 8 g/L, yeast extract 4 g/L, Tween 80 1 mL/L, dipotassium hydrogen phosphate 2 g/L, sodium acetate $3H_2O$ 5 g/L, triammonium citrate 2 g/L, magnesium sulphate $7H_2O$ 0.2 g/L, manganese sulphate $4H_2O$ 0.05 g/L) and with the addition of two different sugars: in one case, 4.5 g/L of glucose, while in the other, 3.75 g/L of ribose were added. In the medium with ribose, according to the observation by McLeod et al. [3], a small amount of glucose was also added (0.2 g/L) in order to stimulate the growth of the microorganism in the initial phase.

The cells grown overnight under statically microaerophilic conditions at 30 °C in modified MRS were collected by centrifugation at 10,000 rpm for 10 min and suspended in a defined medium (DM), containing macro components, vitamins, nucleotides and amino acids. This DM, whose composition is reported in Table 1, is a modification of the medium proposed by Lauret et al. [13] for the growth of *L. sakei*. Amino acids were added at 0.2 g/L. The consequent mM concentration was as follows: alanine (ala) 2.24, arginine (arg) 1.15, asparagine (asg) 1.52, aspartic acid (asp) 1.50, cysteine (cys) 1.65, glutamic acid (glu), glutamine (glm) 1.36, glycine (gly) 2.66, histidine (his) 1.29, isoleucine (ile) 1.52, leucine (leu) 1.52, lysine (lys) 1.37, methionine (met) 1.34, phenylalanine (phe) 1.21, proline (pro) 1.74, serine (ser) 1.90, threonine (thr) 1.68, tryptophan (try) 0.98, tyrosine (tyr) 1.10, valine (val) 1.71.

Table 1. Chemically defined medium (DM) composition.

Compound	Concentration (g/L)
Macro components	
Sodium acetate	2.0
K_2HPO_4	1.75
$MnSO_4$ $4H_2O$	0.012
$MgSO_4$ $7H_2O$	0.2
Tween 80	1 mL
Vitamins	
Thiamine HCl	0.0011
Folic acid	0.0002
Riboflavin	0.001
Calcium pantothenate	0.001
Nicotinic acid	0.001
Pyridoxal-5-phosphate	0.0005
p-amino benzoic acid	0.0004
Nucleotides	
Adenine	0.005
Guanine	0.01
Uracil	0.01

The cells of *L. sakei* grown in the presence of glucose were suspended in DM with 2.5 mM of glucose and subsequently inoculated in the two DM added with glucose 2.5 mM (2.5 G) or 25 mM (25 G), at a cell concentration of about 7 log CFU/mL. The same procedure was applied for *L. sakei* Chr82 cells grown with ribose, but in this case the cells were suspended in DM with ribose 2.5 mM and then inoculated (cell load 7 log CFU/mL) in the two DM added with ribose 2.5 mM (2.5 R) or 25 mM (25 R). In the samples containing ribose, a small amount of glucose (0.1 mM) was added in order to provide the energy needed to activate the ribose metabolism related genes, as indicated by McLeod et al. [3]. The medium was sterilized by filtration at 0.22 µm (Sartorius Lab Instruments GmbH & Co. KG, Göttingen, Germany). The initial pH of the medium was 6.50 ± 0.02.

Inoculated samples were incubated at 30 °C and monitored at different times.

2.3. Growth Modeling and pH Measurement

Growth performances were analyzed by measuring the increase in the optical density at 600 nm (OD_{600}) using the UV–VIS spectrophotometer 6705 UV- Vis (Jenway, Stone, UK). Before each detection, a calibration of the instrument was performed with the blank (non-inoculated medium) of the respective sample.

The results of the optical density were modelled using the STATISTICA program (Statsoft Italia, Vigonza, Italy) through the Gompertz equation [14]:

$$y = k + A \cdot e^{-e^{[(\frac{\mu max \cdot e}{A}) \cdot (\lambda - t) + 1]}}$$

where y is the OD_{600} at time t, A represents the maximum OD_{600} value reached, μ_{max} is the maximum OD_{600} increase rate in the exponential phase and λ is the lag time.

The pH meter Basic 20 (Crison, Modena, Italy) was used for the sample pH measurement in order to monitor the acidification activity in the different conditions.

2.4. Microbiological Analysis

The microbiological counts of *L. sakei* Chr82 were carried out by plate counting in MRS agar (Oxoid, Basingstoke, United Kingdom) incubated aerobically for 48 h at 30 °C.

2.5. Organic Acids Content

The quantification of organic acids was performed using a HPLC instrument (PU-2089 Intelligent HPLC quaternary pump, UV-VIS multiwavelength detector UV 2070 Plus; Jasco Corp., Tokyo, Japan) equipped with a manual Rheodyne injector with a 20 µL loop (Rheodyne, Rohnert Park, CA, USA) and a Bio-Rad Aminex HPX-87H column with a size of 300 × 7.8 mm (Bio-Rad Laboratories, Hertfordshire, UK).

The analysis was performed in isocratic conditions at 65 °C with a rate flow of 0.6 mL/min of mobile phase H_2SO_4 0.005 M. The UV detector was set at 210 nm. Chromatographic peaks were identified by comparing retention times with those of standards (Sigma-Aldrich, St. Louis, MO, USA) and quantification was carried out by using the external standard method.

2.6. Quantification of Amino Acids

To evaluate the variation in the amino acid concentration, samples were analyzed by HPLC (PU-1580 Intelligent HPLC, Intelligent Fluorescence Detector FP-1520 and Intelligent Sampler AS-2055 Plus, with 10 µL loop; Jasco Corp., Tokyo, Japan), after a derivatization using an AccQ-Fluor Reagent kit (Waters Corp., Milford, MA, USA) according to the method described by Montanari et al. [6].

The separation of amino acids was performed using an AccQ-Tag$_{TM}$ column (3.9 × 150 mm; Waters Corp.) at 30 °C using mobile phase A (100 mL of AccQ-Tag Eluent (Waters Corp., Milford, MA, USA), diluted 1:10 with H_2O for chromatography (Sigma-Aldrich, St. Louis, MO, USA) and mobile phase B (60% acetonitrile and 40% H_2O for chromatography (Sigma-Aldrich, St. Louis, MO, USA)) at a flow rate of 1 mL/min. The fluorescent detector was set at an excitation wavelength of 250 nm and emission wavelength of 395 nm. Under the adopted conditions, good separation of the amino acids was obtained with the exception of the couples histidine + glutamine and serine + asparagine, which coeluted in unique peaks. Tryptophan was not detectable with this protocol.

2.7. Flow Cytometric Analysis

Flow cytometry (FCM) was used to monitor the physiological state of *L. sakei* Chr82 cells in each sample. Cell suspensions were analyzed with the flow cytometer Accuri C6 (BD Biosciences, Milan, Italy), using setting parameters, emission filters and thresholds according to Arioli et al. [15].

Before the analysis, where necessary, the samples were diluted in the corresponding DM up to a concentration of 7 log CFU/mL, the optimal cell density for a correct sample staining by fluorochromes.

The cells were stained with SYBR-Green I (1X), propidium iodide (PI) 7.5 μM and $DiBAC_4$ (3) (Bis-(1,3-Dibutylbarbituric Acid) Trimethine Oxonol) 3.0 μM as reported by Tabanelli et al. [16]. The data obtained were analyzed using the BD ACCURITM C6 software version 1.0 (BD Biosciences, Milan, Italy). Before analysis, each aliquot was kept at 37 °C for 15 min in order to let the dye react with the cells.

2.8. Untargeted Metabolomics Analysis by ^{1}H-NMR

For the metabolomics investigation by ^{1}H-NMR, an analysis solution was created, with 3-(trimethylsilyl)-propionic-2,2,3,3-d4 acid sodium salt (TSP) 10 mM in D_2O, set at pH 7.00 ± 0.02 by means of 1 M phosphate buffer. The solution contained also 10 μL of NaN_3 2 mM, to avoid microbial proliferation, while TSP was employed as the ^{1}H NMR chemical-shift reference, as suggested by Zhu et al. [17]. Growth medium samples were prepared for ^{1}H-NMR by thawing and centrifuging 1 mL of each for 15 min at 18,630 g and 4 °C. An amount of 700 μL of supernatant was added to 200 μL of the ^{1}H NMR analysis solution. Finally, each of the so-obtained samples was centrifuged again at the above conditions right before analysis.

^{1}H-NMR spectra were recorded at 298 K with an AVANCE III spectrometer (Bruker, Milan, Italy), operating at a frequency of 600.13 MHz, equipped with the software Topspin 3.5. Following the procedure described by Laghi et al. [12], the HOD residual signal was suppressed by applying the first increment of the nuclear overhauser effect spectroscopy (NOESY) pulse sequence and a spoil gradient. This was done by employing the NOESYGPPR1D sequence, part of the standard pulse sequence library. Each spectrum was acquired by summing up 256 transients using 32 K data points over a 7184 Hz spectral window, with an acquisition time of 2.28 s. The spectra were phase- and baseline-adjusted in Tospin, that was employed also for the calculation of the signal-to-noise ratio. Spectra were elaborated with the R package (R Core Team, 2018, Vienna, Austria) as reported by Zhu et al. [17]. Molecules identification was performed by comparing their signals with those of pure compounds by the Chenomx software ver. 8.3 (Chenomx Inc., Edmonton, AB, Canada) with the Chenomx (ver. 10) and HMDB (release 2) libraries.

3. Results and Discussion

3.1. Determination of Growth Curves and Microbiological Analysis

The DM containing the two different sugars (glucose and ribose) at the two different concentrations (2.5 and 25 mM) were inoculated with approximately 7 log CFU/mL of *L. sakei* Chr82. The experimental data of the growth dynamics monitored by measuring the OD_{600} were modelled with the Gompertz equation [14] and are reported in Supplementary Materials Figure S1. The growth parameters obtained, together with the cell concentrations after 24 and 48 h of incubation, are summarized in Table 2. The amount of sugars influenced the maximum OD_{600}, which reached values of 0.264 and 0.282 in the presence of glucose and ribose at 2.5 mM, respectively, while, under the same conditions, the maximum

OD_{600} predicted when the sugars were added at 25 mM were 1.446 and 1.151. The addition of ribose determined a slightly longer λ time and lower μ_{max}. Regarding the cell concentrations after 24 and 48 h of growth, determined by plate counting, no significant differences were found after 24 h in the different media (Table 2). After 48 h, cell loads showed a drastic decline (1 log unit or more) if compared with the counts at 24 h and the higher survival rate was observed in the sample containing 2.5 mM·R. The rapid beginning of the death phase in culture media for this species has already been observed [18] and it is in contrast with the long survival showed by *L. sakei* in stricter conditions as those characterizing fermented sausages during ripening.

Table 2. Cell load (expressed as log CFU/mL) after 24 and 48 h of incubation of *L. sakei* Chr82 at 30 °C in the different DM and growth parameters obtained modelling the growth dynamics in different DM (measure of OD_{600}) with the Gompertz equation.

Sample	Time	Cell Counts (Log CFU/mL)	Growth Parameters (Gompertz Equation) k	A	μ_{max}	λ
	0 h	6.94 (±0.21)				
2.5 G	24 h 48 h	8.15 (±0.23) 6.67 (±0.11)	0.044	0.264	0.155	11.457
2.5 R	24 h 48 h	8.41 (±0.16) 7.62 (±0.13)	0.046	0.282	0.120	11.570
25 G	24 h 48 h	8.13 (±0.21) 6.98 (±0.13)	0.045	1.446	0.280	10.839
25 R	24 h 48 h	8.10 (±0.11) 6.89 (±0.14)	0.074	1.151	0.275	11.257

3.2. Organic Acid Content and pH

In Table 3, the organic acids accumulated after 24 and 48 h of incubation and detected by HPLC are reported.

Table 3. Organic acid content and pH values of *L. sakei* Chr82 samples incubated at 30 °C in the different DMs. Acetic acid concentration is expressed as a difference with respect to the initial amount added in the media as sodium acetate (24 mM).

Sample	Time	L-Lactic Acid (mM)	D-Lactic Acid (mM)	Acetic Acid (mM)	pH
2.5 G	24 h	4.25 (±0.09)	- *	0.52 (±0.20)	5.78 (±0.29)
	48 h	3.88 (±0.14)	-	0.62 (±0.15)	5.83 (±0.16)
2.5 R	24 h	1.66 (±0.07)	-	2.85 (±0.11)	6.17 (±0.30)
	48 h	1.39 (±0.04)	-	3.15 (±0.09)	6.21 (±0.27)
25 G	24 h	39.60 (±1.98)	0.41 (±0.09)	2.58 (±0.03)	4.15 (±0.17)
	48 h	43.81 (±2.05)	0.39 (±0.10)	2.77 (±0.28)	4.00 (±0.22)
25 R	24 h	15.61 (±0.90)	1.43 (±0.20)	21.00 (±0.43)	4.73 (±0.21)
	48 h	16.02 (±0.63)	1.09 (±0.14)	22.70 (±0.67)	4.37 (±0.24)

* Under the detection limit (0.1 mM).

Considering the lower sugar concentration (2.5 mM), only L-lactate was detected in these samples. In the presence of glucose, its presence (approximately 4 mM) was accompanied by a lower proportion of acetate (0.52 and 0.62 mM after 24 and 48 h, respectively). Higher amounts of acetate (approximately 3 mM) were produced, as expected, in the presence of ribose. In addition, the molar production of the two acids represented more that 90% of the theoretical yield. In the presence of ribose, the ratio acetate/lactate was higher than 1, indicating the activation of pathways alternative to homolactic and heterolactic fermentations [19].

The presence of fermentable carbohydrates at 25 mM determined the production of more than 40 mM of lactate and small amounts (2 mM) of acetate in the medium added with glucose, while the addition of ribose resulted in the accumulation of lactate (more than 17 mM, including L- and D-lactate) and relevant quantities of acetate (more than 21 mM). In this latter case, the quantitative production of acetate was higher than expected (acetate/lactate molar ratio higher than 1) as a consequence of the activity of secondary pathways. As observed by McLeod et al. [9], *L. sakei* alters its pyruvate metabolism when grown in the presence of ribose, generating more ATP per ribose unit up-regulating pyruvate decarboxylases and pyruvate dehydrogenases which can bring to the accumulation of acetate [19].

The presence of glucose compared with ribose always determined a lower pH (Table 3), and after 48 h, values of 5.83 vs. 6.21 were found with the addition of the sugars at 2.5 mM, while at 25 mM, the pH values measured were 4.00 and 4.37 for glucose and ribose, respectively.

3.3. Amino Acids Quantification

It is well known that *L. sakei*, because of its adaptation to meat environments, is auxotrophic for 18 amino acids [8]. The study of the variations in the amino acid content in a defined medium is important to elucidate how this species uses these molecules. In the first instance, they are used to assemble all the proteins (and enzymes) necessary to sustain growth and multiplication. However, it is interesting to evidence amino acid alternative uses in the perspective of explaining the high persistence of *L. sakei* cells in habitats, such as fermented sausages, in which fermentable sugars are rapidly depleted. In a previous work, Montanari et al. [6] described amino acid variations due to the resuspension of resting cells in a defined medium. In the present research, the changes in amino acid concentration are studied after *L. sakei* growth, and the influence of cell metabolism on each amino acid, as revealed by HPLC analyses, is shown in Table 4.

The concentration of many amino acids after 24 and 48 h of incubation showed small variations with respect to the initial level. Aspartate, alanine, valine, lysine and leucine were always consumed in an amount lower than 20% of the initial concentration. Smaller variations were observed for isoleucine, tyrosine, threonine and glycine. Phenylalanine was accumulated (up to 17% of the initial concentration) in all the conditions tested with a trend similar to that observed by McLeod et al. [5]. Glutamic acid, generally consumed in all the other conditions, was accumulated in the presence of glucose 25 mM. The remaining amino acids were subjected to more relevant variations (Figure 1).

The use of arginine to produce ATP through the ADI pathway has been well studied in *L. sakei* [20]. However, different patterns of decrease have been observed among strains [5,6]. In this case, glucose retarded arginine depletion, even if after 48 h, the presence at 2.5 mM of this sugar caused a high consumption of this amino acid (Figure 1a). By contrast, the presence of ribose as a fermentable sugar determined a rapid and massive consumption of this amino acid, independently of the ribose concentration.

The activation of the ADI pathway is considered crucial to allow the survival of this species in meat environments. Some strains possess a second putative ADI pathway, which improves their ability to take advantages from the high amounts of arginine in meats [5]. This activity is confirmed by the accumulation of ornithine (Figure 1b). Ornithine is the final product of the ADI pathway and was produced in higher amounts in the presence of ribose, particularly 2.5 mM. The absence of a correspondence between arginine consumption and ornithine production could be attributed to the ability of this strain to decarboxylate this amino acid. In fact, only *L. sakei* Chr82, among the six strains tested by Montanari et al. [6], was able to produce putrescine from the decarboxylation of ornithine. However, in the conditions adopted in the present study, this biogenic amine was never detected.

Table 4. Amino acid concentrations (mM) detected by HPLC in DM after 24 and 48 h of incubation of *L. sakei* Chr82 at 30 °C in the different DMs. In brackets, the relative variations (as percentage) with respect to the initial concentration are reported.

Sample	Time	Asp	Ser + Asg	Glu	Gly	His + Glm	Arg	Thr	Ala	Pro	Cys	Tyr	Val	Met	Lys	Ile	Leu	Phe
Initial concentration		1.50	3.42	1.36	2.66	2.66	1.15	1.68	2.24	1.74	1.65	1.10	1.71	1.34	1.37	1.52	1.52	1.21
2.5 G	24 h	1.39 (−7.49)	3.15 (−7.94)	1.27 (−6.53)	2.91 (9.53)	2.25 (−15.27)	1.01 (−12.00)	1.79 (6.57)	2.12 (−1.09)	1.92 (10.49)	2.18 (32.02)	1.11 (0.85)	1.84 (7.68)	1.27 (−5.45)	1.36 (−0.45)	1.50 (−1.54)	1.39 (−8.51)	1.28 (6.02)
	48 h	1.27 (−15.53)	2.65 (−22.41)	1.07 (−21.36)	2.50 (−5.97)	1.85 (−30.44)	0.29 (−74.57)	1.80 (7.29)	2.03 (−9.22)	1.80 (3.20)	2.45 (48.56)	1.04 (−5.83)	1.66 (−2.90)	0.69 (−48.88)	1.44 (4.96)	1.53 (0.68)	1.49 (−2.22)	1.41 (16.51)
2.5 R	24 h	1.32 (−12.26)	2.98 (−13.00)	1.18 (−13.50)	2.81 (5.61)	2.13 (−19.94)	0.10 (−91.57)	1.65 (−1.87)	2.21 (−1.38)	1.78 (2.09)	2.01 (21.78)	1.10 (0.12)	1.69 (−0.96)	1.36 (1.18)	1.36 (−0.67)	1.58 (3.63)	1.43 (−6.03)	1.26 (4.53)
	48 h	1.44 (−11.04)	2.58 (−24.42)	1.08 (−20.47)	2.48 (−6.82)	1.79 (−32.85)	0.19 (−83.76)	1.71 (2.08)	2.02 (−10.03)	1.77 (1.75)	2.53 (53.09)	1.14 (3.93)	1.62 (−5.05)	0.94 (−30.22)	1.41 (2.99)	1.53 (0.39)	1.50 (−1.58)	1.34 (10.72)
25 G	24 h	1.30 (−4.00)	1.21 (−64.72)	1.73 (27.56)	2.71 (1.78)	1.45 (−45.33)	0.90 (−21.38)	1.56 (−6.93)	1.92 (−14.09)	1.69 (−3.07)	1.75 (6.31)	1.07 (−2.87)	1.59 (−7.27)	1.31 (−2.42)	1.23 (−9.96)	1.59 (4.93)	1.39 (−8.25)	1.42 (17.71)
	48 h	1.41 (−13.27)	0.89 (−73.95)	1.63 (19.88)	2.33 (−12.43)	1.10 (−58.54)	0.80 (−30.78)	1.36 (−19.13)	1.73 (−22.80)	1.75 (0.81)	2.00 (21.05)	1.02 (−7.05)	1.50 (−12.17)	1.26 (−6.27)	1.10 (−19.61)	1.39 (−8.81)	1.45 (−4.67)	1.32 (9.38)
25 R	24 h	1.28 (−14.55)	2.05 (−40.17)	1.33 (−2.17)	2.62 (−1.41)	1.80 (−32.31)	0.09 (−92.08)	1.77 (5.53)	2.02 (−9.78)	1.75 (0.30)	1.43 (−13.05)	1.01 (−7.88)	1.59 (−6.93)	1.15 (−14.50)	1.21 (−11.82)	1.49 (−2.04)	1.46 (−4.22)	1.35 (11.82)
	48 h	1.20 (−20.06)	1.83 (−46.41)	1.18 (−13.32)	2.35 (−11.60)	1.50 (−43.72)	0.12 (−89.31)	1.64 (−2.14)	1.86 (−17.08)	1.80 (3.60)	1.57 (−5.09)	1.08 (−2.27)	1.52 (−11.15)	1.04 (−22.76)	1.09 (−20.45)	1.34 (−11.72)	1.37 (−9.93)	1.32 (9.45)

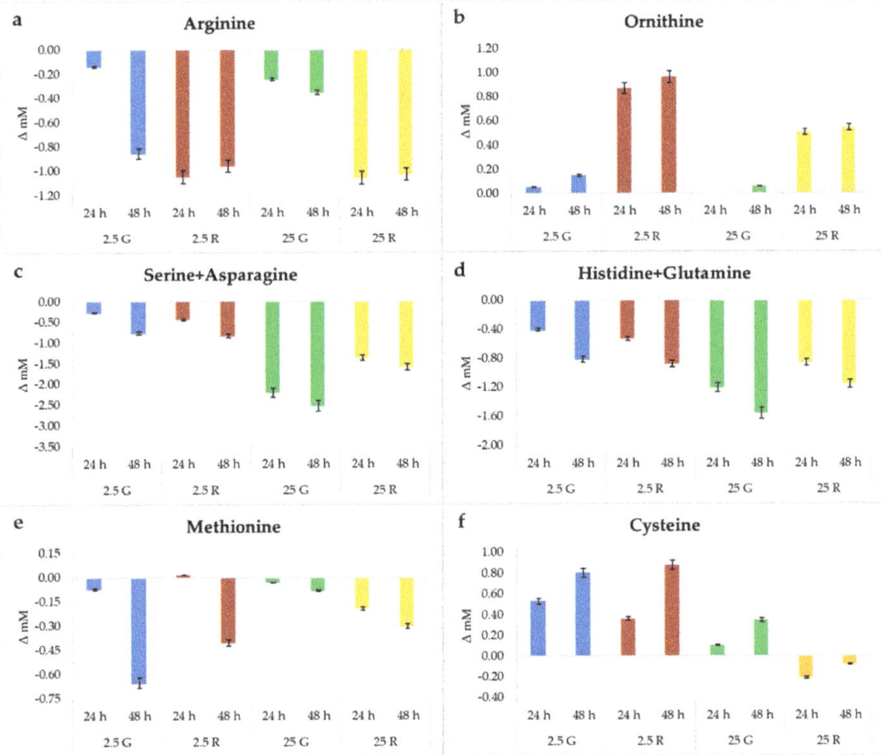

Figure 1. Variation in the amounts of amino acids characterized by the most relevant modification after 24 and 48 h of incubation at 30 °C of *L. sakei* Chr82 with respect to the initial concentration in the DM, and ornithine production. The standard deviations are reported.

The sum of serine and asparagine (not separated under the adopted HPLC analytical protocol) showed a drastic decrease, especially in the samples containing a high sugar concentration and after 48 h of incubation (Figure 1c). Serine can be used to supply pyruvate, which can then be used to produce energy through the PFL or POX pathways [21]. The conversion of serine into pyruvate has been described in *Pediococcus pentosaceus* as the result of the activity of a serine dehydratase [22], while *Lactobacillus plantarum* could metabolize serine with the production of formate, succinate and acetate [23]. McLeod et al. [5] showed a high use of serine and asparagine in *L. sakei* strains grown under glucose limiting conditions. Among the six *L. sakei* strains tested by Montanari et al. [6] under resting conditions, the strain Chr82 was the most efficient in serine + aspargine uptake in the absence of fermentable sugars. The decrease was higher in the sample containing glucose, in contrast to the trend observed for arginine.

Further, the consumption of histidine + glutamine (not separated under the adopted HPLC analytical protocol) was higher in the media containing glucose (Figure 1d). The resuspension of resting cells of the same strain in the DM did not markedly change the concentration of these amino acids. By contrast, growing cells decreased the concentration of these amino acids, especially after 48 h and when glucose was present in the medium. McLeod et al. [5] showed a strong decrease in glutamine during a continuous cultivation in a glucose-limited medium inoculated with two *L. sakei* strains, while the concentration of histidine was scarcely affected.

Finally, the sulfur amino acids methionine and cysteine showed a correlated trend (Figure 1e,f). Relevant diminutions of methionine were observed after 48 h in the samples with 2.5 mM of both sugars. However, these decreases were accompanied by concomitant increases in cysteine. Only the samples containing ribose 25 mM presented a simultaneous decrease in both amino acids.

3.4. Flow Cytometric Analysis

The same samples were also subjected to flow cytometric (FCM) analysis to define some parameters linked to cell viability. Each sample was labeled with SYBR-Green I and propidium iodide (PI) in a 1:1 ratio. This dual staining allowed to discriminate three sub-populations: alive, damaged or dead cells. The results are shown in Figure 2.

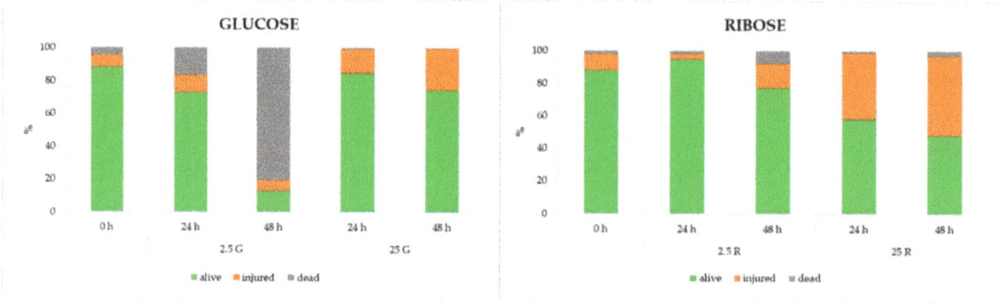

Figure 2. Distribution of alive, damaged and dead cells of *L. sakei* Chr82 after 24 and 48 h of incubation at 30 °C in different DMs. The data are reported as the relative frequency of the total population obtained by flow cytometric (FCM) analysis with dual staining (SYBR-Green I and PI).

In the samples containing glucose, a higher viability in the cells grown in the presence of 25 mM of this sugar was generally observed. The percentage of cells recognized as alive was 84.6% and 74.4%, after 24 and 48 h, respectively. At the same sampling times, the injured cells passed from 7.6% to 14.9% and 25.5% and dead cells from 4.28% to 0.5% and 0.1%, respectively. The presence of a limited amount of glucose (2.5 mM) determined a drastic increase in the dead cells (approximately 18% after 24 h and 80% after 48 h). When the ribose was added at the higher concentration (25 mM), the number of dead cells remained comparable to those observed in the presence of glucose, but the portion of damaged cells was much more relevant, already starting from 24 h of incubation. Conversely, the addition of ribose 2.5 mM resulted, after 48 h, in a higher viability if compared with the sample added with the same amount of glucose.

Regarding the membrane depolarization, expressed as fluorescence of $DiBAC_4$ (3), the results obtained for the sugars at the two concentrations (Figure 3) showed that this parameter is inversely proportional to the media pH. In fact, in the presence of the higher sugar concentrations, and therefore with the lowest media pH values (4 in the presence of glucose and 4.6 in the presence of ribose), cell membrane depolarization was greater. In the 2.5 G and 2.5 R samples, characterized by higher pH values (about 5.8 in the presence of glucose and 6.2 in the presence of ribose), the depolarization degree was lower.

Finally, the results of membrane permeability (Figure 4) showed higher values in the presence of glucose 2.5 mM, while the trend was opposite in the presence of ribose, where the membrane permeability was higher in the presence of the higher concentration of this sugar.

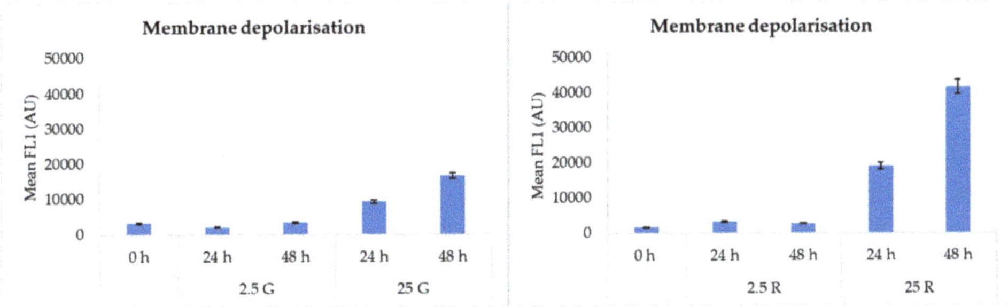

Figure 3. FCM analysis related to membrane depolarization in *L. sakei* Chr82 after 24 and 48 h of incubation at 30 °C in different DMs. Data are reported as mean fluorescence of $DiBAC_4$ (3) dye (arbitrary unit, AU).

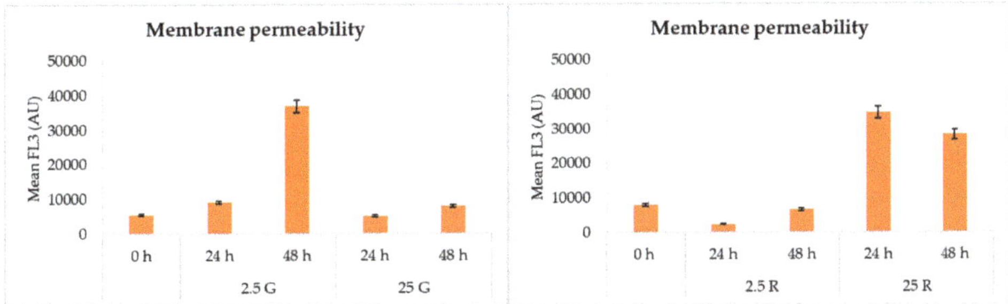

Figure 4. FCM analysis related to membrane permeability in *L. sakei* Chr82 after 24 and 48 h of incubation at 30 °C in different DMs. Data are reported as mean fluorescence of the propidium iodide (PI) (arbitrary unit, AU).

3.5. Untargeted Metabolomics Analysis by ^{1}H-NMR

With the aim to have a deeper insight in the metabolomic responses of the strain *L. sakei* Chr82, the same samples were further analyzed applying a quantification protocol based on ^{1}H-NMR. An example of portions of the ^{1}H-NMR spectrum obtained from one representative sample is reported in Supplementary Materials Figure S2. Regarding the amino acid concentration, the correlation between the NMR and HPLC results was satisfying, as demonstrated by the regression analysis reported in Figure 5, characterized by a high R^2 (0.8702), an intercept of 0.1542 and an angular coefficient close to 1 (1.1372). Previous works supported the suitability of the ^{1}H-NMR approach: for example, Biagioli et al. [24] were able to observe divergent metabolic activities of two batches of the same probiotic preparation; Parolin et al. [25] identified the metabolome traits distinguishing vaginal lactobacilli with different anti-candida activity; and Picone et al. [26] followed the adaptation of *Escherichia coli* 555 to increasing doses of carvacrol.

NMR data were focused, at a first instance, to the evaluation of the amino acids not separated by the HPLC analysis, i.e., serine + asparagine and histidine + glutamine. The results regarding these amino acids, expressed as concentration variation, are reported in Figures 6 and 7.

Taking into consideration serine + asparagine, the total decrease in these two molecules detected with the two methods were comparable. However, the ^{1}H-NMR approach indicated that the main decrease concerned serine (Figure 6), confirming the hypothesis that this amino acid can provide a supply of pyruvate, which can be addressed to alternative metabolic pathways that are important for when available sugars become a limiting factor. Liu et al. [27] proved the central role of pyruvate deriving from serine in *L. plantarum* metabolism, demonstrating that it was involved in the regeneration of NADH and in the production of ATP, acetate, formate, ethanol, acetoin, diacetyl and 2-3-butanediol.

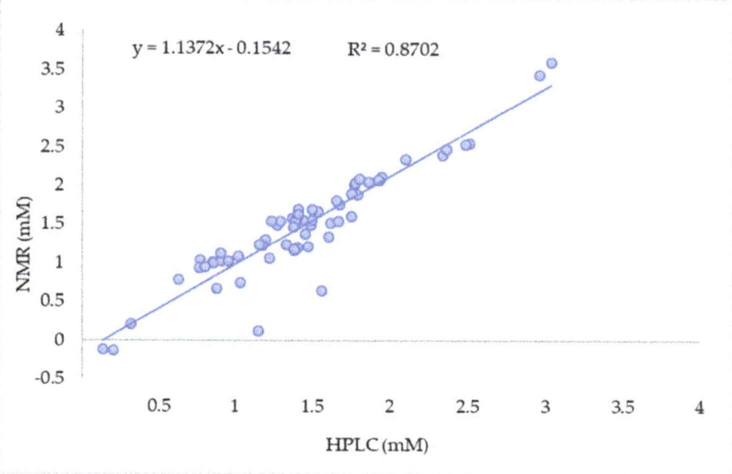

Figure 5. Correlation matrix between ^{1}H-NMR and HPLC results (expressed as mM). In the figure are reported the correlation line between NMR results (y) and HPLC results (x) and the correlation coefficient (R^2).

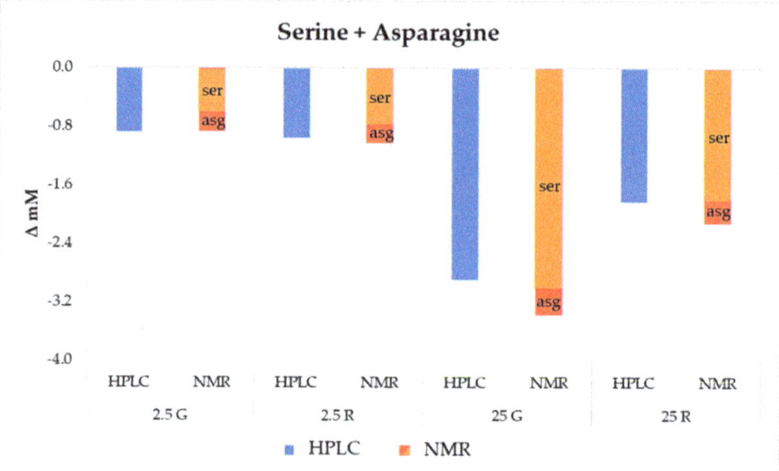

Figure 6. Variation (mM) in serine and asparagine with respect to the initial concentration in the different DMs after 48 h of incubation.

Lower concordance was observed for the data of histidine + glutamine (Figure 7). While in the presence of a low sugar concentration, the ^{1}H-NMR analysis reported lower amounts of amino acids uptake if compared with the HPLC results, and an opposite trend characterized the samples with the higher glucose and ribose concentration. In any case, according to the ^{1}H-NMR results, the consumption of the two amino acids was reduced at a similar rate in the presence of 2.5 mM of the sugars, while, when the sugars were present at high concentration, histidine was depleted much more than glutamine.

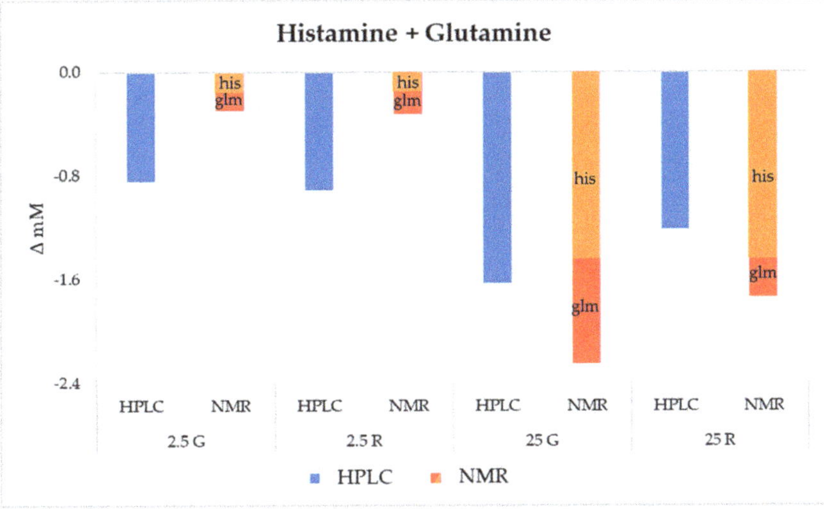

Figure 7. Variation (mM) in histidine and glutamine with respect to the initial concentration in the different DMs after 48 h of incubation.

In addition to the amino acid content, the untargeted ^{1}H-NMR protocol applied allowed the detection of other metabolic compounds (Table 5).

Table 5. Concentration (expressed as mM) of some metabolic compounds detected by ^{1}H-NMR after 48 h of *L. sakei* Chr82 incubation at 30 °C in different DMs.

	Ethanol	Acetoin	2,3-Butanediol	Glucose	Ribose
2.5 G	0.01	0.10	0.01	- *	-
2.5 R	0.01	0.15	0.01	-	-
25 G	0.03	0.14	0.06	0.14	-
25 R	0.01	0.21	0.19	-	0.15

* Under the detection limit (0.01 mM).

Regarding sugars, they resulted in being completely depleted in the media in which they were added at 2.5 mM, while in the samples added with 25 mM, small residual quantities (approximately 0.15 mM) were detected. Further, ethanol was detected in extremely low amounts, indicating that the secondary metabolic pathways activated by *L. sakei* Chr82 were mainly addressed towards the production of acetic acid which allows the production of energy rather than the regeneration of reduced NADH. The production of acetoin and 2,3-butanediol was higher in the samples containing a higher sugar concentration, particularly in the sample with ribose.

4. Conclusions

The results obtained in this work increase the knowledge on the physiological and metabolic responses of *L. sakei* in relation to different sugar amounts. Indeed, the combined use of HPLC and ^{1}H-NMR approaches allowed to better elucidate the consumption of amino acids and the resulting metabolites produced during incubation.

As expected, higher concentrations of glucose or ribose induced higher growth performances, acidification of the growth medium and accumulation of lactic and acetic acids.

The flow cytometric analysis evidenced a different physiological adaptation to the conditions; in fact, even if cells grown on glucose at a high concentration had a high viability, the same sugar in low amounts induced the presence of dead cells, while ribose determined the higher percentage of injured cells, but only few cells were recognized as dead also when this pentose was present at 2.5 mM.

The analysis of amino acids confirmed the rapid depletion of specific amino acids, mainly arginine, whose consumption was higher in the presence of ribose and resulted in the production of ornithine. Other amino acids highly consumed by this strain were serine, asparagine, glutamine and histidine. Since the HPLC protocol adopted was not able to separate these metabolites (co-elution of serine + asparagine and glutamine + histidine), ^{1}H-NMR analysis in this case was helpful to discriminate the single metabolites, allowing to evidence a higher consumption of serine, especially when sugars were present at an optimal concentration. This consumption of serine confirmed other findings reported in the literature for lactic acid bacteria, i.e., the use of this amino acid as a source of pyruvate, which can then be used to produce energy through secondary pathways.

The untargeted ^{1}H-NMR analysis performed on the samples with the aim to set up a fast method to simultaneously quantify amino acids, sugars, organic acids and other molecules was successful. This approach resulted as indeed suitable and very promising to evaluate the metabolic response of *L. sakei* in terms of the consumption and accumulation of specific metabolites.

This information can be helpful to optimize the use of this species as a starter culture for the industrial production of fermented sausages, since stressful conditions can affect the microbial technological performances or induce the activation of specific metabolic pathways, whose final products can have a significant impact on the sensorial features of the fermented sausages obtained.

Supplementary Materials: The following are available online at http://www.mdpi.com/2304-8158/9/6/720/s1, Figure S1: Growth curves of *L. sakei* Chr82 at 30 °C under different conditions. In the square the Gompertz parameters are reported. Points: experimental data; lines: fitting curves as predicted by Gompertz equation, Figure S2. Portions of 1H-NMR spectrum from *L. sakei* Chr82 inoculated in DM with 25 mM of glucose reporting, for the molecules listed in Tables 4 and 5, the signals employed for quantification. The exact extremes of the signals are represented by dashed lines (p. 14).

Author Contributions: Conceptualization, G.T. and F.G.; Literature data collection, G.T. and F.B.; Microbiological, HPLC and FCM analyses, C.M. and F.B.; NMR analyses, L.L.; Writing—original draft preparation, G.T. and F.B.; Writing—review and editing, C.M., F.G. and L.L. All authors have read and agreed to the published version of the manuscript.

Funding: This research received no external funding.

Conflicts of Interest: The authors declare no conflict of interest.

References

1. Cocconcelli, P.S.; Fontana, C. Starter cultures for meat fermentation. In *Handbook of Meat Processing*; Toldrá, F., Ed.; Blackwell Publishing: Ames, IA, USA, 2010; pp. 199–218, ISBN 9780813820897.
2. Chaillou, S.; Lucquin, I.; Najjari, A.; Zagorec, M.; Champomier-Vergès, M.C. Population genetics of *Lactobacillus sakei* reveals three lineages with distinct evolutionary histories. *PLoS ONE* **2013**, *8*, e73253. [CrossRef]
3. McLeod, A.; Nyquistb, O.N.; Snipenb, L.; Naterstada, K.; Axelsson, L. Diversity of *Lactobacillus sakei* strains investigated by phenotypic and genotypic methods. *Syst. Appl. Microbiol.* **2008**, *31*, 393–403. [CrossRef]
4. Rimaux, T.; Vrancken, G.; Vuylsteke, B.; De Vuyst, L.; Leroy, F. The pentose moiety of adenosine and inosine is an important energy source for the fermented-meat starter culture *Lactobacillus sakei* CTC 494. *Appl. Environ. Microbiol.* **2011**, *77*, 6539–6550. [CrossRef]
5. McLeod, A.; Mosleth, E.F.; Rud, I.; Branco dos Santos, F.; Snipen, L.; Liland, K.H.; Axelsson, L. Effects of glucose availability in *Lactobacillus sakei*; metabolic change and regulation of the proteome and transcriptome. *PLoS ONE* **2017**, *12*, e0187542. [CrossRef] [PubMed]
6. Montanari, C.; Barbieri, F.; Magnani, M.; Grazia, L.; Gardini, F.; Tabanelli, G. Phenotypic diversity of *Lactobacillus sakei* strains. *Front. Microbiol.* **2018**, *9*, 2003. [CrossRef] [PubMed]

7. Chaillou, S.; Daty, M.; Baraige, F.; Dudez, A.M.; Anglade, P.; Jones, R.; Alpert, C.A.; Champomier-Vergès, M.C.; Zagorec, M. Intra-species genomic diversity and natural population structure of the meat borne lactic acid bacterium *Lactobacillus sakei*. *Appl. Environ. Microbiol.* **2009**, *75*, 970–980. [CrossRef]
8. Chaillou, S.; Champomier-Vergès, M.C.; Cornet, M.; Crutz-Le Coq, A.M.; Dudez, A.M.; Martin, V.; Beaufi, S.; Darbon-Rongère, E.; Bossy, R.; Loux, R.; et al. The complete genome sequence of the meat-borne lactic acid bacterium *Lactobacillus sakei* 23K. *Nat. Biotechnol.* **2005**, *23*, 1527–1533. [CrossRef]
9. McLeod, A.; Zagorec, M.; Champomier-Vergès, M.C.; Naterstad, K.; Axelsson, L. Primary metabolism in *Lactobacillus sakei* food isolates by proteomic analysis. *BMC Microbiol.* **2010**, *10*, 120. [CrossRef]
10. McLeod, A.; Snipen, L.; Naterstad, K.; Axelsson, L. Global transcriptome response in *Lactobacillus sakei* during growth on ribose. *BMC Microbiol.* **2011**, *11*, 145. [CrossRef]
11. Fiehn, O. Metabolomics – the link between genotypes and phenotypes. *Plant Mol. Biol.* **2002**, *48*, 155–171. [CrossRef]
12. Laghi, L.; Versari, A.; Marcolini, E.; Parpinello, G.P. Metabonomic investigation by ^{1}H-NMR to discriminate between red wines from organic and biodynamic grapes. *Food Nutr. Sci.* **2014**, *5*, 52–59. [CrossRef]
13. Lauret, R.; Morel-Deville, F.; Berthier, F.; Champomier-Vergès, M.C.; Postma, P.; Ehrlich, S.D.; Zagorec, M. Carbohydrate utilization in Lactobacillus sake. *Appl. Environ. Microbiol.* **1996**, *62*, 1922–1927. [CrossRef]
14. Zwietering, M.H.; Jongenburger, I.; Rombouts, F.M.; van't Riet, K. Modeling of the bacterial growth curve. *Appl. Environ. Microbiol.* **1990**, *56*, 1875–1881. [CrossRef]
15. Arioli, S.; Montanari, C.; Magnani, M.; Tabanelli, G.; Patrignani, F.; Lanciotti, R.; Mora, D.; Gardini, F. Modelling of *Listeria monocytogenes* Scott A after a mild heat treatment in the presence of thymol and carvacrol: Effects on culturability and viability. *J. Food Eng.* **2019**, *240*, 73–82. [CrossRef]
16. Tabanelli, G.; Montanari, C.; Arioli, S.; Magnani, M.; Patrignani, F.; Lanciotti, R.; Mora, D.; Gardini, F. Physiological response of *Saccharomyces cerevisiae* to citral combined with thermal treatment. *LWT Food Sci. Technol.* **2019**, *101*, 827–834. [CrossRef]
17. Zhu, C.; Li, C.; Wang, Y.; Laghi, L. Characterization of yak common biofluids metabolome by means of proton nuclear magnetic resonance spectroscopy. *Metabolites* **2019**, *9*, 41. [CrossRef]
18. Fadda, S.; Anglade, P.; Baraige, F.; Zagorec, M.; Talon, R.; Vignolo, G.; Champomier-Vergès, M.C. Adaptive response of *Lactobacillus sakei* 23K during growth in the presence of meat extracts: A proteomic approach. *Int. J. Food Microbiol.* **2010**, *142*, 36–43. [CrossRef]
19. Axelsson, L. Lactic acid bacteria: Classification and physiology. In *Lactic Acid Bacteria: Microbiology and Functional Aspects*, 3rd ed.; Salminen, S., von Wright, A., Ouwehand, A., Eds.; Marcel Dekker: New York, NY, USA, 2004; pp. 1–66, ISBN 0-8247-5332-1.
20. Rimaux, T.; Riviére, A.; Illeghems, K.; Weckx, S.; De Vuyst, L.; Leroy, F. Expression of the arginine deiminase pathway genes in *Lactobacillus sakei* is strain-dependent and is affected by environmental pH. *Appl. Environ. Microbiol.* **2012**, *78*, 4874–4883. [CrossRef]
21. Gänzle, M.G. Lactic metabolism revisited: Metabolism of lactic acid bacteria in food fermentations and food spoilage. *Curr. Opin. Food Sci.* **2015**, *2*, 106–117. [CrossRef]
22. Irmler, S.; Bavan, T.; Oberli, A.; Roetschi, A.; Badertscher, R.; Guggenbühl, B.; Berthoud, H. Catabolism of serine by *Pediococcus acidilactici* and *Pediococcus pentosaceus*. *Appl. Environ. Microbiol.* **2013**, *79*, 1309–1315. [CrossRef]
23. Skeie, S.; Kieronczyk, A.; Næss, R.M.; Østlie, H. *Lactobacillus* adjuncts in cheese: Their influence on the degradation of citrate and serine during ripening of a washed curd cheese. *Int. Dairy J.* **2008**, *18*, 158–168. [CrossRef]
24. Biagioli, M.; Laghi, L.; Carino, A.; Cipriani, S.; Distrutti, E.; Marchianò, S.; Parolin, C.; Scarpelli, P.; Vitali, B.; Fiorucci, S. Metabolic variability of a multispecies probiotic preparation impacts on the anti-inflammatory activity. *Front. Pharmacol.* **2017**, *8*, 505. [CrossRef]
25. Parolin, C.; Marangoni, A.; Laghi, L.; Foschi, C.; Ñahui Palomino, R.A.; Calonghi, N.; Cevenini, R.; Vitali, B. Isolation of vaginal Lactobacilli and characterization of anti-*Candida* activity. *PLoS ONE* **2015**, *10*, e0131220. [CrossRef]

26. Picone, G.; Laghi, L.; Gardini, F.; Lanciotti, R.; Siroli, L.; Capozzi, F. Evaluation of the effect of carvacrol on the *Escherichia coli* 555 metabolome by using ^{1}H-NMR spectroscopy. *Food Chem.* **2013**, *141*, 4367–4374. [CrossRef]
27. Liu, S.Q.; Holland, R.; McJarrow, P.; Crow, V.L. Serine metabolism in *Lactobacillus plantarum*. *Int. J. Food Microbiol.* **2003**, *89*, 265–273. [CrossRef]

foods

MDPI

Article

Technological Characterisation of Probiotic Lactic Acid Bacteria as Starter Cultures for Dry Fermented Sausages

Nadia de L. Agüero [1], Laureano S. Frizzo [2,3,4], Arthur C. Ouwehand [4], Gonzalo Aleu [1,†] and Marcelo R. Rosmini [3,4,*]

1 Faculty of Agricultural Sciences, Catholic University of Córdoba, 5000 Cordoba, Argentina; nadiaaguero@hotmail.com (N.d.L.A.); gonzaloaleu@gmail.com (G.A.)
2 Laboratory of Food Analysis "Rodolfo Oscar DALLA SANTINA", Institute of Veterinary Science (ICiVet Litoral), National University of the Litoral—National Council of Scientific and Technical Research (UNL/CONICET), 3080 Esperanza, Argentina; lfrizzo@fcv.unl.edu.ar
3 Department of Public Health, Faculty of Veterinary Science, National University of Litoral, 3080 Esperanza, Argentina
4 Global Health and Nutrition Sciences, DuPont Nutrition and Biosciences, 02460 Kantvik, Finland; arthur.ouwehand@dupont.com
* Correspondence: mrosmini@unl.edu.ar; Tel.: +54-342-479-7991
† Members of HealthyMeat-Network CYTED (ref. 119RT0568).

Received: 10 April 2020; Accepted: 1 May 2020; Published: 7 May 2020

check for updates

Abstract: The objective of this study was to investigate probiotic microorganisms for use as starter cultures in dry fermented sausages production. A total of eight strains were studied evaluating technological and safety characteristics including the ability to grow, lactic acid production, gas formation, catalase activity, nitrate reductase activity, proteolytic activity, lipolytic activity, hydrogen peroxide production, salt tolerance, performance at low temperatures, decarboxylation of amino acids and antimicrobial activity against pathogens associated with the product. *Lactobacillus rhamnosus* R0011, *L. rhamnosus* Lr-32, *Lactobacillus paracasei* Lpc-37, *Lactobacillus casei* Shirota and *Enterococcus faecium* MXVK29 were good candidates for use as fermented sausages starters cultures because they showed the best technological and safety properties since they did not demonstrate amino acid decarboxylation but showed antimicrobial activity against *Listeria monocytogenes*, *Escherichia coli*, *Salmonella Dublin* and *Staphylococcus aureus*. *L. rhamnosus* Lr-32 was the strain best tolerating the levels of salt, nitrate and low pH during the simulated stages of fermentation and ripening of sausage. The strain was thus the most promising of the tested probiotics as sausage starter culture. The findings warrant studies in a meat matrix, such as that of raw-cured sausage, to evaluate the effects of *L. rhamnosus* Lr-32 under actual conditions.

Keywords: probiotic; dry fermented sausages; healthy meats; lactobacillus

1. Introduction

The demand for safe, innovative and healthy food products has been a stimulus for the development of new fermented meat products. Today's consumer requires that food provides the necessary nutrients and supports health; consequently, meat products with added functional ingredients, or that contain smaller amounts of ingredients that are considered not beneficial for health (salt, fat, etc.) are in greater demand [1,2]. Functional foods are defined as foods that contain some health promoting component beyond traditional nutrients. An example of a functional food is those in which probiotic microorganisms have been incorporated [3].

Probiotics are defined as 'live microbial food supplements that, when administered in adequate amounts confer a health benefit on the host' [4]. Such probiotic microorganisms are capable of contributing to the balance of the host's intestinal microbiota, modulate immune response and can act in various other ways as health promoters [5]. Until now, the most common use of these microorganisms is in fermented dairy products but some authors propose the use of probiotic lactic acid bacteria (LAB) as starter cultures for fermented meat products [2,6–11]. The commercialisation of probiotic meat products and commercial application of probiotic microorganisms in fermented sausages is not common yet [12], and in vitro and industrial scale studies are still scarce. The use of probiotic cultures can positively affect the process of fermented sausages, resulting in new technological properties and a beneficial effect on human health. Much of the research is based on the study of bacteria that are commonly associated with the meat environment and which possess the appropriate physiological requirements and health-promoting properties. Such bacteria can be obtained by screening natural sausage or existing commercial meat starter cultures for probiotic properties. Alternatively, the performance of strains with documented health-promoting properties may be investigated in a fermented meat environment [11,13]. Other researchers have studied the use of a mixed starter combining a food starter with a probiotic microorganism [12,14].

LAB as well as Coagulase-Negative Cocci (CNC) are important microorganisms used as starter cultures in fermented meat products [15]. The fermentation improves the quality, safety and stability of the product, extends its shelf life and provides microbial diversity that result in new sensory properties [7,16,17]. The use of probiotic microorganisms as starter cultures could contribute beneficially to the health of the consumer [17]. The development of probiotic starter cultures requires prior knowledge of the microbiota that participates in the spontaneous process and technologically characterise the strains that are intended to be used; in order to select those that present the best technological properties. The fermented sausage is a complex ecological system in which the natural microbiota, together with some raw materials can present adverse conditions for microorganisms incorporated for technological purposes. Thus, probiotic LAB should tolerate the presence of salts, low pH levels and low temperatures [11,15].

These microorganisms must be well adapted to the ecological environment and processing conditions of sausages and therefore, be able to develop more efficiently and grow rapidly to compete with the natural microbiota present [18,19]. As a result, they dominate the fermentation process in order to carry out the desired metabolic activities and reach levels that enable the display of health-promoting effects [11]. The enzymatic properties of lipases, proteases, catalases, nitrate and nitrite reductases are very important during the manufacturing process. Some strains of LAB can carry out these functions in the meat matrix which makes them of interest in the development of competitive starter cultures adapted to meat matrices [17,20–22]. The tolerance of CNCs, which will be part of the mixed starter culture, should also be studied [17,21,22].

Biogenic amines (BA), organic bases with aliphatic, aromatic or heterocyclic structures, are found in various foods and are mainly produced by microbial decarboxylation of amino acids [23,24]. The accumulation of BA in food requires the presence of precursors (amino acids), microorganisms with amino acid decarboxylase activity and favourable conditions for growth and metabolic activity. Such requirements are met during sausage fermentation [21]. A wide variety of meat and meat product LAB can decarboxylate amino acids [21] and BA concentrations can be high enough to cause food poisoning [24,25]. As a result of this, it is necessary to select LAB that do not show amino decarboxylase activity when they are to be used as starter cultures in sausage preparation [26,27].

LAB found in meat may produce a variety of bacteriocins that are generally active against other LAB (which contribute to the competitiveness of the producing strain) and Gram positive pathogens that are transmitted through food, such as *Listeria monocytogenes*, *Staphylococcus aureus*, *Clostridium perfringens* and *Bacillus cereus* [28]. The production of bacteriocins with a wide inhibition range, especially towards foodborne pathogens, combined with acid production and reduction of the batter pH is therefore highly desirable, as this would ensure the competitiveness of the initiating strain while reducing the development of an undesirable microbiota [21]. Probiotic bacteria could be incorporated into the complex meat system and be potentially successful starter cultures. In the process of selecting these probiotic bacteria, properties such as: growth capacity and production of lactic acid at the desired temperatures, tolerance to low pH and high salt concentrations (nitrite and NaCl) and enzymatic and antagonistic activities should be taken into account.

The first stage in the design of a starter culture for a meat product is to characterise the probiotic LAB strains and select the most appropriate [21]. In meat fermentations, the main function of LAB is to obtain a rapid pH drop of the batter. Furthermore, the strains must be able to grow and survive in the conditions that exist in the sausages.

The objective of this work was to study commercial LAB probiotics and select the most suitable to be used as a starter of a dry fermented sausage, according to its technological characteristics, strain safety and antimicrobial activity against pathogenic microorganisms associated with these foods.

2. Materials and Methods

2.1. Sampling

The microorganisms used in the tests are listed in Table 1. Probiotic strains were obtained from Lallemand (Montreal, Canada), DuPont (Madison, WI, USA), Yakult (Tokyo, Japan), BioGaia (Stockholm, Sweden) and the Autonomous Metropolitan University (Distrito Federal, México). *Pediococcus pentosaceus* PCFF-1 and *Staphylococcus xylosus* DD-34 were purchased from Chr. Hansen (Hørsholm, Denmark). Some trials included strains of *Lactobacillus plantarum* and *Lactobacillus sakei* isolated from the indigenous microbiota of sausage (Córdoba, Argentina). *Staphylococcus aureus* and *Listeria monocytogenes* were kindly provided by the Pediatric Hospital of the Infant Jesus (Córdoba, Argentina). *Escherichia coli* and *Salmonella* Dublin belong to the strain collection of the National University of the Litoral (Department of Public Health, National University of the Litoral, Esperanza, Santa Fe, Argentina). The beneficial effects of probiotic strains have been documented among others in the following studies: *L. rhamnosus* R0011 and *L. helveticus* R0052 by Evans et al. [29]; *L. rhamnosus* Lr-32 by Miyazima et al. [30]; *L. paracasei* Lpc-37 by Ouwehand et al. [31]; *Lactobacillus casei* Shirota by Tripolt et al. [32]; *Enterococcus faecium* MXVK29 by Alvarez-Cisneros et al. [33]; *Lactobacillus reuteri* DSM17938 by Savino et al. [34] and *L. reuteri* DSM17918 by Lin et al. [35].

Table 1. Bacterial strains used in the tests.

Species	Strain	Manufacturer/Origin
Probiotics		
Lactobacillus rhamnosus	R0011	Lallemand
Lactobacillus helveticus	R0052	Lallemand
Lactobacillus rhamnosus	Lr-32	DuPont
Lactobacillus paracasei	Lpc-37	DuPont
Lactobacillus casei	Shirota	Yakult
Enterococcus faecium	MXVK29	Autonomous Metropolitan University, México, Mexico
Lactobacillus reuteri	DSM17918	BioGaia
Lactobacillus reuteri	DSM17938	BioGaia
Sausage starters		
Lactobacillus plantarum	Lp-UCC	Catholic University of Cordoba, Cordoba, Argentina
Lactobacillus sakei	Ls-UCC	Catholic University of Cordoba, Cordoba, Argentina
Pedicoccus pentosaceus	PCFF-1	Chr. Hansen
Staphylococcus xylosus	DD-34	Chr. Hansen
Sausage associated pathogens		
Listeria monocytogenes		Pediatric Hospital of the Infant Jesus, Cordoba, Argentina
Staphylococcus aureus	ATCC 25923	Pediatric Hospital of the Infant Jesus, Cordoba, Argentina
Escherichia coli	ATCC 25922	DSPV-FCV-UNL, Esperanza, Argentina
Salmonella Dublin	DSPV 595T	DSPV-FCV-UNL, Esperanza, Argentina

2.2. Quantification of Strains

The strains were inoculated in de Man, Rogosa, Sharpe (MRS) broth (Britania®, Buenos Aires, Argentina) and incubated in a thermostatic bath with continuous shaking at 37 °C. Optical density was measured with a wide band filter (OD 600 nm) with an automatic turbidimeter (Bioscreen C, Turku, Finland) and parallel plate counts were performed to determine colony forming units (CFU). This was done at the start ($t = 0$), at 2 h (when turbidity was observed in the tube) and over 12 h (measurements approximately every 30–40 min) until 20 measurements were taken in total. The tests were carried out in triplicate and on two different occasions. For each strain, its standard calibration curve was constructed from the OD 600 obtained and the logarithm of the CFU at 37 °C; we assumed no influence of cultivation temperature on the calibration curve. The curves were used to assess the ability of the strains to grow and also to quickly quantify the strains when used in subsequent trials.

2.3. Production of Acid and Gas from Carbohydrates

The pure cultures were inoculated in MRS broth added with 1% glucose, bromocresol green indicator (Sigma-Aldrich®, Buenos Aires, Argentina) 0.05 g/L and Durham bells [36] and incubated at 37 °C for 24 h. The colour change from green to yellow indicated the production of acid and the presence of gas manifested by the bubbles inside the bells were considered as positive reactions.

2.4. Lactic Acid Production

A modified method of Erkkilä et al. [7] was used. Strains were cultivated on MRS agar for 2 days at 37 °C. Three colonies of each strain were picked and inoculated in 10 mL MRS broth. After 48 h of incubation at 22 °C, the broths were purified with strong anion-exchange solid phase extraction cartridges (EFS-SPE). Lactic acid was separated with a reverse-phase column (LiChrosorb® Hibar® RP-18 250 × 4.6 mm, 5 µm) in a high-performance liquid chromatography (HPLC) system. The lactic acid concentration of the samples was determined at a wavelength of 210 nm by using the external standard method with 50, 100, 150 and 200 mM standards. The mobile phase (H_2SO_4 50 mM) was pumped at 1.5 mL/min and the column temperature was 30 °C. The relative retention value for lactic acid was 4.8 min. Standards and each sample was run three times. The pH of the broths were measured at the beginning and after the incubation period. As a control the commercial starter *P. pentosaceus* PCFF-1 was used.

2.5. Effect of Temperature on the Growth of Probiotic LAB

LAB growth at different temperatures was observed in MRS broth after incubation at 7 and 15 °C for 14 days and at 22, 30, 37 and 43 °C for 7 days [37]. The first three values correspond to temperatures used commonly for the fermentation of raw-cured sausages; while 43 °C is the temperature used in the recovery tests for microorganisms from raw-cured sausages and is of interest for future studies. The cultures were inoculated at 1%. Optical density determinations at 600 nm (DO_{600}) were performed with an automatic turbidimeter (Bioscreen C) and correlated with colony forming units (CFU) at the beginning (day 0) and after the incubation period.

2.6. Effect of pH, Sodium Nitrite and Sodium Chloride on Probiotic LAB Survival and Growth

To study in vitro the conditions of fermentation and maturation that occur during the preparation of a sausages and the combined effect of pH, sodium nitrite and sodium chloride on the strains, a modified method of Korkeala et al. [38] was used. The cultures were grown in 2 different media; fermentation medium with pH levels 4.5, 100 mg/L $NaNO_2$ and 2% NaCl (*w/v*); and ripening medium with pH levels 5.5, 100 mg/L $NaNO_2$ and 2% NaCl (*w/v*). Both media were prepared with MRS. MRS broth adjusted to pH 6.5 without $NaNO_2$ and NaCl was used as a control. The wells of the microplates were seeded with 240 µL of medium and inoculated with 2.4 µL of 18 h culture. The determinations were made in triplicate. Growth was monitored using a Synergy™ HT (BioTek®, Winooski, VT, USA)

Multi-Modal microplate reader. The microplates were incubated at 37 °C for 24 h and the optical density was measured every 30 min with a wide band filter (OD 620 nm). The delta absorbance (difference between the first and last absorbance reading), the gradient (slope of the logarithmic growth phase) and the lag phase (time when the bacterial growth started) were collected, analysed and exported with the Gen5 Data Analysis Software and finally used to characterise the growth. If no bacterial growth was observed the lag phase was assessed as 24 h.

2.7. Enzymatic Characterisation

The isolates were further characterised taking into account the following biochemical tests:

2.7.1. Hydrogen Peroxide Production

The capacity of the strains to produce H_2O_2 was determined following the methodology described by McLean and Rosenstein [39]. Bacterial cultures were streaked on 15 mL of MRS agar (Britania®) supplemented with 2.5 mg/mL Tetramethylbenzidine (TMB) (Sigma-Aldrich®) and 0.01 mg/mL Peroxidase from radish (HRP) (Sigma-Aldrich®). The plates were incubated at 37 °C for 48 h under anaerobic conditions. After the incubation period, the plates were opened and exposed to atmospheric air for 30 min. After exposure, the colonies of the strains that produce H_2O_2 turned in blue. In the presence of H_2O_2 the HRP enzyme oxidises the TMB (colourless) to give rise to the formation of a blue pigment. When interpreting the results, it was considered that the hue of the blue colour was a semiquantitative measure of the amount of H_2O_2 produced and released into the medium by the strain [40]. From this, the hue of the blue colour was differentiated into the following categories: light blue (lower production) or dark blue (higher production). In the trial, *L. plantarum* Lp-UCC and *L. sakei* Ls-UCC strains were used as positive controls [41] and the commercial initiator *P. pentosaceus* PCFF-1 to evaluate its behaviour.

2.7.2. Catalase Activity and Hydrogen Peroxide Hydrolysis

To demonstrate the presence of the enzyme catalase (an enzyme capable of decomposing hydrogen peroxide in water and oxygen), the techniques described by Harrigan [42] were followed and tests were carried out on plates and tubes. The pure cultures were grown on MRS agar (Britania®), MRS broth (Britania®) and Nutritive broth (Britania®) at 37 °C for 24 to 48 h. After the incubation period, a few drops of 3% H_2O_2 were placed on the colonies grown in plates and 1 mL of the same solution was added to the cultures grown in tubes. The appearance of bubbles in the plates and in the broths caused by the release of oxygen, were considered positive tests. The tests were carried out in triplicate and *S. aureus* ATCC 25923 was used as a positive catalase control strain.

2.7.3. Nitrate and Nitrite Reduction

The technique described in MacFaddin [43] was used to determine the ability of microorganisms to reduce nitrates to nitrites or free nitrogen gas. Nutritive broth (Britania®) with potassium nitrate (Sigma-Aldrich®) 1 g/L were made. Two reagents were used to reveal the presence of nitrites in the medium: reagent A (sulfanilic acid (Sigma-Aldrich®) at 0.8% in 5 N acetic acid) and reagent B (α-naphthylamine (Sigma-Aldrich®) 0.5% in acetic acid (Sigma-Aldrich®) 5 N). Nitrated broths were inoculated with pure cultures and incubated for 48 h at 37 °C. Assays were performed in triplicate and *S. aureus* ATCC 25923 and *S. xylosus* DD34 were used as control strains. After the incubation period 0.5 mL of reagent A was added and then 0.5 mL of reagent B. The appearance of red colour in the medium (after approximately 30 s) indicated the presence of nitrites in the medium.

The reduction of nitrates can give other more reduced products such as ammonia, molecular nitrogen, nitric oxide, nitrous oxide or hydroxylamine, in this case, after the addition of the reagents, there is no colour change in the medium. To confirm that the process was negative, zinc powder (Sigma-Aldrich®) was added, which, if nitrates existed, reduced them to nitrites, turning the medium red and consequently indicating that previously nitrates had produced the reaction. If after the addition of zinc powder the colour did not turn red, this indicated that there was a reduction and that the product obtained was not nitrite but a smaller product.

2.7.4. Milk Casein Hydrolysis

To examine protease activity, the methodology described by Harrigan [42] was followed with some modifications. Two culture media were prepared. (1) Skim milk agar: 5% skim milk powder and 1.3% agar (Britania®). (2) Skim milk agar with peptone: 2% skim milk powder in distilled water and agar (Britania®) at a concentration of 1.3% in peptone water (Britania®) were prepared separately; they were sterilised, allowed to cool to 45 °C and then homogenised and distributed in Petri dishes. Plates with 6 mm diameter perforations were prepared that were filled with 20 µL of fresh cultures and plates without perforations that were streaked with the same fresh cultures. *S. aureus* ATCC 25923 was used as a control. The plates were incubated at 37 °C for 48 to 72 h. The presence of a transparent halo around the grown colonies and wells was considered as a positive casein proteolysis reaction. The test was carried out in triplicate and on two different occasions.

2.7.5. Lecithinase and Lipase Activity

To examine the activity of microorganisms to produce the enzyme lecithinase and lipase, the methodology described in MacFaddin [43] was followed. Two culture media were prepared: Soy Tryptin agar (Britania®) and Brain Heart Infusion agar (Britania®) enriched with 10% sterile egg yolk. Plates with 6 mm diameter perforations were prepared which were filled with 20 µL of fresh cultures and plates without perforations were streaked with the same fresh cultures. *Staphylococcus aureus* ATCC 25923 was used as a positive control. The plates were incubated at 37 °C for 48 to 72 h. The hydrolysis of lecithin releases phosphorus and choline in stages, and the precipitation of insoluble fats (diglycerides) produces the opalescence of the medium. An opaque halo surrounding the colonies was considered as a positive lecithinase test, on the other hand, the enzyme lipase, catalyses the hydrolysis of triglycerides and diglycerides to fatty acids and glycerol, this, which is evidenced by an oily, iridescent shine on the colonies and around them, it was considered as positive lipase activity.

2.8. Antagonism against Staphylococcus Xylosus

To evaluate the antagonism of target microbes against *S. xylosus*, a sausage starter strain, a modified method of the agar spot test [44] and the agar diffusion technique [45] was used. Petri dishes were filled with MRS agar. A 2.6 µL amount of fresh culture of LAB strains was spotted onto and incubated at 37 °C for 24 h. The plates were overlaid with 15 mL of brain heart infusion (BHI)-1.5% agar (Britania®) inoculated with con 1.5 µL of fresh culture of *S. xylosus* and incubated at 37 °C for 24 h. The results were reported as positive or negative considering the presence or absence of clear areas of inhibition. In the agar diffusion technique, the production of inhibitory substances in cell-free extracts (CFE) of LAB strains was evaluated. LAB were grown in MRS broth for 24 h at 37 °C. They were then centrifuged at 5000× *g* in a 4 °C refrigerated centrifuge (IEC Multi RF Thermo, Spain) for 10 min. Subsequently, an aliquot of the supernatant was adjusted to pH 6.5 using NaOH as neutraliser to avoid acid inhibition (neutralised CFE). Another aliquot was used directly as a CFE without neutralising. Both of the supernatants were filter-sterilised, 0.22 µm pore diameter (Millipore, Merck, Germany). On a Petri dish with 15 mL of BHI solid agar (1.5% agar) were covered with 9 mL of semi-solid agar BHI (0.8% agar) inoculated with con 1.5 µL of fresh culture of *S. xylosus*. On the upper agar layer, various numbers of holes (6 mm) were punched out of the agar and were filled with 20 µL of neutralised and non-neutralised CFE of target microbes. The inoculated dishes were incubated at 4 °C for 1 h to allow diffusion of the extracts and then incubated at 37 °C for 24 h and examined for clear inhibition halos around the wells. The results were reported as positive or negative considering the presence or absence of inhibition zones, respectively.

2.9. Amino Acid Decarboxylation

To evaluate the amino decarboxylase activity [46] a qualitative method was used. A culture medium was prepared with the following composition: Casein peptone (Sigma-Aldrich®) 0.5%, Yeast extract (Oxoid®, Basingstoke, United Kingdom) 0.3%, D (+) Glucose (Biopack®,) 0.1% and bromocresol purple (Sigma-Aldrich®) 0.0016%. Then, 0.5% of the amino acid to be tested was added to 0.9% of the base medium, in this case L-Arginine (Anedra®, Buenos Aires, Argentina), L-Histidine (Anedra®), L-Lysine (Anedra®), L-Tyrosine (Anedra®) and L-Tryptophan (Anedra®). The pH of the medium was then adjusted to 6.7 ± 0.1 at 25 °C. The tubes were inoculated with 10 μL of 24 h culture and incubated 72 h at 37 °C in under aerobic conditions. The colour of the medium at the time of inoculation was purple. In addition to the probiotic strains, the *P. pentosaceus* PCFF-1, *L. plantarum* Lp-UCC and *L. sakei* Ls-UCC strains were used in the assay to observe how they behaved with respect to the amino acids tested. In cases where the strain did not decarboxylate the amino acid, the colour of the medium remained yellow until the end of the test and the test was considered negative. On the contrary, when the strain decarboxylated the amino acid the colour turned purple and the test was considered positive.

2.10. Antagonism against Pathogens Associated with Sausages

The antagonism of the probiotic strains against the pathogens *E. coli*, *L. monocytogenes*, *Salmonella* Dublin and *S. aureus* was studied according to the method of the agar spot test described by Lewus et al. [44], the agar diffusion technique described by Tagg and McGiven [45] indicated above and the microplate antimicrobial activity described by Ruíz-Moyano et al. [47] with some modifications.

In the antimicrobial activity in microplates, inhibitory effects of strains on selected potential harmful microorganisms were studied by following the ability of the target microbes to grow in a medium containing sterilised neutralised supernatants from probiotic strain cultures. The ability of fresh selected harmful microorganisms to grow in broth (BHI) supplemented with of filter-sterilised supernatant was evaluated by following the microbial growth at 37 °C for 24 h with an automated turbidimeter Synergy™ HT (BioTek®) Multi-Modal microplate reader. Each well of the microplates was seeded with 120 μL of double concentration BHI broth supplemented with 120 μL of neutralised CFE of the probiotic strain under study. The wells were inoculated with 2.4 μL of 18 h culture of target pathogen. Simple concentration BHI broth, pH 6.5 without supplementation, inoculated with each of the pathogenic strains, was used as a positive control. The sowing was done in triplicate. The optical density was measured with a wide band filter (OD 620 nm) every 30 min. The delta absorbance (difference between the first and last absorbance reading), the gradient (slope of the logarithmic growth phase) and the lag phase (time when the bacterial growth started) were collected, analysed and exported with the Gen5 Data Analysis Software (BioTek®) and finally used to characterise the growth. If no bacterial growth was observed the lag phase was assessed as 24 h. The inhibitory effect of probiotic culture supernatants was determined by comparing the maximum growth with respect to the positive control.

2.11. Statistical Analysis

To quantify the strains, a logarithmic regression model was used. The model was built with OD600 data and the decimal logarithm of microbial growth at 37 °C [48]. Lactic acid production data, growth in fermentation and maturation conditions and antimicrobial activity represented by continuous variables were analysed using an ANOVA. All the experiments in the present study were conducted in triplicate and the values presented in results are mean values from each triplicate. The experimental data obtained after mean from each triplicate set was statistically evaluated at a significance level of $p \leq 0.05$. The IBM SPSS Statistics 19 program was used. When the differences were significant ($p \leq 0.05$), the Duncan multiple means comparisons test was applied.

3. Results

The strains studied showed different technological capabilities, compared to the different conditions in which the tests were performed. These results are presented in Tables 2–4 and Table S1. To be able to select the most suitable strains, we performed a ranking with a numerical categorisation, assigning positive and negative values according to the results obtained for the different parameters measured (Table 5).

Table 2. Technological and safety features of commercial probiotic bacteria.

	L. rhamnosus R0011	*L. helveticus* R0052	*L. rhamnosus* Lr-32	*L. paracasei* Lpc-37	*L. casei* Shirota	*E. faecium* MXVK29	*L. reuteri* DSM17918	*L. reuteri* DSM17938
				Biomass generation (log_{10} CFU/mL)				
	9.4 ± 0.08	8.3 ± 0.10	9.3 ± 0.13	8.6 ± 0.07	9.7 ± 0.10	9.6 ± 0.10	9.2 ± 0.10	9.3 ± 0.08
				Growth at (log_{10} CFU/mL)				
7 °C	8.9 ± 0.13	ng	8.3 ± 0.10	8.5 ± 0.11	8.8 ± 0.19	9.4 ± 0.36	ng	ng
15 °C	9.3 ± 0.18	ng	9 ± 0.10	8.6 ± 0.37	9.6 ± 0.33	9.5 ± 0.31	ng	ng
22 °C	9.3 ± 033	7.5 ± 0.32	9.1 ± 0.16	8.6 ± 0.24	9.7 ± 0.10	9.6 ± 0.80	ng	ng
30 °C	9.4 ± 0.40	8.3 ± 0.25	9.3 ± 0.35	8.6 ± 0.36	9.7 ± 0.22	9.6 ± 0.13	9.2 ± 0.40	9.3 ± 0.14
37 °C	9.4 ± 0.50	8.3 ± 0.30	9.3 ± 0.33	8.6 ± 0.33	9.7 ± 0.50	9.6 ± 0.51	9.2 ± 0.45	9.3 ± 0.07
43 °C	9.2 ± 0.20	8.3 ± 0.53	9.3 ± 0.25	ng	ng	9.5 ± 0.09	9.2 ± 0.40	9.3 ± 0.35
				Gas production				
	nd	nd	nd	nd	nd	nd	nd	nd
				Acid production at (nmol/L) *				
22 °C	215.94 ^{A} ± 0.52	28.27 ^{D} ± 0.39	177.44 ^{B} ± 0.15	198.29 ^{A} ± 0.42	201.52 ^{A} ± 0.59	147.85 ^{C} ± 0.91	7.5 ^{E} ± 0.41	7.06 ^{E} ± 0.54
				Growth (OD620) simulated sausages conditions *				
Fermentation	0.57 ^{D,E} ± 0.06	0.77 ^{B,C,D} ± 0.06	0.90 ^{A,B,C} ± 0.17	0.63 ^{C,D} ± 0.12	0.30 ^{E,F} ± 0.17	0.00 ^{F} ± 0.00	1.07 ^{A,B} ± 0.04	1.17 ^{A} ± 0.06
Ripening	1.10 ^{A,B} ± 0.17	0.60 ^{C,D} ± 0.10	1.27 ^{A,B} ± 0.12	1.23 ^{A,B} ± 0.23	0.93 ^{B,C} ± 0.15	0.33 ^{D} ± 0.29	1.33 ^{A} ± 0.12	1.07 ^{A,B} ± 0.32
				Catalase activity				
	nd	nd	nd	nd	nd	nd	nd	nd
				H_2O_2 production				
	nd	d	nd	nd	d	nd	d	d
				Nitroreductase activity				
	nd	nd	nd	nd	nd	nd	nd	nd
				Proteolytic activity				
	d	nd	d	d	d	d	nd	nd
				Lipolytic activity				
	d	nd	d	d	d	d	nd	nd
				Antagonism against *S. xylosus*; coagulase negative cocci (CNC)				
	nd	nd	nd	nd	nd	nd	nd	nd
				Antagonism against pathogens				
S. aureus	d	d	d	d	d	d	d	d
L. monocytogenes	nd	nd	d	d	d	d	d	d
S. dublin	d	d	d	d	d	d	d	d
E. coli	d	d	d	d	d	d	d	d

(ng): no growth; (d): detected; (nd): not detected; (*, $^{A–F}$): Different letters indicate significant differences ($p \leq 0.05$).

Table 3. Antimicrobial activity of probiotic strains against pathogens associated with fermented sausages.

	Inhibitory Strain															
	Colony Diameter (mm)								Clearing Zone Diameter (mm)							
Target Strain	**R0011**	**R0052**	**Lr-32**	**Lpc-37**	**Shirota**	**MXVK29**	**DSM17918**	**DSM17938**	**R0011**	**R0052**	**Lr-32**	**Lpc-37**	**Shirota**	**MXVK29**	**DSM17918**	**DSM17938**
S. aureus	7	6.5	8	7.5	7	6.5	8	7.5	14	4	12	14	9	7	10	11
L. monocytogenes	7	6	9	8	7	7	9	8	15	5	12	15	13	15	7	8
Salmonella Dublin	5	5	7	7	7	5	7	7	12	nd	16	14	16	12	11	10
E. coli	5	5	6.5	7	6	6	7	6.5	12	nd	12	17	11	8	8	10

Table 4. Antimicrobial activity of probiotic strains against pathogens associated with fermented sausages.

Target Strain *	**Growth (OD_{620})**								
	Control	**R0011**	**R0052**	**Lr-32**	**Lpc-37**	**Shirota**	**MXVK29**	**DSM17918**	**DSM17938**
S. aureus	1.13 ^{C}	1.0 ^{B}	1.03 ^{B}	1.0 ^{B}	1.0 ^{B}	0.9 ^{A}	1.0 ^{B}	1.03 ^{B}	1.0 ^{B}
L. monocytogenes	1.03 ^{D}	1.0 ^{D}	1.03 ^{D}	0.90 ^{C,D}	0.70 ^{B}	0.80 ^{B,C}	0.70 ^{B}	0.80 ^{B,C}	0.40 ^{A}
S. Dublin	1.40 ^{E}	1.20 ^{D}	1.20 ^{D}	1.10 ^{C}	0.97 ^{A,B}	1.10 ^{C}	0.90 ^{A}	0.90 ^{A}	1.0 ^{B}
E. coli	1.57 ^{E}	1.27 ^{D}	1.23 ^{D}	1.10 ^{C}	1.10 ^{C}	1.10 ^{C}	0.93 ^{A}	1.03 ^{B,C}	1.0 ^{A,B}

(*, $^{A–E}$): Different letters indicate significant ($p \leq 0.05$) differences.

Table 5. Strain selection criteria according to results obtained.

	L. rhamnosus Lr-32	*L. rhamnosus* R0011	*L. paracasei* Lpc-37	*E. faecium* MXVK29	*L. casei* Shirota	*L. reuteri* DSM17918	*L. reuteri* DSM17938	*L. helveticus* R0052
Biomass generation	1	1	1	1	1	1	1	1
Growth at all different temperatures tested	1	1	0	1	0	0	0	0
No gas production	1	1	1	1	1	1	1	1
Acid production	1	1	1	1	1	0	0	0
Growth at fermentation condition	1	0	0	0	0	1	1	0
Growth at ripening condition	1	1	1	0	0	1	1	0
Catalase activity	0	0	0	0	0	0	0	0
No H_2O_2 production	1	1	1	1	0	0	0	0
Nitroreductase activity	0	0	0	0	0	0	0	0
Proteolytic activity	1	1	1	1	1	0	0	0
Lipolytic activity	1	1	1	1	1	0	0	0
Antagonism against *S. xylosus*	1	1	1	1	1	1	1	1
Antagonism against all pathogens	1	0	1	1	1	1	1	0
No amino acid decarboxylation	1	1	1	0	0	0	0	0
Total number of positive reaction results	12	10	10	9	7	6	6	3

(0): negative result to the desired reaction; (1): positive result to the desired reaction.

In general, they showed very good capacity to generate biomass in a short period of time (37 °C; 12 h; Table 2). All reached more than 8 $\log_{10}$ CFU/mL in the time tested. The logarithmic model used adjusted significantly ($p \leq 0.01$) and allowed quantifying the strains studied (Figure S1). The strains with the highest number of cells per unit volume were *L. casei* Shirota, *E. faecium* MXVK29, *L. rhamnosus* R0011 and *L. rhamnosus* Lr-32.

All tested strains fermented glucose, producing lactic acid without gas generation, indicating that their metabolism is homofermentative (Table 2).

The amount of lactic acid produced by each strain is shown in Table 2. The highest concentration was generated by the commercial initiator *P. pentosaceus* PCFF-1 ($p \leq 0.05$), while *L. reuteri* DSM 17918, *L. reuteri* DSM 17938 and *L. helveticus* R0052 had the lowest production. This was related to their low capacity to grow at the incubation temperature of the trials (22 °C), which was chosen because it is the usual fermentation temperature of sausages in which this strain is intended to be used in the future. The strains *L. rhamnosus* R0011, *L. rhamnosus* Lr-32, *L. paracasei* Lpc-37, *L. casei* Shirota and *E. faecium* MXVK29 generated lactic acid in the range of 216 to 148 mM (Table 2).

The results in Table 2 show that at 7 °C *L. helveticus* R0052, *L. reuteri* DSM17918 and *L. reuteri* DSM17938 did not increase their biomass; *L. rhamnosus* Lr-32 had slow development but the strain remained viable and an increase was observed on day 15 of the incubation period. Between 30 and 37 °C, the most characteristic temperature of the fermentation of central European meat products, all strains had good growth, which indicates that they are mesophilic and that it corresponds to their optimum growth temperature. At 43 °C the strains *L. rhamnosus* R0011, *L. rhamnosus* Lr-32, *E. faecium* MXVK29, *L. helveticus* R0052, *L. reuteri* DSM17918 and *L. reuteri* DSM17938 had good growth. The first three of the above strains showed a wide optimum range of growth temperatures. For *L. helveticus* R0052, *L. reuteri* DSM17918 and *L. reuteri* DSM17938 the range was narrower and established only at the upper levels. *L. paracasei* Lpc-37 and *L. casei* Shirota did not grow at high temperature (Table 2).

The strains *L. rhamnosus* Lr-32, *L. reuteri* DSM17918 and *L. reuteri* DSM17938 were best adapted to the simulated fermentation conditions (pH 4.5, NaCl 2% and $NaNO_2$ 100 mg/L). The Nmax values of these strains were similar and, in turn, higher and different from the rest of the strains studied ($p \leq 0.05$). On the contrary, it was observed that *E. faecium* MXVK29 failed to develop under the conditions studied (Table 2). On the other hand, in the ripening stage (pH 5.5, 3% NaCl, 200 mg/L $NaNO_2$) it can be seen that *L. rhamnosus* Lr-32 and *L. rhamnosus* R0011, *L. paracasei* Lpc-37, *L. reuteri* DSM17918 and *L. reuteri* DSM17938 had the highest Nmax values ($p \leq 0.05$) (Table 2).

The strains that produce H_2O_2 showed that their white colonies turn to light blue colour (in the case of *P. pentosaceus* PCFF-1, *L. plantarum* Lp-UCC, *L. sakei* Ls-UCC and *L. casei* Shirota) and intense blue, indicating a high production of hydrogen peroxide, (for *L. helveticus* R0052, *L. reuteri* DSM17918 and *L. reuteri* DSM17938) (Table 2). All strains were negative for catalase activity and were not observed to breakdown H_2O_2. Nor were any of the tested trains found to reduce nitrates (Table 2).

The probiotics strains studied showed no inhibition halos against *S. xylosus* DD-34 (a CNC strain commonly used in the meat industry) (Table 2).

Probiotics *L. rhamnosus* R0011, *L. rhamnosus* Lr-32, *L. paracasei* Lpc-37, *L. casei* Shirota and *E. faecium* MXVK29 hydrolysed milk casein generating a transparent halo around the colonies. These strains also hydrolysed triglycerides, showing an oily shine on the colonies and around them (Table 2).

Lactobacillus rhamnosus R0011, *L. rhamnosus* Lr-32 and *L. paracasei* Lpc-37 strains did not decarboxylate any of the amino acids tested. However, the other tested strain decarboxylated at least one amino acid after 24 or 36 h (Table S1).

Table 3 shows the results of the agar spot test. Here it can be seen that all probiotic strains tested generated halos of inhibition against the pathogens *L. monocytogenes* and *S. aureus*. Except for *L. helveticus* R0052, probiotics formed inhibition halos against *E. coli* and *Salmonella* Dublin. In the agar diffusion technique, only the *E. faecium* MXVK29 strain was able to inhibit *L. monocytogenes* in both non-neutralised and neutralised CFE (Figure S2).

Table 4 shows the results of microbial antagonism in microplate. The CFE of the different probiotics produced lower growth of *E. coli*, *Salmonella* Dublin and *S. aureus* ($p \leq 0.05$). The *L. casei* Shirota strain had the greatest antagonistic effect against *S. aureus*. The *L. paracasei* Lpc-37, *E. faecium* MXVK29 and *L. reuteri* DSM17918 strains strongly inhibited the growth of *Salmonella* Dublin while the *E. faecium* MXVK29 and *L. reuteri* DSM17938 were the strains with the bigger inhibition against *E. coli*. There was no significant difference ($p \leq 0.05$) in the growth of *L. monocytogenes* in the control medium with respect to media containing CFE of *L. rhamnosus* R0011 and *L. helveticus* R0052. On the contrary, there were, with the CFE of the other probiotics tested and a pronounced inhibition effect was evidenced in the medium containing CFE of the *L. reuteri* DSM17938 strain.

Table 5 shows the results of a strain selection ranking. There it emerged that the most suitable strains to be selected as starter are *L. rhamnosus* Lr-32, *L. rhamnosus* R0011, *L. paracasei* Lpc-37, *E. Faecium* MXVK29 and *L. casei* Shirota. Of the aforementioned strains, the one that best tolerated the fermentation and maturation conditions of sausages was *L. rhamnosus* Lr-32.

4. Discussion

The growth shown by the strains studied, all above 8 $\log_{10}$ CFU/mL in 12 h was important. This characteristic is related to the adaptive power of microorganisms to the environment and its ability to develop rapidly and is essential when producing probiotic products as they should be able to deliver an adequate amount of live organisms that enable the health-promoting effects [11]. Further, when microorganisms are incorporated into sausage batter, they must grow rapidly and reach high numbers to become dominant and competitive against the indigenous microbiota present in raw materials [18,19].

For the fermentation process to occur, the participation of LAB is necessary [15]. The ability of the strains studied to produce only lactic acid from carbohydrates was very important. Heterofermentative LAB are not suitable for sausage production due to the formation of carbon dioxide, which accumulates in the matrix and generates holes of different sizes in the product [26]. In addition, these LAB produce concentrations of acetic acid that cause a strong off flavour.

The aptitude of starter strains to fast acidification is an important characteristic as it has an impact on taste, safety, aroma and bacteriostatic or bactericidal properties [49,50]. The main function of LAB is to reduce the pH of the matrix through production of lactic acid from the fermentation of carbohydrates. The production of acids in dry sausage depends on the type and concentration of sugars added to the meat mixture, the diameter of the dry sausage and the LAB microbiota [11]. A reduction in pH is necessary for fibrillar proteins to coagulate, resulting in improved firmness and cohesiveness of the final product, facilitating slicing [51,52]. They also promote the spontaneous reduction of nitrites to nitric oxide, which reacts with myoglobin to form nitrosomyoglobin, the compound responsible for the typical red colour of cured sausages [53]. In addition, they contribute to the flavour of the final product through the formation of its typical acidic taste. The acidity reached by the meat matrix also contributes to increasing the activity of cathepsin D, which is responsible for muscle proteolysis [54]. The production of organic acids is undoubtedly the determining factor on which the product's shelf life and safety depends. The inhibition of pathogens and spoilage microbiota also depends on a rapid and adequate production of organic acids and the associated reduction in pH [15]. Finally, it has been reported that the rapid decrease in pH caused by negative aminodecarboxylase cultures decreases the production of biogenic amines in sausage [55]. The immediate and rapid formation of acid at the beginning of the fermentation process, and the production of sufficient quantity of organic acids that allow to reach a pH below 5.1 is an essential requirement of the starter LAB. However, excessive acid formation is often associated with colour defects (due to inhibition of CNCs) and sometimes with gas formation (in the case of heterofermentative bacteria), one of the most important problems in sausage processing [26]. Comparing the results of this work with those obtained by Erkkilä et al. [7] it can be assumed that the studied strains produced enough lactic acid to act as the main initiating microorganism in sausage production. However, a sausage test should be performed to confirm this hypothesis. The acid taste of fermented meat products, which is correlated to the acid content, is appreciated in some countries whereas it is undesirable in others [11]. Therefore, the acid production is an important factor for selection of strains as starter cultures in fermented sausages also considering the characteristics of the product to be produced.

The manufacturing temperature of fermented meat sausages ranges between 4 and 7 °C when the mixture is prepared, rises between 18 and 24 °C during the fermentation period and is reduced from 12 to 15 °C during the drying and ripening period [21,56]. Therefore, it is necessary to select isolates of LAB which are capable of developing over a wide range of temperatures [15]. In the temperature range of the typical Mediterranean fermentation (15–22 °C), which is also generally used in the Argentinean meat industry, the strains *L. rhamnosus* R0011, *L. rhamnosus* Lr-32, *L. paracasei* Lpc-37, *L. casei* Shirota and *E. faecium* MXVK29 were the ones that best adapted and developed.

The growth of the LAB strains under the conditions of fermentation and maturation of the sausages is decisive so that they can be considered as a potential starter. The strains studied showed a different adaptive response to the fermentation and drying conditions studied. Three of the strains studied (*L. reuteri* DSM 17938, *L. reuteri* DSM 17918 and *L. rhamnosus* Lr-32) showed the best performance in both conditions. The concentration of salt added to raw-cured sausage batter, approximately 2% depending on each product, can reach up to 2.5–3.5 *w*/*w* in the final product [57]. The initial pH of the mixture, which is generally around 6, decreases during fermentation and reaches values between 4.6–5.1. Subsequently, yeasts can increase the pH of the product, reaching final values, ranging from 5.1–5.5 [21,57]. Bacterial growth in food matrices may be affected by intrinsic and extrinsic factors. Therefore, it is necessary to select isolates of LAB which are capable of developing tolerating adverse conditions including the presence of sodium chloride, sodium nitrite and acidic pH, it is essential that they adapt properly to the conditions of the food matrix. Thus, the above strains will present a greater capacity to persist during the fermentation of food [15]. The development of a new probiotic fermented meat product requires the application of probiotic bacteria that are resistant to salt, nitrite and acidic pH, so that they are able to activate and grow rapidly during fermentation and maturation [16]. This will allow competition with the natural microbiota and result in a successful fermentation. Further, when producing functional foods with probiotic cultures, it is necessary that the product contains a high enough number of viable cells at the time of consumption, which must exceed the minimum suggested dose for a health benefit. This is a prerequisite to exert beneficial effects on the host [57]. Tolerance to high salt concentration and fast acidification constitute key functions for a significant role of the organisms in meat fermentation [11].

Most lactobacilli are capable of forming hydrogen peroxide through the enzyme lactate oxidase during sausage fermentation. Hydrogen peroxide can interfere with the organoleptic properties of fermented meat products by increasing the rancidity due to lipid oxidation [15] and discolouration of the final product [21]. Authors' trials have reported that *L. plantarum* and *L. sakei* are capable of producing H_2O_2, which is an oxidising agent, and it has been suggested that such strains could be related to defects in colour and flavour of the sausage [16,41]. Taking into account the aforementioned, *L. helveticus* R0052, *L. casei* Shirota, *L. reuteri* DSM17918 and *L. reuteri* DSM17938 would not be the most suitable strains to be used as sausage starters. However, it should also be considered that the LAB strains will be part of a mixed starter culture and that the CNCs are capable of lysing the H_2O_2 produced [15].

Some LAB strains involved in the fermentation of meat such as *L. sakei*, *L. plantarum*, *L. pentosus* and *P. acidilactici* have haemo-dependent catalase activity that is active in meat products, since these substrates contain abundant hemin [21]. Although all the probiotic strains studied were catalase negative and its activity of LAB used as starter cultures in meats, it is a desirable property [21], it is not essential since this function is performed by the CNCs [15] that are part of the starter culture. CNCs can neutralise pro-oxidant molecules, limiting the oxidative processes based on their superoxide dismutase (SOD) and catalase (CAT) [15].

Although several authors have reported that some meat LABs have nitrate reductases [21,22,58] the probiotic strains tested did not reduce nitrates but Gram-positive cocci species contribute to the formation of the typical characteristics of fermented meat products, such as colour. Nitrate reductase confers a characteristic colour to the product [15]. The probiotic LAB will be part of a mixed culture along with a micrococcus.

Starter cultures affect the aroma and taste of fermented meat products and the use of their enzymatic pattern as selection criterion could be of interest [16]. Five of the strains studied showed their lipolytic and proteolytic capacity, which is interesting as a technological property. This is a desirable characteristic, which could contribute to the catabolism of proteins and fats by generating precursors of sausage flavour compounds [21]. These activities are associated with flavour development [59,60].

As mentioned above, the starter cultures of meat products are mainly a mixture of LAB and CNC strains, and it has been used as a commercial starter for many years for sausage making in different countries worldwide [11]. To carry out the expected functions it is necessary that the starter LAB strain show tolerance or even synergism with the CNC that is part of starter culture [21]. Therefore, the results obtained indicate that the probiotic strains studied could be combined with *S. xylosus* DD-34 thus forming a mixed starter culture to be used in the preparation of probiotic sausage. It has been shown that some of these coagulase negative staphylococci could affect both growth and proteolytic activities of LAB strains. Such interactions are therefore of interest and should be considered for selection of starters in order to improve the organoleptic properties of fermented sausages [61].

Biogenic amines (BA) are organic bases of low molecular weight, polar or semi-polar compounds, resulting from the decarboxylation of amino acids [62]. These compounds are usually non-toxic, but when ingested in high amounts by people with gastrointestinal diseases, or incapable of detoxification due to genetic problems and/or combined with the ingestion of alcohol, they can cause nausea, diarrheal and hyperdilation of blood vessels [62]. The most prevalent BAs in food and beverages are histamine, tyramine, putrescine, cadaverine and β-phenylethylamine [62]. Residual nitrite present in fermented sausages can react with amines and form carcinogenic nitrosamines [63]. The results indicate a low capacity of the microorganisms studied to generate biogenic amines, especially the probiotic strains, which did not decarboxylate any of the amino acids tested. Previous studies have reported deamination of arginine by strains of *L. sakei* but not by strains of *L. plantarum* [16,64]. Other authors reported the formation of biogenic amines by both *L. plantarum* and *L. sakei* strains [47]. The absence of biogenic amines formation has been proposed as a selection criterion for new strains used as starter cultures for meats [21,26,47]. The rapid growth and production of acid will further prevent the development of an indigenous microbiota that produces amines [21]. In addition, other studies indicate that the use of amino decarboxylase-negative starter cultures significantly reduces the levels of biogenic amines formed in sausage [8,65]. This activity has already been described in some LABs for fermentation of sausage, such as *L. plantarum* and *L. casei* [66]. The most promising microorganisms utilised as starter cultures are identified microorganisms, previously characterised as safe and exhibiting desired metabolic activity [67]. Based on these concepts, the probiotics studied *L. rhamnosus* R0011, *L. rhamnosus* Lr-32 and *L. paracasei* Lpc-37 would be the main candidates to be used as a starter culture in the preparation of a healthy and safe sausage.

LAB are known to promote food safety and quality because they present antagonistic activity against spoilage and pathogenic microorganisms [68]. In addition to the use as a starter microorganism, based mainly on acidification capacities, the antibacterial capacity demonstrated by the studied strains could help to develop biopreservation properties in dry fermented meat products. These approaches appear promising for future development of biopreservation for enhanced shelf life and safety of meat products [50]. Antagonistic activity may occur through several mechanisms, including competition for nutrients and adhesion sites, as well as production of bactericidal compounds such as organic acids (lactic, acetic and propionic), carbon dioxide, diacetyl, hydrogen peroxide, reuterin and bacteriocins [68]. The main antimicrobial effect of LAB responsible for biopreservation of foods is their acidification capacity. However, bacteriocins produced by LAB can provide additional control against pathogens in meat sausage [15]. The use of a starter culture capable of reducing or inhibiting the growth of pathogenic microorganisms associated with sausage would contribute to ensuring product safety. The ability to inhibit pathogens capable of producing foodborne illnesses is a quality of interest for the starters used in the manufacture of raw-cured meat products [26]. Although this effect is usually restricted to a limited number of microorganisms, a broad inhibitory effect is expected in strains already characterised as those used in this work. Further on this characteristic, focus was on pathogen inactivation, as probiotic strains with additional food safety properties could confer added value to healthy fermented meat products.

According to the data obtained, *L. rhamnosus* Lr-32, *L. rhamnosus* R0011, *L. paracasei* Lpc-37, *E. faecium* MXVK29 and *L. casei* Shirota strains are the main candidates to be used as sausages starters. These results coincide with those obtained by Rebucci et al. [69] who proposed the strains of *L. casei* and *L. rhamnosus* isolated from sausages are the best potential functional starter cultures in meat products. Accordingly, recent investigations have shown that strains of *L. rhamnosus*, *L. fermentum* and *L. paracasei* of human intestinal origin were able to survive the dry sausage manufacturing process, being detected in high numbers in the final product [7–12,14,70].

Considering that the adaptability and growth of the LAB strains under the fermentation and maturation conditions of the raw-cured sausages is a decisive selection criterion, *L. rhamnosus* Lr-32 is considered the best of the strains studied to be tested as an initiator in the preparation of such sausages. A probiotic starter culture adapted to the ecology of the meat fermentation, with those potential properties and antimicrobial activity could provide significant health benefits and contribute to enhance the hygienic quality of the products [16]. We emphasised that strain *L. rhamnosus* Lr-32 is of interest for further tests in sausages according to the results obtained. Although the strain had excellent results in vitro, its performance as an initiator in sausages will be dependent on its ability to survive and fulfil its functions reliant on the raw materials, the conditions of preparation of the sausages and the existing indigenous microbiota communities.

5. Conclusions

The objective of this study was to evaluate technological and safety properties of eight commercial probiotic strains to select those that were most suitable to act as a sausages starter microorganism. In addition to contributing to human health, probiotic fermented meats need to be of sufficient commercial value. According to the data obtained *L. rhamnosus* Lr-32, *L. rhamnosus* R0011, *L. paracasei* Lpc-37, *E. faecium* MXVK29 and *L. casei* Shirota strains are the main candidates to be used as sausages starters because they showed very good capacity to grow, a high production of lactic acid without gas formation, generate good growth at the low temperatures at which this product is manufactured, demonstrated to have proteolytic and lipolytic capacity that could contribute to flavour and did not show antagonism against the CNC, that are usually part of the starter culture mix. These strains did not exhibit amino acid decarboxylase activity, a key requirement for the selection of a new safe LAB starter culture. In addition, they could prevent the growth of the native biogenic amine-producing microbiota, helping to produce a safer product. The strains showed antimicrobial activity against harmful microorganisms frequently associated with sausage, which could contribute to improving the hygienic quality of the product and preventing possible diseases.

The theoretical basis for the selection of a new starter culture includes technological aspects (growth in food base, sensory properties, stability and viability) and safety as mentioned above. The beneficial effects of LAB have been attributed to their ability to suppress the growth of pathogens, probably by secretion of antibacterial substances such as lactic acid and bacteriocins. Of the aforementioned strains, *L. rhamnosus* Lr-32 was the strain that best tolerated the levels of salt, nitrate and the low pH during the simulated stages of fermentation and maturation of sausage.

The novelty of this work was to study existing probiotics strains as starter and to show that the probiotic strain could be used without being combined with a commercial starter culture. It is expected that the strain fulfils both functions being a starter at the same time as a beneficial microorganism.

The great advance that could be achieved in the science of the meat using probiotic microorganism as a starter would be a functional meat product with improvements in the safety aspect of the product, potential benefits to the health of the consumer and it could improve the negative image that some consumers have about sausages.

It is necessary to carry out new studies in manufacturing conditions of the raw-cured sausages in order to verify if probiotic *L. rhamnosus* Lr-32 maintains the characteristics identified in the in vitro work presented here.

Supplementary Materials: The following are available online at http://www.mdpi.com/2304-8158/9/5/596/s1, Figure S1: Standard calibration curves used to quantify probiotic LAB; Figure S2. Inhibition of *Listeria monocytogenes* by *Enterococcus faecium* MXVK29 in agar well diffusion assay; Table S1. Amino acid decarboxylation assay.

Author Contributions: Conceptualisation, M.R.R. and L.S.F.; methodology, N.d.L.A. and L.S.F.; validation, L.S.F.; formal analysis, N.d.L.A. and G.A.; investigation, N.d.L.A. and G.A.; resources, M.R.R. and L.S.F.; data curation, L.S.F.; writing—original draft preparation, N.d.L.A., L.S.F. and M.R.R.; writing—review and editing, A.C.O. and N.d.L.A.; visualisation, A.C.O., N.d.L.A. and M.R.R.; supervision, M.R.R.; project administration, M.R.R.; funding acquisition, M.R.R. All authors have read and agreed to the published version of the manuscript.

Funding: This research was funded by Catholic University of Córdoba (Nadia de Lourdes Agüero PhD Scholarship) and National University of Litoral (Argentina), CAID 50120110100198.

Acknowledgments: The companies that generously provided the investigated strains are gratefully acknowledged.

Conflicts of Interest: A.C.O. is an employee of DuPont Nutrition and Biosciences. DuPont Nutrition and Biosciences manufactures and markets probiotics; two of which were investigated in the present study. The other authors declare no conflict of interest. The funders had no role in the design of the study; in the collection, analyses, or interpretation of data; in the writing of the manuscript, or in the decision to publish the results.

References

1. Rosmini, M.R.; Frizzo, L.S.; Zogbi, A.P. Meat products with low sodium content: Processing and properties. In *Technological Strategies for Functional Meat Products Development*; Juana, L.F., Ed.; Transworld Research Network: Kerala, India, 2008; pp. 87–108.
2. Zhang, W.; Xiao, S.; Samaraweera, H.; Lee, E.J.; Ahn, D.U. Improving functional value of meat products. *Meat Sci.* **2010**, *86*, 15–31. [CrossRef]
3. Soccol, C.R.; Vandenberghe, L.P.d.S.; Spier, M.R.; Medeiros, A.B.P.; Yamaguishi, C.T.; Lindner, J.d.D.; Ashok, P.; Thomaz-Soccol, V. The potential of probiotics: A review. *Food Technol. Biotechnol.* **2010**, *48*, 413–434.
4. Hill, C.; Guarner, F.; Reid, G.; Gibson, G.R.; Merenstein, D.J.; Pot, B.; Morelli, L.; Canani, R.B.; Flint, H.J.; Salminen, S. Expert consensus document: The International Scientific Association for Probiotics and Prebiotics consensus statement on the scope and appropriate use of the term probiotic. *Nat. Rev. Gastroenterol. Hepatol.* **2014**, *11*, 506. [CrossRef] [PubMed]
5. Trush, E.A.; Poluektova, E.A.; Beniashvilli, A.G.; Shifrin, O.S.; Poluektov, Y.M.; Ivashkin, V.T. The Evolution of Human Probiotics: Challenges and Prospects. *Probiotics Antimicrob. Proteins* **2020**. [CrossRef] [PubMed]
6. Rivera-Espinoza, Y.; Gallardo-Navarro, Y. Non-dairy probiotic products. *Food Microbiol.* **2010**, *27*, 1–11. [CrossRef] [PubMed]

7. Erkkilä, S.; Petäjä, E.; Eerola, S.; Lilleberg, L.; Mattila-Sandholm, T.; Suihko, M.L. Flavour profiles of dry sausages fermented by selected novel meat starter cultures. *Meat Sci.* **2001**, *58*, 111–116. [CrossRef]
8. Erkkilä, S.; Suihko, M.L.; Eerola, S.; Petäjä, E.; Mattila-Sandholm, T. Dry sausage fermented by Lactobacillus rhamnosus strains. *Int. J. Food Microbiol.* **2001**, *64*, 205–210. [CrossRef]
9. Ruiz-Moyano, S.; Martín, A.; Benito, M.J.; Hernández, A.; Casquete, R.; de Guia Córdoba, M. Application of Lactobacillus fermentum HL57 and Pediococcus acidilactici SP979 as potential probiotics in the manufacture of traditional Iberian dry-fermented sausages. *Food Microbiol.* **2011**, *28*, 839–847. [CrossRef]
10. Rubio, R.; Aymerich, T.; Bover-Cid, S.; Guàrdia, M.D.; Arnau, J.; Garriga, M. Probiotic strains Lactobacillus plantarum 299V and Lactobacillus rhamnosus GG as starter cultures for fermented sausages. *LWT-Food Sci. Technol.* **2013**, *54*, 51–56. [CrossRef]
11. Rubio, R.; Jofré, A.; Martín, B.; Aymerich, T.; Garriga, M. Characterization of lactic acid bacteria isolated from infant faeces as potential probiotic starter cultures for fermented sausages. *Food Microbiol.* **2014**, *38*, 303–311. [CrossRef]
12. Mati, M.; Magala, M.; Karovičová, J.; Staruch, L. The influence of Lactobacillus paracasei LPC-37 on selected properties of fermented sausages. *Potravin. Slovak J. Food Sci.* **2015**, *9*, 58–65. [CrossRef]
13. De Vuyst, L.; Falony, G.; Leroy, F. Probiotics in fermented sausages. *Meat Sci.* **2008**, *80*, 75–78. [CrossRef] [PubMed]
14. Pidcock, K.; Heard, G.M.; Henriksson, A. Application of nontraditional meat starter cultures in production of Hungarian salami. *Int. J. Food Microbiol.* **2002**, *76*, 75–81. [CrossRef]
15. dos Santos Cruxen, C.E.; Funck, G.D.; Haubert, L.; da Silva Dannenberg, G.; de Lima Marques, J.; Chaves, F.C.; da Silva, W.P.; Fiorentini, Â.M. Selection of native bacterial starter culture in the production of fermented meat sausages: Application potential, safety aspects, and emerging technologies. *Food Res. Int.* **2019**. [CrossRef]
16. Papamanoli, E.; Tzanetakis, N.; Litopoulou-Tzanetaki, E.; Kotzekidou, P. Characterization of lactic acid bacteria isolated from a Greek dry-fermented sausage in respect of their technological and probiotic properties. *Meat Sci.* **2003**, *65*, 859–867. [CrossRef]
17. Vinderola, G.; Ouwehand, A.; Salminen, S.; von Wright, A. *Lactic Acid Bacteria: Microbiological and Functional Aspects*; CRC Press: Boca Raton, FL, USA, 2019. [CrossRef]
18. Drosinos, E.; Mataragas, M.; Xiraphi, N.; Moschonas, G.; Gaitis, F.; Metaxopoulos, J. Characterization of the microbial flora from a traditional Greek fermented sausage. *Meat Sci.* **2005**, *69*, 307–317. [CrossRef] [PubMed]
19. Landeta, G.; Curiel, J.A.; Carrascosa, A.V.; Muñoz, R.; De Las Rivas, B. Technological and safety properties of lactic acid bacteria isolated from Spanish dry-cured sausages. *Meat Sci.* **2013**, *95*, 272–280. [CrossRef] [PubMed]
20. Cocolin, L.; Dolci, P.; Rantsiou, K. Biodiversity and dynamics of meat fermentations: The contribution of molecular methods for a better comprehension of a complex ecosystem. *Meat Sci.* **2011**, *89*, 296–302. [CrossRef]
21. Ammor, M.S.; Mayo, B. Selection criteria for lactic acid bacteria to be used as functional starter cultures in dry sausage production: An update. *Meat Sci.* **2007**, *76*, 138–146. [CrossRef]
22. Hammes, W.P.; Bantleon, A.; Min, S. Lactic acid bacteria in meat fermentation. *FEMS Microbiol. Rev.* **1990**, *7*, 165–173. [CrossRef]
23. Silla Santos, M.H. Biogenic amines: Their importance in foods. *Int. J. Food Microbiol.* **1996**, *29*, 213–231. [CrossRef]
24. Ruiz-Capillas, C.; Herrero, A.M. Impact of biogenic amines on food quality and safety. *Foods* **2019**, *8*, 62. [CrossRef] [PubMed]
25. Shalaby, A.R. Significance of biogenic amines to food safety and human health. *Food Res. Int.* **1996**, *29*, 675–690. [CrossRef]
26. Buckenhüskes, H.J. Selection criteria for lactic acid bacteria to be used as starter cultures for various food commodities. *FEMS Microbiol. Rev.* **1993**, *12*, 253–271. [CrossRef]
27. Barbieri, F.; Montanari, C.; Gardini, F.; Tabanelli, G. Biogenic amine production by lactic acid bacteria: A review. *Foods* **2019**, *8*, 17. [CrossRef]
28. Reis, J.; Paula, A.; Casarotti, S.; Penna, A. Lactic acid bacteria antimicrobial compounds: Characteristics and applications. *Food Eng. Rev.* **2012**, *4*, 124–140. [CrossRef]

29. Evans, M.; Salewski, R.; Christman, M.; Girard, S.; Tompkins, T. Effectiveness of Lactobacillus helveticus and Lactobacillus rhamnosus for the management of antibiotic-associated diarrhoea in healthy adults: A randomised, double-blind, placebo-controlled trial. *Br. J. Nutr.* **2016**. [CrossRef]
30. Miyazima, T.; Ishikawa, K.; Mayer, M.; Saad, S.; Nakamae, A. Cheese supplemented with probiotics reduced the Candida levels in denture wearers—RCT. *Oral Dis.* **2017**, *23*, 919–925. [CrossRef]
31. Ouwehand, A.C.; DongLian, C.; Weijian, X.; Stewart, M.; Ni, J.; Stewart, T.; Miller, L.E. Probiotics reduce symptoms of antibiotic use in a hospital setting: A randomized dose response study. *Vaccine* **2014**, *32*, 458–463. [CrossRef]
32. Tripolt, N.J.; Leber, B.; Triebl, A.; Kofeler, H.; Stadlbauer, V.; Sourij, H. Effect of Lactobacillus casei Shirota supplementation on trimethylamine-N-oxide levels in patients with metabolic syndrome: An open-label, randomized study. *Atherosclerosis* **2015**, *242*, 141–144. [CrossRef]
33. Alvarez-Cisneros, Y.M.; Fernández, F.J.; Sainz-Espuñez, T.; Ponce-Alquicira, E. Assessment of virulence factors, antibiotic resistance and amino-decarboxylase activity in Enterococcus faecium MXVK29 isolated from Mexican chorizo. *Lett. Appl. Microbiol.* **2017**, *64*, 171–176. [CrossRef] [PubMed]
34. Savino, F.; Garro, M.; Montanari, P.; Galliano, I.; Bergallo, M. Crying Time and RORγ/FOXP3 Expression in Lactobacillus reuteri DSM17938-Treated Infants with Colic: A Randomized Trial. *J. Pediatrics* **2018**, *192*, 171–177. [CrossRef] [PubMed]
35. Lin, J.; Zhang, Y.; He, C.; Dai, J. Probiotics supplementation in children with asthma: A systematic review and meta-analysis. *J. Paediatr. Child Health* **2018**, *54*, 953–961. [CrossRef] [PubMed]
36. Rodríguez Haro, I.; Salazar Castillo, M.; Villalobos Infante, E. Lactobacillus spp. del tracto intestinal de Gallus gallus con potencial probiótico. *Rev. REBIOL* **2012**, *32*, 62–72.
37. Cai, Y.; Benno, Y.; Nakase, T.; Oh, T.K. Specific probiotic characterization of Weissella hellenica DS-12 isolated from flounder intestine. *J. Gen. Appl. Microbiol.* **1998**, *44*, 311–316. [CrossRef]
38. Korkeala, H.; Alanko, T.; Tiusanen, T. Effect of sodium nitrite and sodium chloride on growth of lactic acid bacteria. *Acta Vet Scand* **1992**, *33*, 27–32.
39. McLean, N.W.; Rosenstein, I.J. Characterisation and selection of a Lactobacillus species to re-colonise the vagina of women with recurrent bacterial vaginosis. *J. Med Microbiol.* **2000**, *49*, 543–552. [CrossRef]
40. Rosenstein, I.J.; Fontaine, E.A.; Morgan, D.J.; Sheehan, M.; Lamont, R.F.; Taylor-Robinson, D. Relationship between hydrogen peroxide-producing strains of lactobacilli and vaginosis-associated bacterial species in pregnant women. *Eur. J. Clin. Microbiol. Infect. Dis.* **1997**, *16*, 517–522. [CrossRef]
41. Talon, R.; Walter, D.; Montel, M.C. Growth and effect of staphylococci and lactic acid bacteria on unsaturated free fatty acids. *Meat Sci.* **2000**, *54*, 41–47. [CrossRef]
42. Harrigan, W.F. *Laboratory Methods in Food Microbiology*; Academic Press: Cambridge, MA, USA, 1998.
43. MacFaddin, J.F. *Pruebas Bioquímicas Para la Identificación de Bacterias de Importancia Clínica*, 3rd ed.; Médica Panamericana: Buenos Aires, Argentina, 2003.
44. Lewus, C.B.; Kaiser, A.; Montville, T.J. Inhibition of food-borne bacterial pathogens by bacteriocins from lactic acid bacteria isolated from meat. *Appl. Environ. Microbiol.* **1991**, *57*, 1683–1688. [CrossRef]
45. Tagg, J.R.; McGiven, A.R. Assay system for bacteriocins. *Appl. Microbiol.* **1971**, *21*, 943. [CrossRef] [PubMed]
46. Schillinger, U.; Lücke, F.K. Identification of lactobacilli from meat and meat products. *Food Microbiol.* **1987**, *4*, 199–208. [CrossRef]
47. Ruiz-Moyano, S.; Martín, A.; Benito, M.J.; Casquete, R.; Serradilla, M.J.; Córdoba, M.d.G. Safety and functional aspects of pre-selected lactobacilli for probiotic use in Iberian dry-fermented sausages. *Meat Sci.* **2009**, *83*, 460–467. [CrossRef] [PubMed]
48. Frizzo, L.S.; Soto, L.P.; Bertozzi, I.; Sequeira, G.; Marti, L.E.; Rosmini, M.R. Evaluación in vitro de las capacidades probióticas microbianas orientadas al diseño de inóculos probióticos multiespecie para ser utilizados en la crianza de terneros. *Rev. FAVE-Cienc. Vet.* **2006**, *5*, 69–80. [CrossRef]
49. Leroy, F.; Verluyten, J.; De Vuyst, L. Functional meat starter cultures for improved sausage fermentation. *Int. J. Food Microbiol.* **2006**, *106*, 270–285. [CrossRef]
50. Zagorec, M.; Champomier-Vergès, M.C. Lactobacillus sakei: A starter for sausage fermentation, a protective culture for meat products. *Microorganisms* **2017**, *5*, 56. [CrossRef]
51. Ordoñez, J.A.; Hierro, E.M.; Bruna, J.M.; de la Hoz, L. Changes in the components of dry-fermented sausages during ripening. *Crit. Rev. Food Sci. Nutr.* **1999**, *39*, 329–367. [CrossRef]

52. Drosinos, E.H.; Paramithiotis, S.; Kolovos, G.; Tsikouras, I.; Metaxopoulos, I. Phenotypic and technological diversity of lactic acid bacteria and staphylococci isolated from traditionally fermented sausages in Southern Greece. *Food Microbiol.* **2007**, *24*, 260–270. [CrossRef]
53. Hugas, M.; Monfort, J.M. Bacterial starter cultures for meat fermentation. *Food Chem.* **1997**, *59*, 547–554. [CrossRef]
54. Molly, K.; Demeyer, D.; Johansson, G.; Raemaekers, M.; Ghistelinck, M.; Geenen, I. The importance of meat enzymes in ripening and flavour generation in dry fermented sausages. First results of a European project. *Food Chem.* **1997**, *59*, 539–545. [CrossRef]
55. Maijala, R.L.; Eerola, S.H.; Aho, M.A.; Hirn, J.A. The effect of GDL-induced pH decrease on the formation of biogenic amines in meat. *J. Food Prot.* **1993**, *56*, 125–129. [CrossRef] [PubMed]
56. Montel, M.C. Fermented Meat Products. In *Encyclopedia of Food Microbiology*; Robinson, R.K., Ed.; Elsevier: Oxford, UK, 1999; pp. 744–753.
57. Ruiz-Moyano, S.; Martín, A.; Benito, M.J.; Nevado, F.P.; de Guía Córdoba, M. Screening of lactic acid bacteria and bifidobacteria for potential probiotic use in Iberian dry fermented sausages. *Meat Sci.* **2008**, *80*, 715–721. [CrossRef] [PubMed]
58. Wolf, G.; Arendt, E.K.; Pfähler, U.; Hammes, W.P. Heme-dependent and heme-independent nitrite reduction by lactic acid bacteria results in different N-containing products. *Int. J. Food Microbiol.* **1990**, *10*, 323–329. [CrossRef]
59. Chen, Q.; Kong, B.; Han, Q.; Xia, X.; Xu, L. The role of bacterial fermentation in lipolysis and lipid oxidation in Harbin dry sausages and its flavour development. *LWT* **2017**, *77*, 389–396. [CrossRef]
60. Chen, Q.; Liu, Q.; Sun, Q.; Kong, B.; Xiong, Y. Flavour formation from hydrolysis of pork sarcoplasmic protein extract by a unique LAB culture isolated from Harbin dry sausage. *Meat Sci.* **2015**, *100*, 110–117. [CrossRef]
61. Tremonte, P.; Reale, A.; Di Renzo, T.; Tipaldi, L.; Di Luccia, A.; Coppola, R.; Sorrentino, E.; Succi, M. Interactions between Lactobacillus sakei and CNC (Staphylococcus xylosus and Kocuria varians) and their influence on proteolytic activity. *Lett. Appl. Microbiol.* **2010**, *51*, 586–594. [CrossRef]
62. Papageorgiou, M.; Lambropoulou, D.; Morrison, C.; Kłodzińska, E.; Namieśnik, J.; Płotka-Wasylka, J. Literature update of analytical methods for biogenic amines determination in food and beverages. *TrAC Trends Anal. Chem.* **2018**, *98*, 128–142. [CrossRef]
63. De Mey, E.; De Klerck, K.; De Maere, H.; Dewulf, L.; Derdelinckx, G.; Peeters, M.C.; Fraeye, I.; Vander Heyden, Y.; Paelinck, H. The occurrence of N-nitrosamines, residual nitrite and biogenic amines in commercial dry fermented sausages and evaluation of their occasional relation. *Meat Sci.* **2014**, *96*, 821–828. [CrossRef]
64. Hugas, M.; Garriga, M.; Aymerich, T.; Monfort, J.M. Biochemical characterization of lactobacilli from dry fermented sausages. *Int. J. Food Microbiol.* **1993**, *18*, 107–113. [CrossRef]
65. Maijala, R.; Eerola, S.; Lievonen, S.; Hill, P.; Hirvi, T. Formation of Biogenic Amines during Ripening of Dry Sausages as Affected by Starter Culture and Thawing Time of Raw Materials. *J. Food Sci.* **1995**, *60*, 1187–1190. [CrossRef]
66. Fadda, S.; Vignolo, G.; Oliver, G. Tyramine degradation and tyramine/histamine production by lactic acid bacteria and Kocuria strains. *Biotechnol. Lett.* **2001**, *23*, 2015–2019. [CrossRef]
67. Leroy, F.; De Vuyst, L. Lactic acid bacteria as functional starter cultures for the food fermentation industry. *Trends Food Sci. Technol.* **2004**, *15*, 67–78. [CrossRef]
68. da Costa, W.K.A.; de Souza, G.T.; Brandão, L.R.; de Lima, R.C.; Garcia, E.F.; dos Santos Lima, M.; de Souza, E.L.; Saarela, M.; Magnani, M. Exploiting antagonistic activity of fruit-derived Lactobacillus to control pathogenic bacteria in fresh cheese and chicken meat. *Food Res. Int.* **2018**, *108*, 172–182. [CrossRef]
69. Rebucci, R.; Sangalli, L.; Fava, M.; Bersani, C.; Cantoni, C.; Baldi, A. Evaluation of functional aspects in Lactobacillus strains isolated from dry fermented sausages. *J. Food Qual.* **2007**, *30*, 187–201. [CrossRef]
70. Sidira, M.; Galanis, A.; Nikolaou, A.; Kanellaki, M.; Kourkoutas, Y. Evaluation of Lactobacillus casei ATCC 393 protective effect against spoilage of probiotic dry-fermented sausages. *Food Control* **2014**, *42*, 315–320. [CrossRef]

foods

MDPI

Article

Biochemical Changes during the Manufacture of Galician Chorizo Sausage as Affected by the Addition of Autochthonous Starter Cultures

Miriam Rodríguez-González [1,2], Sonia Fonseca [1], Juan A. Centeno [1,2] and Javier Carballo [1,2,*]

1 Food Technology Area, Faculty of Sciences, University of Vigo, 32004 Ourense, Spain; mirrodriguez@uvigo.es (M.R.-G.); soniafonseca@uvigo.es (S.F.); jcenteno@uvigo.es (J.A.C.)
2 CITACA, Agri-Food Research and Transfer Cluster, Campus da Auga, University of Vigo, 32004 Ourense, Spain
* Correspondence: carbatec@uvigo.es

Received: 30 October 2020; Accepted: 1 December 2020; Published: 7 December 2020

check for updates

Abstract: In this work, the effect of the use of two autochthonous starter cultures (*Lactobacillus sakei* LS131 + *Staphylococcus equorum* SA25 (EQU), or *L. sakei* LS131 + *Staphylococcus saprophyticus* SB12 (SAP)) on the physicochemical, microbiological, proteolytic and lipolytic changes taking place during the manufacture of Galician chorizo, a traditional Spanish sausage, was studied. Three different batches (control (CNT), EQU and SAP) were manufactured in triplicate and analysed during the manufacturing process (samples were taken and analysed at 0, 2, 5, 9, 14, 21 and 30 days of ripening) for proximate composition, pH, a_w, colour parameters, nitrogen fractions, free amino acids, biogenic amines, fat parameters and free fatty acids. The use of either of these two starter cultures slightly but significantly reduced the pH values during the fermentation and increased the percentage of transformation to nitrosyl-heme pigments as well as the a* and b* values in the final products. The two starters significantly decreased the *Enterobacteriaceae* counts in the final product, but without this microbial group completely disappearing. Both starter cultures significantly increased the α-amino acidic nitrogen and the total basic volatile nitrogen fractions during manufacturing, also increasing the free amino acid content and reducing the total biogenic amine content by approximately 20%. The SAP starter enhanced the lipolytic processes, increasing the free fatty acid content. Due to their performances, these two starter cultures seem to be suitable for increasing the quality and safety of the Galician chorizo sausage.

Keywords: Galician chorizo; starter cultures; *Staphylococcus equorum*; *Staphylococcus saprophyticus*; *Lactobacillus sakei*; physicochemical characteristics; free amino acids; free fatty acids; biogenic amines

1. Introduction

In 2018, 1,429,000 Mt of meat products were manufactured in Spain [1]. Cooked sausages were the major products (430,000 Mt), followed by dry-cured hams and forelegs (306,000 Mt), dry-fermented sausages (214,000 Mt), fresh and marinated products (200,000 Mt), cooked hams and forelegs (175,000 Mt) and pre-prepared dishes (104,000 Mt). Despite the strong domestic demand, 63,103 Mt of dry-fermented sausages and 49,138 Mt of dry-cured hams were exported to the international market in 2019. The dry-fermented sausages were the meat products whose production experienced a greater increase in the last five years (from 186,000 Mt in 2013 to 214,000 Mt in 2018), and also the product that registered the greatest increase in exports (from 40,218 Mt in 2013 to 63,103 Mt in 2019) [1]. Spanish dry-fermented sausages are very diverse in size, appearance and organoleptic and nutritional characteristics, reflecting the traditions as well as the diversity of preferences and climatic conditions in the different regions and areas [2].

Galician chorizo is the typical traditional sausage of Galicia (NW of Spain), the one that enjoys the greatest acceptance and the most widely produced and consumed in this region, being also abundantly consumed in other regions of Spain and foreign countries. Its physicochemical [3–6] and microbiological [7,8] characteristics have been reasonably studied and described. It is both artisanal and industrially produced. Industries have adapted the traditional manufacturing procedures, incorporating the modern technologies in distinct steps of the production process. However, although the manufacturing method is nowadays perfectly standardized, batches of Galician chorizo present in the markets are very diverse in organoleptic characteristics, reflecting, above all, the diversity of the quality of the raw materials used. This variability diminishes the acceptance by the consumers, limits its demand and hinders its expansion in national and international markets.

It is widely known that the organoleptic characteristics of dry-fermented sausages are the result of a series of biochemical changes that take place during maturation [9], promoted by the meat and fat's autochthonous enzymes and by those from the microorganisms that grow during the fermentation and maturation processes [10,11]. Both enzymatic activities are modulated by the rest of the ingredients (salt, spices, additives, etc.) and by the environmental conditions (temperature and relative humidity) present during the production process. Therefore, the diversity of the organoleptic characteristics of the dry-fermented sausages mainly reflects the diversity of the microorganisms acting during the maturation process.

Taking into account all these considerations, the use of an appropriate starter culture seems to be the most feasible solution for the heterogeneity of the organoleptic characteristics of Galician chorizo. The use of starter cultures, generally composed by a lactic acid bacteria and a coagulase-negative staphylococci (CNS), is a common and effective practice in the manufacture of fermented sausages in order to improve the colour and flavour development, ensure safety and extend their shelf-life [12–16]. However, the use of a commercial non-autochthonous starter culture could have a negative impact on the sensory characteristics of the sausages, resulting in losses of the desirable particular organoleptic properties that characterize each type of sausage [17,18].

With the aim of developing a specific and appropriate starter culture for the Galician chorizo sausage, bacterial strains isolated from Galician traditional artisanal sausages were isolated and adequately characterized regarding their technological and safety properties [19–21]. The most suitable lactic acid bacteria and *Staphylococcaceae* strains were then selected, and mixed cultures of *Lactobacillus sakei* and diverse species of *Staphylococcus* were developed and monitored throughout the ripening of this sausage using molecular methods [22]. In this last work, it was possible to verify that these cultures are capable of being implanted in the sausages, dominating the spontaneous flora.

As one of the final steps of this whole work, the aim of the present study was to evaluate the effect of two of these autochthonous starter cultures, consisting of a combination of one strain of *Lactobacillus sakei* and a strain of *Staphylococcus equorum* or *Staphylococcus saprophyticus*, on the biochemical changes that take place during the sausage ripening and that were responsible for the organoleptic characteristics of the final product.

Apart from the novelty of developing and testing a specific starter culture for an economically relevant dry-fermented sausage, the main novelty of the present work is the use of *Staphylococcus equorum* and *Staphylococcus saprophyticus* as starter cultures. *Staphylococcus xylosus* and *S. carnosus* are the only *Staphylococcus* species assayed as starter culture until now [12,14,15,23–26]. Regarding the lactic acid bacteria, the use of *Lactobacillus sakei* as a starter culture is common [13,24,25,27,28], although *Lb. plantarum* [10,11,15,26], *Lactobacillus curvatus* [12,23,29] and *Pediococcus pentosaceus* [14,24] were in many other cases preferred.

2. Materials and Methods

2.1. Preparation of the Starter Cultures

In order to inoculate the Galician chorizo batches, one *Lactobacillus* strain (*L. sakei* LS131-CECT 8335-) and two *Staphylococcus* strains (*S. equorum* SA25-CECT 8337- and *S. saprophyticus* SB12-CECT 8336-) were used as starter cultures. These strains were previously isolated from artisanal Androlla and Botillo, two traditional sausages made in Galicia (NW of Spain), and appropriately identified in our laboratory. These strains were chosen from a large set of isolates after testing their suitable technological and safety properties [19–21]. In brief, the strain *Lactobacillus sakei* LS 131 has a mild acidifying activity. The strain *Staphylococcus equorum* SA25 is slightly lipolytic; it has a medium proteolytic activity on the sarcoplasmic proteins and a lack of hydrolytic activity on the myofibrillar proteins. The strain *Staphylococcus saprophyticus* SB12 has high lipolytic activity, high proteolytic activity on the myofibrillar proteins and a lack of activity on the sarcoplasmic proteins [21,30]. The *Lactobacillus* strain was subcultured on MRS broth (Oxoid Ltd., Basingstoke, UK) to a final volume of 500 mL with a concentration of 10^8 CFU/mL, whereas the *Staphylococcus* strains were subcultured on BHI broth (Oxoid) to a final volume of 1000 mL with a concentration of 10^8 CFU/mL. The cell concentration was assessed by interpolation into the correspondent growth curve of the values of absorbance measured at 600 nm. Next, cells were obtained (centrifugation at 4000× *g* for 5 min at 4 °C) and washed (with 0.85% NaCl sterile solution). The pellets of cells were finally resuspended in 40 mL of sterile distilled water before addition to the sausage batches.

2.2. Production of Sausages and Sampling

Following the traditional procedure, three different batches of Galician chorizo were manufactured in triplicate. Batches were designed according to the starter culture added: CNT batch (control not inoculated), EQU batch (inoculated with *L. sakei* CECT 8335 + *S. equorum* CECT 8337) and SAP batch (inoculated with *L. sakei* CECT 8335 + *S. saprophyticus* CECT 8336). *L. sakei* CECT 8335 was inoculated in the sausage mix in a concentration of 10^6 CFU/g, while *S. equorum* CECT 8337 and *S. saprophyticus* CECT 8336 were added in an amount of 10^7 CFU/g. The mix of sausages was formulated according to traditional procedures, including lean pork shoulder (80%), pork back fat (20%), sweet paprika (22 g/kg)), spicy paprika (1 g/kg)), garlic (4 g/kg)), salt (15 g/kg)) and water (40 mL/kg). Lean and back fat were firstly ground using a 10-mm diameter mincing plate and next mixed together with the other ingredients for 3 min under vacuum. The resulting mix was allowed to stand at 4 °C for 24 h and then stuffed into porcine gut of 36–38 mm in diameter. Sausages were initially fermented for 9 days (6 °C and 80% relative humidity) and then dry-ripened for another 21 days (12 °C and 75% RH). From each replicate of each batch, samples were taken for subsequent analysis at 0 (mix before stuffing), 2, 5, 9, 14, 21 and 30 days of ripening.

2.3. Microbial Analysis

Ten grams of sample were taken in triplicate from the mix before stuffing or from the inner of the sausages at the different sampling times. Samples were aseptically added to 40 mL of a sterile solution containing 0.1% peptone (Oxoid), 0.85% NaCl (Oxoid) and 1% Tween 80 (Panreac Química SLU, Barcelona, Spain), and then homogenized in a Masticator Classic blender (IUL Instruments, Barcelona, Spain) for 2 min at room temperature. Serial decimal dilutions in sterile peptone water (0.1% (*w/v*)) were prepared and poured or spread in the corresponding agar media. Total mesophilic aerobic bacteria were enumerated in standard plate count agar (SPCA) (Oxoid) after incubation at 30 °C for 72 h; staphylococci on mannitol salt agar (MSA) (Oxoid) incubated at 30 °C for 48 h; lactic acid bacteria (LAB) in pH 5.7 de Man, Rogosa, Sharpe (MRS) agar (Merck GmbH, Darmstadt, Germany), overlaid and incubated at 30 °C for 5 days; and *Enterobacteriaceae* in violet red bile glucose agar (VRBGA) (Oxoid), overlaid and incubated at 37 °C for 24 h. Counts were expressed as log CFU/g.

2.4. Determination of the Proximate Composition and Physico-Chemical Parameters

Moisture, fat, protein (Kjeldahl nitrogen × 6.25), ash and NaCl contents were assessed following the standards ISO 1442:1997 [31], ISO 1443:1973 [32], ISO 937:1978 [33], ISO 936:1998 [34] and ISO 1841-1:1996 [35], respectively. Water activity was measured using a Fast-lab device (GBX, Bourg-de-Péage, France). The pH values were measured with a pH meter GLP21 (Crison Instruments, S.A., Barcelona, Spain) after mixing 10 g of sample with 90 mL of distilled water. Titratable acidity, nitrosyl-heme pigments, total heme pigments and percentage of conversion to cured meat pigments were determined according to the procedures described by Zaika et al. [36]. Colour parameters were measured using a portable CR-400 colorimeter (Konica Minolta Sensing Inc., Osaka, Japan). The results were expressed in the CIELAB space [37] as lightness (L*), redness (a*) and yellowness (b*).

2.5. Determination of the Nitrogen Fractions, Free Amino Acids and Biogenic Amines

The total non-protein nitrogen (NPN), α-amino acidic nitrogen (NH_2-N) and total basic volatile nitrogen (TBVN) were quantified following the methods of Johnson [38], Moore and Stein [39] and Pearson [40], respectively, after precipitation of the proteins with 0.6 N $HClO_4$, according to the procedure described by De Ketelaere et al. [41].

The extraction of free amino acids was performed as described by Alonso et al. [42]. The identification and quantification were carried out by HPLC techniques, using the conditions described by Alonso et al. [42], with some minor modifications. The liquid chromatography equipment consisted of a SpectraSystem module (Thermo Finnigan, San José, CA, USA) equipped with a SCM1000 vacuum membrane degasser, a P4000 pump, an AS3000 automatic injector, a UV6000LP photodiode array detector and ChromQuest Chromatography Workstation software. Separation was made in a reversed phase C18 Ultrasphere 5-ODS, 4.6 mm × 250 mm column (Hichrom Ltd., Theale, Berkshire, UK). The temperature of the column was maintained at 50 ± 1 °C with a column heater (SpectraSystem 3000) and the wavelength of the detector was at 254 nm. The standards of the 22 amino acids were supplied by Sigma Chemical Co. (St Louis, MO, USA). All the samples and standards were injected at least in duplicate. Repeatability tests were carried out by injecting a sample and a standard six times consecutively in a day. Reproducibility tests were also performed by injecting the sample and the standard two times per day during three consecutive days under the same experimental conditions. No significant differences ($p < 0.05$) were observed among the results obtained in these tests. Data were expressed as mg/100 g of total solids (TS).

The extraction of the biogenic amines was performed following the method described by Eerola et al. [43]. The separation, identification and quantification were carried out by HPLC techniques also following the procedure described by Eerola et al. [43], using the HPLC equipment already described. The separation was carried out in a reversed phase C18 mod. Kromasil 100 column (25 cm, 4 mm ID) (Teknokroma S. Coop. C. Ltda., San Cugat del Vallés, Barcelona, Spain). The temperature of the column was set at 40 ± 1 °C and the wavelength of the detector at 254 nm. The chromatographic conditions used were those described by Lorenzo et al. [44]. A standard containing appropriate amounts of histamine, tyramine, tryptamine, 2-phenylethylamine, putrescine, cadaverine, spermidine, spermine and 1,7-diaminoheptane (acting this later as internal standard) was used for identification and quantification. All the samples and standards were injected at least in duplicate. Repeatability and reproducibility tests were also carried out as indicated for the free amino acid analysis and significant differences ($p < 0.05$) were also not found between the results obtained in these tests. The contents of each biogenic amine were expressed as mg/kg of TS. From the values of the individual biogenic amines, the biogenic amine index (BAI) and the total vasoactive biogenic amine content (TVBA) were calculated as indicated in the foot of the table that shows the amine content in the results section.

2.6. Determination of Fat Indexes and Free Fatty Acids

After the fat extraction following the procedure of Folch et al. [45], the fat acidity and the peroxide values were determined following the Spanish Official Standards UNE 50.011 and UNE 55.023, respectively [46]. The TBA (thiobarbituric acid) value was measured following the method of Tarladgis et al. [47], with some modifications. All parameters were measured at least in duplicate in each fat sample.

The separation of the free fatty acids from the total fat was carried out in NH_2-aminopropyl mini-columns, according to the method described by Kaluzny et al. [48]. The fatty acid methylation was carried out following the procedure described by Shehata et al. [49], with some modifications. The separation, identification and quantification of the fatty acid methyl esters were performed by gas chromatography techniques in a Trace GC chromatograph (Thermo Finnigan, Austin, TX, USA) equipped with a split/splitless, an AI 3000 autoinjector and a flame ionisation detector. The samples were injected in split mode. The separation of the different fatty acids was carried out on an Innowax column (length 30 m, ID25 mm, film thickness 0.25 mm) (Agilent Technologies, Santa Clara, CA, USA). The temperature of the detector was set at 250 °C and that of the injector at 230 °C. The gases used were hydrogen (35 mL/min), air (350 mL/min) and helium (carrier gas) (30 mL/min). The chromatographic conditions and the procedures for identification and quantification of the individual free fatty acids were those described by Méndez-Cid et al. [50]. All samples and standards were injected at least in duplicate. Repeatability and reproducibility tests were also carried out as previously indicated for the free amino acid determination. The free fatty acid contents were expressed as mg/100 g of fat.

2.7. Statistical Analysis

In order to analyse significant differences among batches and ripening times in the parameters studied, an analysis of variance (ANOVA) was performed using the General Linear Model (GLM) procedure of the SPSS package, version 23.0 (IBM SPSS, Chicago, IL, USA). The analysis of each parameter and significance was given as $p < 0.05$, $p < 0.01$ and $p < 0.001$. To determine the correlations between variables, Pearson's linear coefficient was used, employing the same SPSS package.

3. Results and Discussion

3.1. Effect on Physicochemical and Microbial Changes during the Manufacturing Process

Table 1 shows the values of the proximate composition of the sausage batches along the manufacturing process. The evolution of the a_w values is summarized in Figure 1B. The trends in moisture loss and the a_w decrease during the manufacture of the three sausage groups are very similar than those reported in the literature for other similar dry-fermented sausages [51–53] and are basically determined by the size of the sausages and by the environmental conditions (temperature and relative humidity) during the process. The protein, fat, ash and NaCl contents expressed as percentage of the total solids are within the wide range of values reported for similar sausage types and reflect the proportions of lean and fat used and the quantities of salt added in the mix preparation. None of these compositional parameters were significantly affected by the addition of starter cultures. The titratable acidity increased significantly from values of 0.15, 0.17 and 0.15 g of lactic acid/100 g of TS in the mix for the CNT, EQU and SAP batches, respectively, to values of 0.52, 0.69 and 0.71 g of lactic acid/100 g of TS (CNT, EQU and SAP batches, respectively) after 9 days of ripening, and then decreasing until reaching final values of 0.32, 0.59 and 0.67 g of lactic acid/100 g of TS, respectively. Significant differences ($p < 0.001$) were observed between the non-inoculated and inoculated batches. These values of titratable acidity are quite low, reflecting a moderate acidification during the manufacture of this sausage type. Information on the evolution of this parameter throughout the maturation of raw-cured sausages is not abundant in the literature, nor is there any discussion of the phenomena involved in such an evolution. The values in the present study are in accordance with that reported in previous works for similar sausages [52,53] and the trends in this parameter also agree with those indicated by Salgado et al. [53] in another variety of chorizo sausage. The decrease after day 14 of ripening is probably due to the consumption of organic acids by the microorganisms present, above all, moulds and yeasts. The increase in titratable acidity during the first 9 days of manufacturing and also the values reached were significantly ($p < 0.001$) higher in the batches manufactured using starter cultures than in the control, reflecting the acidifying capacity of the strain of *L. sakei* added.

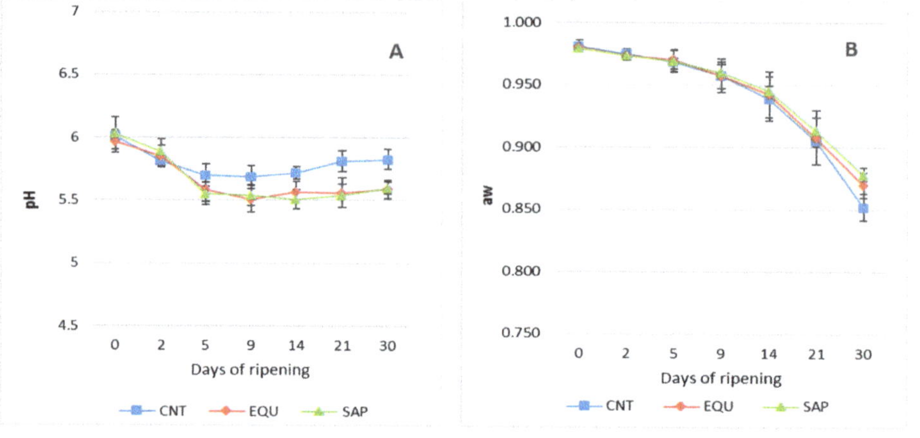

Figure 1. Evolution of the pH (**A**) and a_w (**B**) values along the manufacturing process of Galician sausage made without and with additives (plotted values are means ± standard deviations of three replicates in each sausage group). CNT: Non-inoculated control batches; EQU: Batches inoculated with *L. sakei* + *S. equorum*; SAP: Batches inoculated with *L. sakei* + *S. saprophyticus*.

The evolution of the pH values is shown in Figure 1A. Initial acidification plays an important role in the microbiological, biochemical and sensory characteristics of the fermented foods. In the case of the fermented sausages, the acidification until the pI of the muscle proteins causes denaturation of these proteins and determines the cohesiveness of the mass, the firmness and the sliceability of the final products [54]. The pH values reached, in addition, modulate the activity of the meat enzymes responsible for the ripening and flavour generation [55], and avoids the survival and growth of undesirable spoiling and pathogenic microorganisms.

In the present study, the decrease in pH values during the fermentation phase was very moderate (mean values from 6.02, 5.95 and 6.03 in the mix before stuffing to 5.69, 5.51 and 5.54 after 9 days of manufacturing, for the CNT, EQU and SAP batches, respectively). The pH decrease was significantly ($p < 0.01$) higher in batches of sausages inoculated with starter cultures, not observing significant differences between the two starter cultures used. From day 9 of manufacture, a slight and constant increase, more marked in the control sausages, was observed, reaching final average values of 5.81, 5.60 and 5.59 for the CNT, EQU and SAP batches, respectively. This pH increase in the last stages of the ripening process was already described by other authors in different sausages [15,52] and seems to be due to an increase of basic nitrogen compounds as a result of the proteolytic processes and also to the consumption of lactic acid by the microorganisms.

The pH decrease during fermentation is highly variable in the different sausages and depends on the quantity of fermentable sugars in the mix, the environmental temperature and the activity of the lactic acid bacteria present in the sausages. Therefore, the final pH values of the ripened sausages show a high variability, ranging along the values reported in the literature, from 4.15 [29] to 6.52 [56]. In agreement with some other previous observations [14], the use of starter cultures in the present study decreased the pH values during manufacturing by only a little. This contrasts with the great decrease observed in other studies where *Lactobacillus* strains with a greater acidifying capacity were used as starter cultures [29].

Colour is a very important sensory attribute in this type of food and colour deficiencies are likely to cause rejection even if sausages have a good taste and texture [57,58]. Table 2 shows the evolution of the colour parameters during the manufacture of the three batches of sausage. The percentage of conversion of pigments (from heme to nitrosyl-heme) had initial values of 35.16%, 39.63% and 37.56% for CNT, EQU and SAP sausages, respectively, with no significant ($p > 0.05$) differences between the three batches. As indicated in Table 2, the pigment transformation percentages showed a significant ($p < 0.001$) upward trend throughout the entire ripening process, reaching final values of 81,88% in the CNT batch and 82.02% and 86.26% in the EQU and SAP batches, respectively. The final value in the SAP batch was significantly ($p < 0.05$) higher than in the control batch. The values of percentage of transformation to nitrosyl-heme pigments in the present study are within the wide range of values reported in the literature [59] and reflect a high transformation of pigments. According to Zaika et al. [36], the percentage of pigment conversion is as high as the pH value is low, since the low pH values favour the formation of NO from nitrates that then reacts with myoglobin to form nitrosyl-myoglobin. Results in the present work seem to corroborate this appreciation, since in the present case the inoculated batches having lower pH values show higher conversion percentages. The nitrate reductase activity of the staphylococci strains added as starter cultures in the present work could also have some responsibility in the higher percentage of transformation in the inoculated batches.

Regarding the changes in the CIELAB colour coordinates (L*, a* and b*) throughout the drying–ripening process, the use of starter cultures did not have a significant effect on the luminosity (L*) of the sausages that decreased significantly ($p < 0.05$) during the whole drying–ripening process (values from 46–49 to 31–32), both in the control and in the inoculated batches. The decrease of this parameter during ripening seems to occur as a result of moisture loss [60–62], thus becoming a darker product. The evolution of the a* parameter (red coloration) is in line with the data described by Gómez et al. [58], with an increase up to 5 days of maturation, reaching a maximum value of 31.94 in the SAP batch, then decreasing to the end of the process with final values of 17.47, 20.70 and 20.25 in CNT, EQU and SAP, respectively. Significant differences were observed in this parameter both during ripening ($p < 0.001$) and due to the use of starter cultures ($p < 0.05$). The initial increase could be due to the formation of nitrosyl-myoglobin [63]. The initial redness is also influenced by the use of paprika in the mix formulation with a high colouring power, just as the oxidation of the carotenoids present in this ingredient contribute to the loss of coloration [58]. Finally, the parameter b * undergoes an evolution similar to that observed for redness. The initial increase in this parameter could be related to lipid oxidation processes. Again, we observed significant differences both during ripening ($p < 0.001$) and due to the use of starter cultures ($p < 0.05$).

Despite the fact that the correct implantation of these two starter cultures and their dominance over the indigenous microbiota was already observed and demonstrated using molecular methods [22], the main microbial groups were counted along the manufacture of the three sausage batches using classic culture-dependent procedures. Microbial counts are shown in Table 3. Counts of the microbial groups (5.48–6.57 log CFU/g for the total aerobic mesophilic bacteria, 4.82–6.42 log CFU/g for the total staphylococci, 4.12–5.47 log CFU/g for the lactic acid bacteria and 2.97–3.09 log CFU/g for the *Enterobacteriaceae* in the mix before stuffing) increased until the day 14 of ripening and then remained relatively constant or decreased in the case of the *Enterobacteriaceae* until the end of production. Counts and trends of the different microbial groups basically agree with previous data reported in the literature for similar sausages [12,64,65]. In each sampling time, counts of total aerobic bacteria, staphylococci, and lactic acid bacteria were always significantly ($p < 0.001$) higher in the inoculated batches than in the control batch, which confirmed the correct implantation of the starter cultures added. Higher counts of lactic acid bacteria and staphylococci in inoculated compared to control batches were also reported by Essid and Hassouna [15], using *Staphylococcus xylosus* and *Lactobacillus plantarum* as the starter cultures. Also, from day 14 of ripening, counts of the *Enterobacteriaceae* were significantly ($p < 0.01$) lower in the inoculated than in control sausages. This phenomenon seems to be due to the lower pH values reached in the inoculated sausages, taking into account the acid-sensitivity of the enterobacteria. In all three batches (CNT, EQU and SAP), the enterobacteria did not completely disappear at the end of maturation. In this sense, there is some discrepancies among the data reported in the literature. While some authors reported the total disappearance of the enterobacteria at the end of the ripening process in inoculated sausages having high pH values (5.63–5.76) [14], some others reported the survival of this microbial group in sausages reaching very low pH values (4.15) [15]. The different sources of contamination and the different nature and acid-resistance of the enterobacteria species present in the different sausage types could explain this discrepancy.

3.2. *Effect on Proteolytic Changes during the Manufacturing Process*

Hydrolysis of the meat proteins, both sarcoplasmic and myofibrillar, is considered one of the main degrading processes during the ripening of meat products playing a determinant role not only in the development of their final aroma and taste but also of the texture properties. Despite the fact that the electrophoretic methods, both the SDS-PAGE techniques [12,66,67] and miniaturized procedures [68], are the most exhaustive way of study the changes undergone by the proteins during the maturation processes, the quantification of the classical nitrogen fractions is a very satisfactory alternative.

Table 4 shows the values of these nitrogen fractions during the manufacture of the three batches of sausages. The non-protein nitrogen (126.23–133 mg/100 g of TS in the mix before stuffing) significantly increased ($p < 0.001$) during the manufacturing process, reaching final values of 167.45, 176.21 and 217.55 mg/100 g TS for the CNT, EQU and SAP sausages, respectively. At the end of the manufacture, values in the SAP batch were significantly ($p < 0.001$) higher than in the CNT and EQU batches. The increase in NPN is a common event that takes place in all the ripened sausage types during manufacturing, although in different rates and in unequal proportion during the different steps of the process, as indicated and discussed by Salgado et al. [53]. The low NPN contents in the mix before stuffing and the moderate increase observed in the present study (from 1.32 fold in CNT and EQU batches to 1.68 fold in the SAP batch) when compared to other dry-fermented sausages [53] indicate that proteolysis is only moderate in this type of sausage.

The α-aminoacidic nitrogen also increased significantly ($p < 0.001$) during the manufacturing, from 31.91, 44.48, and 42.71 to 109.73, 131.97 and 166.92 mg/100 g of TS for the CNT, EQU and SAP batches, respectively. Regarding the total basic volatile nitrogen, the increase was again significant ($p < 0.001$) from 13 mg/100 g of TS in the mix before stuffing to 83.05, 91.81 and 95.17 mg/100 g of TS for the CNT, EQU and SAP batches, respectively, at the end of the process. The values observed in the present study for these two nitrogen fractions are in the range of those reported in the literature for other similar sausage types [53,59].

Differences in NPN content among batches were not significant during the first 14 days of manufacture and only in the two last sampling times (21 and 30 days) the contents in the SAP batch were significantly ($p < 0.001$) higher than in the CNT and EQU batches. However, the contents in α-aminoacidic nitrogen were significantly ($p < 0.001$) higher in the inoculated than in the control batches in all the sampling times, and the same occurred for the total basic volatile nitrogen contents from day 2 of manufacture.

Although there is not a total consensus regarding the importance of the tissue and microbial enzymes in the protein degradation during sausage ripening, it seems evident that proteolysis takes place in two different phases. The initial protein degradation seems to be the responsibility of the muscle calpains and cathepsins, and at a later stage the bacterial enzymes further degrade the protein fragments and polypeptides initially formed [55]. The fact that in the present study the use of starter cultures had a greater effect on the α-aminoacidic and total basic volatile nitrogen fractions than on the non-protein nitrogen content seems to corroborate this hypothesis.

Table 1. Evolution of the proximate composition and titratable acidity along the manufacturing process of Galician chorizo made without and with starter cultures (means of three replicates in each sausage group).

Days of Ripening	0			2			5			9			14			21			30			SEM
	CNT	EQU	SAP	CNT	EQU	SAP	CNT	EQU	SAP	CNT	EQU	SAP	CNT	EQU	SAP	CNT	EQU	SAP	CNT	EQU	SAP	
TS #	38.02 ^{a,1}	38.08 ^{a,1}	36.52 ^{a,1}	42.72 ^{b,1}	41.49 ^{b,1}	40.81 ^{b,1}	49.41 ^{c,1}	48.10 ^{c,1}	46.03 ^{c,1}	53.14 ^{d,1}	51.35 ^{d,1}	51.14 ^{d,1}	60.65 ^{e,1}	59.10 ^{e,1}	57.65 ^{e,2}	68.04 ^{f,1}	66.25 ^{f,1}	67.42 ^{f,1}	73.85 ^{g,1}	71.80 ^{g,1}	73.06 ^{g,1}	2.14
Protein (N × 6.25) *	49.22 ^{a,1}	49.67 ^{a,1}	48.34 ^{a,1}	48.67 ^{a,1}	49.08 ^{a,1}	48.13 ^{a,1}	48.47 ^{a,1}	46.96 ^{a,1}	48.63 ^{a,1}	49.40 ^{a,1}	48.39 ^{a,1}	50.11 ^{a,1}	48.92 ^{a,1}	49.47 ^{a,1}	50.42 ^{a,1}	48.79 ^{a,1}	50.59 ^{a,1}	51.08 ^{a,1}	50.75 ^{a,1}	49.09 ^{a,1}	51.22 ^{a,1}	0.99
Fat *	41.22 ^{a,1}	41.97 ^{a,1}	42.96 ^{a,1}	41.88 ^{a,1}	42.94 ^{a,1}	42.42 ^{a,1}	41.46 ^{a,1}	43.12 ^{a,1}	42.53 ^{a,1}	41.65 ^{a1}	42.13 ^{a1}	42.87 ^{a,1}	42.04 ^{a,1}	43.25 ^{a,1}	42.66 ^{a,1}	42.92 ^{a,1}	42.01 ^{a,1}	42.13 ^{a,1}	41.91 ^{a,1}	42.30 ^{a,1}	41.27 ^{a,1}	0.36
Ash *	6.13 ^{a,1}	5.76 ^{a,1}	6.09 ^{a,1}	5.84 ^{a,1}	6.01 ^{a,1}	5.83 ^{a,1}	5.58 ^{a,1}	6.04 ^{a,1}	5.87 ^{a,1}	5.37 ^{a,1}	6.82 ^{b,2}	5.42 ^{a,1}	6.15 ^{a,1}	6.12 ^{a,1}	5.91 ^{a,1}	5.40 ^{a,1}	6.41 ^{a,1}	5.84 ^{a,1}	5.54 ^{a,1}	6.94 ^{b,2}	5.85 ^{a,1}	0.057
NaCl *	3.24 ^{a,1}	3.35 ^{a,1}	3.27 ^{a,1}	3.56 ^{a,1}	3.64 ^{a,1}	3.42 ^{a,1}	3.64 ^{a,1}	3.26 ^{a,1}	3.74 ^{a,1}	3.48 ^{a,1}	3.59 ^{a,1}	3.07 ^{a,1}	3.59 ^{a,1}	3.52 ^{a,1}	3.37 ^{a,1}	3.37 ^{a,1}	3.52 ^{a,1}	3.54 ^{a,1}	3.48 ^{a,1}	3.41 ^{a,1}	3.50 ^{a,1}	0.05
Titratable acidity §	0.15 ^{a,1}	0.17 ^{a,1}	0.15 ^{a,1}	0.32 ^{b,1}	0.47 ^{b,2}	0.49 ^{b,2}	0.45 ^{c,1}	0.67 ^{c,2}	0.69 ^{c,d,2}	0.52 ^{d,1}	0.69 ^{c,2}	0.71 ^{d,2}	0.43 ^{c,1}	0.72 ^{d,2}	0.68 ^{c,3}	0.45 ^{c,1}	0.69 ^{e,2}	0.67 ^{c,2}	0.32 ^{b,1}	0.59 ^{e,2}	0.67 ^{c,3}	0.02

Total solids expressed as g/100g; * Expresed as g/100g of TS.; § Expressed as g of lactic acid/100g of TS. CNT: Non-inoculated control batches; EQU: Batches inoculated with *L. sakei* + *S. equorum*; SAP: Batches inoculated with *L. sakei* + *S. saprophyticus*. $^{a-g}$ Means in the same row and sausage group (CNT, EQU or SAP) not followed by a common letter differ significantly ($p < 0.05$) (differences associated to the ripening time). $^{1-3}$ Means in the same row and ripening time not followed by a common number differ significantly ($p < 0.05$) (differences associated to the use of starter cultures). SEM: standard error of the mean.

Table 2. Evolution of the colour parameters along the manufacturing process of Galician chorizo made without and with starter cultures (means of three replicates in each sausage group).

Days of Ripening	0			2			5			9			14			21			30			SEM
	CNT	EQU	SAP	CNT	EQU	SAP	CNT	EQU	SAP	CNT	EQU	SAP	CNT	EQU	SAP	CNT	EQU	SAP	CNT	EQU	SAP	
Nitrosyl-heme pigments #	23.89 ^{a,1}	26.78 ^{a,2}	25.29 ^{a,3}	31.81 ^{b,1}	37.21 ^{b,2}	30.63 ^{b,3}	41.04 ^{c,1}	54.04 ^{c,2}	54.54 ^{c,2}	60.96 ^{d,1}	69.73 ^{d,2}	70.75 ^{d,2}	71.48 ^{e,1}	79.96 ^{e,2}	91.86 ^{e,3}	101.67 ^{f,1}	104.98 ^{f,2}	120.18 ^{f,3}	111.67 ^{g,1}	120.75 ^{g,2}	128.16 ^{g,3}	3.08
Total heme pigments #	77.90 ^{a,1}	70.70 ^{a,2}	60.98 ^{a,3}	83.05 ^{b,1}	85.28 ^{b,2}	69.42 ^{b,3}	90.39 ^{c,1}	107.95 ^{c,2}	86.54 ^{c,3}	115.62 ^{d,1}	118.85 ^{d,2}	104.47 ^{d,3}	125.63 ^{e,1}	128.53 ^{e,2}	126.15 ^{e,1}	138.54 ^{f,1}	140.44 ^{f,2}	159.74 ^{f,3}	140.22 ^{g,1}	146.66 ^{g,2}	143.97 ^{g,3}	2.65
P.C. §	35.16 ^{a,1}	39.63 ^{a,2}	37.56 ^{a,3}	38.46 ^{b,1}	42.37 ^{b,2}	43.97 ^{b,3}	51.58 ^{c,1}	51.15 ^{c,1}	58.15 ^{c,2}	62.82 ^{d,1}	59.03 ^{d,2}	58.68 ^{d,2}	74.54 ^{e,1}	66.57 ^{e,2}	69.69 ^{e,3}	75.34 ^{e,1}	76.83 ^{f,2}	73.27 ^{f,3}	81.88 ^{g,1}	82.02 ^{g,1,2}	86.26 ^{g,2}	1.45
L*	48.22 ^{a,1}	49.86 ^{a,2}	46.30 ^{a,3}	50.68 ^{b,1}	46.92 ^{b,2}	52.05 ^{b,1}	45.92 ^{c,1}	45.60 ^{c,1}	49.45 ^{c,2}	44.54 ^{d,1}	42.36 ^{d,2}	43.39 ^{d,1,2}	41.86 ^{e,1}	43.16 ^{e,2}	41.91 ^{e,1}	36.26 ^{f,1}	35.87 ^{f,1}	35.68 ^{f,1}	32.21 ^{g,1}	31.39 ^{g,1}	32.00 ^{g,1}	0.62
a*	29.17 ^{a,1}	29.00 ^{a,1}	24.61 ^{a,2}	27.67 ^{b,1}	29.82 ^{b,2}	31.01 ^{b,3}	30.18 ^{c,1}	30.16 ^{c,1}	27.94 ^{c,2}	30.18 ^{c,1}	28.03 ^{d,2}	25.18 ^{d,3}	26.49 ^{d,1}	29.91 ^{e,2}	27.43 ^{c,3}	21.44 ^{e,1}	23.80 ^{f,2}	24.03 ^{f,2}	17.47 ^{f,1}	20.70 ^{g,2}	20.25 ^{g,2}	0.38
b*	38.80 ^{a,1,2}	39.05 ^{a,2}	38.55 ^{a,1}	41.17 ^{b,1}	36.53 ^{b,2}	44.35 ^{b,3}	36.51 ^{c,1}	40.10 ^{c,2}	40.55 ^{c,2}	39.44 ^{d,1}	35.51 ^{d,2}	35.35 ^{d,2}	28.60 ^{e,1}	36.65 ^{e,2}	30.72 ^{e,3}	21.53 ^{f,1}	27.54 ^{f,2}	27.68 ^{f,2}	19.91 ^{g,1}	21.15 ^{g,2}	20.48 ^{g,3}	0.70

Expressed as ppm; § Percentage of transformation into nitrosylheme pigments. CNT: Non-inoculated control batches; EQU: Batches inoculated with *L. sakei* + *S. equorum*; SAP: Batches inoculated with *L. sakei* + *S. saprophyticus*. $^{a-g}$ Means in the same row and sausage group (CNT, EQU or SAP) not followed by a common letter differ significantly ($p < 0.05$) (differences associated to the ripening time). $^{1-3}$ Means in the same row and ripening time not followed by a common number differ significantly ($p < 0.05$) (differences associated to the use of starter cultures). SEM: standard error of the mean.

Table 3. Evolution of the plate counts (log CFU/g) of total aerobic mesophilic bacteria (SPCA), staphylococci (MSA), lactic acid bacteria (MRS) and *Enterobacteriaceae* (VRBGA) along the manufacturing process of Galician chorizo made without and with starter cultures (means of three replicates in each sausage group).

Days of Ripening	0			2			5			9			14			21			30			SEM
	CNT	EQU	SAP	CNT	EQU	SAP	CNT	EQU	SAP	CNT	EQU	SAP	CNT	EQU	SAP	CNT	EQU	SAP	CNT	EQU	SAP	
Total aerobic mesophilic bacteria	5.48 ^{a,1}	6.54 ^{a,2}	6.57 ^{a,2}	5.93 ^{b,1}	6.75 ^{b,2}	7.24 ^{b,3}	6.89 ^{c,1}	7.93 ^{c,2}	7.88 ^{c,2}	8.37 ^{d,1}	8.78 ^{d,2}	8.99 ^{d,3}	8.62 ^{e,1}	8.97 ^{e,2}	9.08 ^{e,2}	8.78 ^{f,1}	9.06 ^{f,2}	9.19 ^{f,2}	9.00 ^{g,1}	9.12 ^{g,2}	9.19 ^{f,2}	0.16
Staphylococci	4.82 ^{a,1}	6.26 ^{a,2}	6.42 ^{a,2}	4.71 ^{b,1}	6.31 ^{a,2}	6.75 ^{b,3}	5.86 ^{c,1}	6.90 ^{b,2}	6.59 ^{a,b,3}	6.40 ^{d,1}	7.63 ^{c,2}	7.53 ^{c,2}	6.33 ^{e,1}	8.05 ^{d,2}	7.81 ^{d,3}	6.24 ^{f,1}	8.13 ^{d,2}	7.95 ^{d,2}	6.29 ^{g,1}	8.08 ^{d,2}	7.89 ^{d,2}	0.15
Lactic acid bacteria	4.12 ^{a,1}	5.47 ^{a,2}	5.30 ^{a,3}	4.96 ^{b,1}	6.15 ^{b,2}	5.80 ^{b,3}	5.42 ^{c,1}	6.36 ^{c,2}	7.12 ^{c,3}	7.35 ^{d,1}	7.68 ^{d,2}	7.86 ^{d,3}	8.40 ^{e,1}	8.98 ^{e,2}	9.00 ^{e,2}	8.37 ^{f,1}	9.01 ^{f,2}	9.15 ^{f,3}	8.60 ^{g,1}	9.14 ^{g,2}	9.23 ^{f,2}	0.14
Enterobacteriaceae	2.98 ^{a,1}	2.97 ^{a,1}	3.09 ^{a,2}	3.56 ^{b,1}	3.02 ^{a,2}	3.28 ^{b,3}	4.77 ^{c,1}	4.91 ^{b,2}	4.84 ^{c,1,2}	5.42 ^{d,1}	5.69 ^{c,2}	5.36 ^{d,1}	5.78 ^{e,1}	5.47 ^{d,2}	5.52 ^{e,2}	5.00 ^{f,1}	4.47 ^{e,2}	4.62 ^{f,3}	3.93 ^{g,1}	3.59 ^{f,2}	3.37 ^{b,3}	0.15

CNT: Non-inoculated control batches; EQU: Batches inoculated with *L. sakei* + *S. equorum*; SAP: Batches inoculated with *L. sakei* + *S. saprophyticus*. $^{a–g}$ Means in the same row and sausage group (CNT, EQU or SAP) not followed by a common letter differ significantly ($p < 0.05$) (differences associated to the ripening time). $^{1–3}$ Means in the same row and ripening time not followed by a common number differ significantly ($p < 0.05$) (differences associated to the use of starter cultures). SEM: standard error of the mean.

Table 4. Evolution of the nitrogen fractions (expressed as mg N/100 g TS) along the manufacturing process of Galician chorizo made without and with starter cultures (means of three replicates in each sausage group).

Days of Ripening	0			2			5			9			14			21			30			SEM
	CNT	EQU	SAP	CNT	EQU	SAP	CNT	EQU	SAP	CNT	EQU	SAP	CNT	EQU	SAP	CNT	EQU	SAP	CNT	EQU	SAP	
Non-protein nitrogen (NPN)	126.23 ^{a,1}	133.00 ^{a,1}	129.45 ^{a,1}	133.38 ^{a,b,1}	150.57 ^{b,1}	140.39 ^{b,1}	148.81 ^{c,1}	162.61 ^{c,1}	139.41 ^{b,1}	156.26 ^{c,1}	163.30 ^{c,1}	154.46 ^{c,1}	140.31 ^{b,1}	166.05 ^{c,1}	167.86 ^{d,1}	156.41 ^{c,1}	163.61 ^{c,1}	193.02 ^{e,2}	167.45 ^{d,1}	176.21 ^{d,1}	217.55 ^{f,2}	7.95
α-Aminoacidic nitrogen (NH_2-N)	31.91 ^{a,1}	44.48 ^{a,2}	42.71 ^{a,3}	40.50 ^{b,1}	50.17 ^{b,2}	54.82 ^{b,3}	46.46 ^{c,1}	63.56 ^{c,2}	66.07 ^{c,3}	55.49 ^{d,1}	76.64 ^{d,2}	96.51 ^{d,3}	59.74 ^{e,1}	89.33 ^{e,2}	117.83 ^{e,3}	84.09 ^{f,1}	106.47 ^{f,2}	148.06 ^{f,3}	109.73 ^{g,1}	131.97 ^{g,2}	166.92 ^{g,3}	3.39
Total basic volatile nitrogen (TBVN)	13.33 ^{a,1}	13.99 ^{a,1}	13.40 ^{a,1}	13.58 ^{b,1}	17.75 ^{b,2}	18.62 ^{b,2}	23.50 ^{c,1}	39.77 ^{c,2}	32.90 ^{c,3}	36.64 ^{d,1}	58.27 ^{d,2}	48.41 ^{d,3}	53.79 ^{e,1}	70.74 ^{e,2}	66.08 ^{e,3}	69.62 ^{f,1}	79.74 ^{f,2}	77.92 ^{f,3}	83.05 ^{g,1}	91.81 ^{g,2}	95.17 ^{g,3}	2.51

CNT: Non-inoculated control batches; EQU: Batches inoculated with *L. sakei* + *S. equorum*; SAP: Batches inoculated with *L. sakei* + *S. saprophyticus*. $^{a–g}$ Means in the same row and sausage group (CNT, EQU or SAP) not followed by a common letter differ significantly ($p < 0.05$) (differences associated to the ripening time). $^{1–3}$ Means in the same row and ripening time not followed by a common number differ significantly ($p < 0.05$) (differences associated to the use of starter cultures). SEM: standard error of the mean.

Release of free amino acids during sausage ripening is a very important event since some amino acids have a particular taste and some others are precursors of taste and odour compounds when degraded following several well-known biochemical pathways [61]. The contents of the free amino acids in the CNT, EQU and SAP sausage batches during manufacturing are shown in Table 5. Free amino acids in the mix before stuffing ranged from 428.12 in CNT batch to 444.53 mg/100 g TS in the SAP batch and no significant differences ($p > 0.05$) were observed among batches. A significant increase ($p < 0.001$) during manufacturing was observed in the three sausage groups reaching final values of 1381.66, 1450.02 and 1593.07 mg/100 g TS for the CNT, EQU and SAP batches, respectively. Significant differences ($p < 0.05$) among batches were observed in the final total free amino acid content. Therefore, from the initial and final values, an increase in the free amino acid content of 3.22, 3.33 and 3.58 times for the CNT, EQU and SAP batches can be observed, respectively. Similar increases were observed by other authors during the ripening of other sausage types [26,66]. Increases during ripening described in the literature are highly variable (around 1.2 times [12,27,69], around 1.5 times [65], around 2.5 times [23,61,70] or even more than 4 times [71]). There are, however, studies in which no increase [23] or a small reduction [12] was observed in control batches prepared in studies to determine the effect of the addition of starter cultures. As occurred in the present study, several authors reported that the addition of starter cultures always increased the release of free amino acids during the sausage ripening [14,15,23,26,71], which undoubtedly proves the participation of peptidases of microbial origin in the release of amino acids during the maturation of sausages.

The free amino acid profile in the mix before stuffing hardly varied among the batches. Arg was the most abundant FAA (73.21, 73.37 and 72.88 mg/100 g TS for the CNT, EQU and SAP batches, respectively), followed, in a decreasing order, by Tau, Ala, Pro, Lys and Leu, the sum of these six amino acids accounting for 61.09, 60.64 and 60.51% of the total FAA in the CNT, EQU and SAP batches, respectively. The free amino acid profile of the sausage mixes widely varies according to the information reported in the literature [12,23,61,65,66,72,73], reflecting the diversity of the operating microorganisms, environmental conditions and ingredients. However, in agreement with our observations, some other works [12,23,61,65,66,72,74] pointed out the abundance of Arg, Tau and Ala in the mix of the sausages. Indeed, Glu, which was reported as the main free amino acid in some studies [26,61,71], was the seventh or even the eighth free amino acid of quantitative importance in the mix before stuffing in the present study.

The individual FAA increased with a different intensity along the manufacturing process (from 2 times in the case of Tau, Arg or Ala to 6 times of Trp, or even 8–10 times of Cys). The most abundant FAA after 30 days of ripening was again Arg (151.09, 152.07 and 167.92 mg/100 g TS for the CNT, EQU and SAP batches, respectively), followed, in a decreasing order, by Ala, Tau, Glu, Pro and Leu. These six FAA accounted for 50.14, 49.65 and 50.13% of the total FAA in 30-day-old sausages in the CNT, EQU and SAP batches, respectively. The FAA profile observed in the mix basically remains in the ripened sausages, with the exception of Glu that increased its abundance and the Lys content that decreased. Again, the FAA profile of the ripened sausages is very variable in the literature [12,14,23,61,66,70,72]. However, in agreement with our results, the predominance of the Arg [61,74] and the abundance of Arg, Ala and Tau [12,23,65,66] was reported in other works.

Table 5. Evolution of the free amino acids (mg/100g of TS) along the manufacturing process of Galician chorizo made without and with starter cultures (means of three replicates in each sausage group).

Days of Ripening	0			2			5			9			14			21			30			SEM
	CNT	EQU	SAP	CNT	EQU	SAP	CNT	EQU	SAP	CNT	EQU	SAP	CNT	EQU	SAP	CNT	EQU	SAP	CNT	EQU	SAP	
Asp	9.05 ^{a,1}	9.34 ^{a,1}	9.21 ^{a,1}	13.12 ^{b,1}	15.36 ^{b,2}	17.72 ^{b,3}	15.94 ^{c,1}	18.88 ^{c,2}	20.47 ^{c,3}	16.43 ^{d,1}	20.60 ^{d,2}	22.29 ^{d,3}	18.12 ^{e,1}	21.83 ^{e,2}	24.39 ^{e,3}	21.49 ^{f,1}	23.70 ^{f,2}	28.41 ^{f,3}	28.02 ^{g,1}	29.27 ^{g,3}	30.97 ^{g,3}	0.55
Glu	19.03 ^{a,1}	19.26 ^{a,1}	19.80 ^{a,1}	30.66 ^{b,1}	37.34 ^{b,2}	43.75 ^{b,3}	53.86 ^{c,1}	54.15 ^{c,1}	68.55 ^{c,2}	80.47 ^{d,1}	91.52 ^{d,2}	97.32 ^{d,3}	85.92 ^{e,1}	101.20 ^{e,2}	104.38 ^{e,3}	100.18 ^{f,1}	113.17 ^{f,2}	118.70 ^{f,3}	108.12 ^{g,1}	118.52 ^{g,3}	126.47 ^{g,3}	3.15
OH-Pro	4.87 ^{a,1}	4.72 ^{a,2}	4.74 ^{a,2}	6.57 ^{b,1}	7.63 ^{b,2}	7.07 ^{b,3}	9.20 ^{c,1}	10.51 ^{c,2}	14.17 ^{c,3}	11.46 ^{d,1}	14.97 ^{d,2}	16.63 ^{d,3}	13.99 ^{e,1}	18.10 ^{e,2}	19.27 ^{e,3}	18.62 ^{f,1}	20.90 ^{f,2}	21.35 ^{f,3}	23.03 ^{g,1}	23.79 ^{g,3}	25.76 ^{g,3}	0.59
Asn	8.32 ^{a,1}	9.37 ^{a,2}	9.43 ^{a,2}	11.62 ^{b,1}	16.12 ^{b,2}	18.99 ^{b,3}	16.23 ^{c,1}	21.26 ^{c,2}	26.35 ^{c,3}	18.48 ^{d,1}	25.58 ^{d,2}	28.92 ^{d,3}	23.54 ^{e,1}	31.34 ^{e,2}	34.81 ^{e,3}	29.93 ^{f,1}	35.84 ^{f,2}	39.03 ^{f,3}	36.32 ^{g,1}	41.73 ^{g,3}	44.08 ^{g,3}	0.96
Ser	11.12 ^{a,1}	11.66 ^{a,2}	11.60 ^{a2}	14.18 ^{b,1}	20.46 ^{b,2}	26.00 ^{b,3}	20.90 ^{c,1}	35.47 ^{c,2}	37.49 ^{c,3}	33.70 ^{d,1}	41.09 ^{d,2}	45.43 ^{d,3}	38.56 ^{e,1}	49.23 ^{e,2}	56.12 ^{e,3}	47.28 ^{f,1}	54.72 ^{f,2}	56.97 ^{f,3}	56.45 ^{g,1}	63.05 ^{g,3}	65.16 ^{g,3}	1.54
Gln	7.89 ^{a,1}	8.20 ^{a,2}	8.34 ^{a,2}	11.36 ^{b,1}	13.11 ^{b,2}	14.28 ^{b,3}	15.20 ^{c,1}	22.21 ^{c,2}	24.15 ^{c,3}	17.68 ^{d,1}	27.79 ^{d,2}	29.46 ^{d,3}	20.71 ^{e,1}	31.07 ^{e,2}	33.90 ^{e,3}	25.27 ^{f,1}	36.02 ^{f,2}	37.72 ^{f,3}	29.38 ^{g,1}	41.06 ^{g,3}	44.29 ^{g,3}	0.98
Gly	15.30 ^{a,1}	14.47 ^{a,2}	15.51 ^{a,1}	23.62 ^{b,1}	26.85 ^{b,2}	30.10 ^{b,3}	27.48 ^{c,1}	45.27 ^{c,2}	51.03 ^{c,3}	39.28 ^{d,1}	51.66 ^{d,2}	55.73 ^{d,3}	45.82 ^{e,1}	57.30 ^{e,2}	42.30 ^{e,3}	54.11 ^{f,1}	67.65 ^{f,2}	72.26 ^{f,3}	66.18 ^{g,1}	73.76 ^{g,3}	73.83 ^{g,3}	1.73
His	8.15 ^{a,1}	8.31 ^{a,1}	8.22 ^{a,1}	11.76 ^{b,1}	14.25 ^{b,2}	18.38 ^{b,3}	17.33 ^{c,1}	21.41 ^{c,2}	28.64 ^{c,3}	20.14 ^{d,1}	27.85 ^{d,2}	35.21 ^{d,3}	23.56 ^{e,1}	34.87 ^{e,2}	37.68 ^{e,3}	28.90 ^{f,1}	38.13 ^{f,2}	42.07 ^{f,3}	37.45 ^{g,1}	42.85 ^{g,3}	45.39 ^{g,3}	1.04
Tau	64.18 ^{a,1}	65.87 ^{a,2}	63.22 ^{a3}	77.16 ^{b,1}	79.10 ^{b,2}	89.08 ^{b,3}	82.03 ^{c,1}	91.16 ^{c,2}	100.51 ^{c,3}	95.54 ^{d,1}	106.81 ^{d,2}	109.13 ^{d,3}	99.60 ^{e,1}	111.19 ^{e,2}	113.40 ^{e,3}	112.84 ^{f,1}	116.16 ^{f,2}	122.63 ^{f,3}	121.80 ^{g,1}	120.12 ^{g,3}	146.44 ^{g,3}	1.66
Arg	73.21 ^{a,1}	73.37 ^{a,1}	72.88 ^{a,1}	93.27 ^{b,1}	104.04 ^{b,2}	117.43 ^{b,3}	114.22 ^{c,1}	125.07 ^{c,2}	129.14 ^{c,3}	120.07 ^{d,1}	129.19 ^{d,2}	135.98 ^{d,3}	127.79 ^{e,1}	131.61 ^{e,2}	140.88 ^{e,3}	138.56 ^{f,1}	143.96 ^{f,2}	148.39 ^{f,3}	151.09 ^{g,1}	152.07 ^{g,1}	167.92 ^{g,3}	2.15
Thr	9.25 ^{a,1}	9.70 ^{a,2}	9.87 ^{a,2}	15.69 ^{b,1}	19.93 ^{b,2}	21.94 ^{b,3}	16.84 ^{c,1}	25.56 ^{c,2}	30.44 ^{c,3}	21.87 ^{d,1}	32.50 ^{d,2}	39.20 ^{d,3}	23.23 ^{e,1}	34.79 ^{e,2}	40.91 ^{e,3}	29.64 ^{f,1}	39.23 ^{f,2}	42.55 ^{f,3}	34.07 ^{g,1}	41.89 ^{g,3}	45.04 ^{g,3}	0.99
Ala	59.68 ^{a,1}	60.08 ^{a,1}	61.80 ^{a,2}	68.46 ^{b,1}	70.22 ^{b,2}	74.98 ^{b,3}	74.71 ^{c,1}	95.72 ^{c,2}	94.99 ^{c,2}	86.20 ^{d,1}	98.10 ^{d,2}	102.84 ^{d,3}	110.90 ^{e,1}	115.89 ^{e,2}	125.01 ^{e,3}	122.45 ^{f,1}	122.91 ^{f,1}	134.34 ^{f,3}	132.01 ^{g,1}	135.77 ^{g,3}	153.46 ^{g,3}	2.42
Pro	21.92 ^{a,1}	20.17 ^{a,2}	24.27 ^{a,3}	33.29 ^{b,1}	41.31 ^{b,2}	43.86 ^{b,3}	48.31 ^{c,1}	60.26 ^{c,2}	65.79 ^{c,3}	74.11 ^{d,1}	90.25 ^{d,2}	93.00 ^{d,3}	82.60 ^{e,1}	92.09 ^{e,2}	95.53 ^{e,3}	99.03 ^{f,1}	102.52 ^{f,2}	109.01 ^{f,3}	102.54 ^{g,1}	111.44 ^{g,3}	117.57 ^{g,3}	2.76
Tyr	9.02 ^{a,1}	9.45 ^{a,1}	9.40 ^{a,1}	12.05 ^{b,1}	14.13 ^{b,2}	16.57 ^{b,3}	16.14 ^{c,1}	22.27 ^{c,2}	26.77 ^{c,3}	21.53 ^{d,1}	29.08 ^{d,2}	35.01 ^{d,3}	25.83 ^{e,1}	35.43 ^{e,2}	37.41 ^{e,3}	34.52 ^{f,1}	37.96 ^{f,2}	40.53 ^{f,3}	47.23 ^{g,1}	41.24 ^{g,3}	52.24 ^{g,3}	1.19
Val	19.13 ^{a,1}	18.21 ^{a,2}	18.88 ^{a,12}	25.93 ^{b,1}	31.88 ^{b,2}	35.73 ^{b,3}	37.84 ^{c,1}	41.51 ^{c,2}	44.16 ^{c,3}	45.36 ^{d,1}	46.76 ^{d,2}	48.18 ^{d,3}	46.35 ^{e,1}	51.84 ^{e,2}	54.12 ^{e,3}	55.81 ^{f,1}	57.74 ^{f,2}	60.27 ^{f,3}	69.83 ^{g,1}	59.28 ^{g,3}	66.88 ^{g,3}	1.27
Met	9.32 ^{a,1}	11.30 ^{a,2}	12.49 ^{a,3}	11.61 ^{b,1}	19.99 ^{b,2}	26.35 ^{b,3}	14.66 ^{c,1}	27.60 ^{c,2}	35.46 ^{c,3}	20.45 ^{d,1}	31.02 ^{d,2}	38.63 ^{d,3}	26.89 ^{e,1}	37.03 ^{e,2}	42.89 ^{e,3}	32.51 ^{f,1}	41.63 ^{f,2}	46.79 ^{f,3}	41.55 ^{g,1}	46.69 ^{g,3}	55.43 ^{g,3}	1.16
Cys	1.66 ^{a,1}	1.74 ^{a,1}	1.79 ^{a,1}	4.35 ^{b,1}	5.21 ^{b,2}	6.19 ^{b,3}	6.45 ^{c,1}	10.80 ^{c,2}	11.20 ^{c,2}	8.07 ^{d,1}	12.89 ^{d,2}	15.69 ^{d,3}	9.29 ^{e,1}	13.57 ^{e,2}	14.86 ^{e,3}	11.59 ^{f,1}	15.04 ^{f,2}	16.18 ^{f,3}	13.48 ^{g,1}	15.75 ^{g,3}	18.39 ^{g,3}	0.47
Ile	13.27 ^{a,1}	13.23 ^{a,1}	11.88 ^{a,2}	21.60 ^{b,1}	22.06 ^{b,1}	26.98 ^{b,3}	25.95 ^{c,1}	33.35 ^{c,2}	36.44 ^{c,3}	32.06 ^{d,1}	36.23 ^{d,2}	40.87 ^{d,3}	37.09 ^{e,1}	40.04 ^{e,2}	45.45 ^{e,3}	45.33 ^{f,1}	45.55 ^{f,1}	51.64 ^{f,2}	55.82 ^{g,1}	48.99 ^{g,3}	53.96 ^{g,3}	1.11
Leu	20.87 ^{a,1}	20.56 ^{a,1}	22.11 ^{a,2}	39.38 ^{b,1}	46.65 ^{b,2}	46.86 ^{b,2}	46.00 ^{c,1}	58.73 ^{c,2}	60.69 ^{c,3}	53.53 ^{d,1}	66.04 ^{d,2}	65.50 ^{d,2}	60.38 ^{e,1}	70.60 ^{e,2}	74.14 ^{e,3}	66.66 ^{f,1}	75.84 ^{f,2}	84.66 ^{f,3}	77.25 ^{g,1}	82.14 ^{g,3}	86.90 ^{g,3}	1.68
Phe	15.80 ^{a,1}	16.54 ^{a,1}	18.18 ^{a,2}	23.04 ^{b,1}	25.19 ^{b,2}	26.19 ^{b,3}	27.06 ^{c,1}	30.33 ^{c,2}	34.32 ^{c,3}	32.26 ^{d,1}	36.21 ^{d,2}	41.37 ^{d,3}	33.68 ^{e,1}	41.89 ^{e,2}	45.80 ^{e,3}	39.74 ^{f,1}	45.93 ^{f,2}	49.36 ^{f,3}	46.05 ^{g,1}	50.95 ^{g,3}	55.64 ^{g,3}	1.01
Trp	5.37 ^{a,1}	5.69 ^{a,12}	6.17 ^{a,2}	6.77 ^{b,1}	8.31 ^{b,2}	12.36 ^{b,3}	10.22 ^{c,1}	12.28 ^{c,2}	15.14 ^{c,3}	22.18 ^{d,1}	24.58 ^{d,2}	26.76 ^{d,3}	25.08 ^{e,1}	26.24 ^{e,2}	30.72 ^{e,3}	29.45 ^{f,1}	29.93 ^{f,1}	34.65 ^{f,2}	33.79 ^{g,1}	34.87 ^{g,3}	36.31 ^{g,3}	0.97
Lys	21.71 ^{a,1}	23.76 ^{a,2}	24.74 ^{a,3}	29.25 ^{b,1}	32.87 ^{b,2}	38.39 ^{b,3}	37.40 ^{c,1}	43.59 ^{c,2}	47.86 ^{c,3}	48.38 ^{d,1}	52.30 ^{d,2}	58.20 ^{d,3}	53.66 ^{e,1}	61.15 ^{e,2}	69.24 ^{e,3}	58.38 ^{f,1}	66.22 ^{f,2}	71.91 ^{f,3}	70.24 ^{g,1}	74.80 ^{g,3}	80.93 ^{g,3}	1.58
Total FAA	428.12 ^{a,1}	435.00 ^{a,1}	444.53 ^{a,1}	584.74 ^{b,1}	672.02 ^{b,2}	759.21 ^{b,3}	733.99 ^{c,1}	907.39 ^{c,2}	1003.77 ^{c,3}	919.26 ^{d,1}	1093.03 ^{d,2}	1181.34 ^{d,3}	1032.59 ^{e,1}	1208.27 ^{e,2}	1283.22 ^{e,3}	1202.29 ^{f,1}	1330.74 ^{f,2}	1429.42 ^{f,3}	1381.66 ^{g,1}	1450.02 ^{g,3}	1593.07 ^{g,3}	30.01
∑ Sweet	117.26 ^{a,1}	116.08 ^{a,1}	123.05 ^{a,2}	155.23 ^{b,1}	178.77 ^{b,2}	196.88 ^{b,3}	188.24 ^{c,1}	262.27 ^{c,2}	279.75 ^{c,3}	255.16 ^{d,1}	313.61 ^{d,2}	336.19 ^{d,3}	301.11 ^{e,1}	349.29 ^{e,2}	359.87 ^{e,3}	352.50 ^{f,1}	387.03 ^{f,2}	415.13 ^{f,3}	391.24 ^{g,1}	425.91 ^{g,3}	455.06 ^{g,3}	9.28
∑ Bitter	78.39 ^{a,1}	79.84 ^{a,1}	83.54 ^{a,2}	121.57 ^{b,1}	145.76 ^{b,2}	162.11 ^{b,3}	151.50 ^{c,1}	191.53 ^{c,2}	211.07 ^{c,3}	183.66 ^{d,1}	216.26 ^{d,2}	234.54 ^{d,3}	204.39 ^{e,1}	241.39 ^{e,2}	262.40 ^{e,3}	240.05 ^{f,1}	266.69 ^{f,2}	292.72 ^{f,3}	290.49 ^{g,1}	288.04 ^{g,3}	318.81 ^{g,3}	6.07
∑ Acid	36.23 ^{a,1}	36.91 ^{a,1}	37.23 ^{a,1}	55.55 ^{b,1}	69.96 ^{b,2}	79.85 ^{b,3}	87.19 ^{c,1}	94.44 ^{c,2}	117.66 ^{c,3}	117.04 ^{d,1}	139.97 ^{d,2}	154.82 ^{d,3}	127.59 ^{e,1}	157.90 ^{e,2}	166.45 ^{e,3}	150.58 ^{f,1}	175.00 ^{f,2}	189.18 ^{f,3}	173.58 ^{g,1}	190.64 ^{g,3}	202.83 ^{g,3}	4.66
∑ Aged	39.70 ^{a,1}	42.55 ^{a,2}	43.56 ^{a,2}	54.43 ^{b,1}	62.36 ^{b,2}	72.68 ^{b,3}	69.48 ^{c,1}	84.74 ^{c,2}	95.10 ^{c,3}	86.34 ^{d,1}	101.98 ^{d,2}	115.50 ^{d,3}	97.61 ^{e,1}	118.41 ^{e,2}	131.04 ^{e,3}	114.39 ^{f,1}	127.88 ^{f,2}	140.85 ^{f,3}	145.48 ^{g,1}	145.31 ^{g,3}	164.14 ^{g,3}	3.19

∑ Sweet: sum of the sweet amino acids (Ala, Gly, Thr, Ser and Pro); ∑ Bitter: sum of the bitter amino acids (Leu, Val, Ile, Met and Phe); ∑ Acid: sum of the acid amino acids (Glu, Asp and His); ∑ Aged: sum of the amino acids responsible for the taste "aged" (Asp, Tyr and Lys). CNT: Non-inoculated control batches; EQU: Batches inoculated with *L. sakei* + *S. equorum*; SAP: Batches inoculated *with L. sakei* + *S. saprophyticus*. $^{a–g}$ Means in the same row and sausage group (CNT, EQU or SAP) not followed by a common letter differ significantly ($p < 0.05$) (differences associated to the ripening time). $^{1–3}$ Means in the same row and ripening time not followed by a common number differ significantly ($p < 0.05$) (differences associated to the use of starter cultures). SEM: standard error of the mean.

The taste of some amino acids and their sensory thresholds were well stablished [75,76]. The amino acids Ala, Gly, Thr, Ser and Pro have a sweet taste. Leu, Val, Ile, Met and Phe are bitter. Glu, Asp and His have an acid taste and, in addition, Glu and Asp cause a pleasantly fresh sensation. Moreover, Asp, Tyr and Lys have been considered as responsible for an "aged" taste in the ripened meat products. On the basis of this knowledge, the FAA were grouped according their tastes (Table 5). In the ripened sausages (30 days) and when compared to the contents observed in mixes before stuffing, sweet FAA increased 3.33, 3.66 and 3.69 times in the CNT, EQU and SAP batches, respectively. Bitter FAA increased 3.70, 3.60 and 3.81 times in the CNT, EQU and SAP batches, respectively. Acid FAA experienced a more marked increase (4.79, 5.16 and 5.44 times, respectively), while "aged" FAA increased 3.66, 3.41 and 3.76 times in the CNT, EQU and SAP batches, respectively. In view of these data, it can be concluded that the use of *Lactobacillus sakei* and *Staphylococcus equorum* as starter cultures increased the sweet and acid tastes and decreased the bitter taste in the final product when compared with the non-inoculated control sausages. In the same way, the use of *Lactobacillus sakei* and *Staphylococcus saprophyticus* as starter cultures increased the four tastes when compared with the non-inoculated control sausages. Taking into account that, according to the information reported by Zhu [75], all these FAA are in the final sausages in concentrations higher than their respective sensory thresholds, the use of the starter cultures assayed in the present work could have some effect on the taste of the manufactured sausages.

Due to their importance for consumer health, derived from their physiological activities, and also due to their effects on the food quality, the biogenic amines are the products of the proteolysis in sausages that demanded more attention in the literature in the last two decades. Fermented sausages offer very favourable conditions for biogenic amine formation because of the high microbial activity during the fermentation process, the high presence of free amino acids (the biogenic amine precursors) as products of the proteolytic processes and the low acidic conditions that favour amino acid decarboxylation [77].

Table 6 shows the evolution of the main biogenic amines during the manufacture of the three Galician chorizo batches. In the present study, tyramine and spermine were the main biogenic amines in the mix, followed by tryptamine, 2-phenylethilamine, putrescine and cadaverine, with spermidine and histamine being the less abundant ones. The contents of most of these biogenic amines significantly ($p < 0.001$) increased during production but in an unequal way, with putrescine (that increased 4.79, 4.72 and 5.24 times in the CNT, EQU and SAP batches, respectively), histamine (4.46, 3.99 and 3.98 times) and cadaverine (4.07, 2.33 and 3.42 times) being the biogenic amines that underwent the greatest increases. Conversely, spermine (whose content did not experience a significant increase), spermidine (that increased 1.71, 1.76 and 1.63 times in the CNT, EQU and SAP batches, respectively) and 2-phenylethylamine (2.06, 1.81 and 1.97 times) were the amines that experienced the lowest increases. As a consequence of the individual amine increase, the total biogenic amine content increased 2.97, 2.48 and 2.53 times in the CNT, EQU and SAP batches, respectively, until reaching final total biogenic amine contents of 289.71, 241.47 and 235.32 mg/kg of TS in the CNT, EQU and SAP batches, respectively. These final content of total biogenic amines are in line with the contents described by other authors [13,68,74], although very much lower (61.71 mg/kg; [14]) and very much higher (1962.1 mg/kg; [72]) contents have been reported in the literature. In the final sausages, the main biogenic amine was putrescine, followed by tyramine, tryptamine and cadaverine. In general, tyramine, putrescine and cadaverine were reported as the main biogenic amines in meat products [24,78–81], with the concentration of cadaverine being the most variable [81]. Spermine and spermidine are the only biogenic amines present at significant levels in fresh meat [82]. According to the information reviewed by Suzzi and Gardini [79], several authors reported that strains of the genus *Lactococcus*, *Leuconostc* and *Lactobacillus* are able to produce tyramine, and therefore the generation of tyramine in sausages could be attributable to the decarboxylase activity of the lactic acid bacteria. Cadaverine and putrescine are associated with the activity of *Enterobacteriaceae* [79,83], and high quantities of these amines indicate poor hygienic practices and high microbial contamination of the raw materials [84]. In the present study, significant correlation coefficients ($r = 0.692$; $p < 0.01$) were observed between the counts in MRS agar and the tyramine contents, which seems to corroborate the responsibility of the lactic acid bacteria in the production of this amine. In the same way, a significant positive correlation ($r = 0.727$; $p < 0.01$) was observed between the counts in MRS agar and the tryptamine contents. The high significant correlations we observed between the counts in VRBGA and cadaverine ($r = 0.635$; $p < 0.01$) and putrescine ($r + 0.560$; $p < 0.01$) contents also suggests some implication of the *Enterobacteriaceae* in the generation of these two biogenic amines.

In the present study, the final content of total biogenic amines was significantly ($p < 0.001$) lower in the inoculated batches than in the control batch, and no significant differences ($p > 0.05$) were observed between the two inoculated batches. The use of starter cultures significantly reduced the total biogenic amine content, with the percentage of reduction being 16.65% in the EQU batch and 18.77% in the SAP batch. Reductions were unequal for the different biogenic amines, being cadaverine (45.03% and 36.26% of reduction for the EQU and SAP batches, respectively), tyramine (12.64% and 21.27%), 2-phenylethylamine (16.57% and 19.23%) and putrescine (12.47% and 17.51%) the amines that underwent the major reductions. In accordance with results of other previous studies [13], the spermine contents remained practically unaltered during the manufacturing process in the three sausage batches.

As indicated by Lorenzo et al. [81], several studies have demonstrated that the use of starter cultures reduce the biogenic amine formation during the sausage fermentation and ripening due to their inhibiting effect on the spoilage bacteria via acidification. However, some authors reported an increase in the biogenic amine content in the ripened sausages when some starter cultures were added [14,85]. This undesirable effect could be due to the increase of proteolysis and subsequent generation of free amino acids (precursors of the biogenic amines), although this possibility was questioned by the results of some studies [14,86]. Rather, it could be that the starter cultures favour the production of biogenic amines via a slight acidification that facilitates the decarboxylation reactions or by the direct production of biogenic amines by the strains integrating the starter cultures. In this sense, recent reports [87] indicated that the species *Staphylococcus xylosus*, largely used as starter culture in fermented sausages, is an effective producer of tyramine. In any case, some authors have indicated that a reduction in the formation of biogenic amines through acidification is only real for some biogenic amines, such as putrescine [85], and that a significant decrease in pH is necessary for this effect to occur. These authors indicated that the acidifying activity of the starter cultures did not reduce the tyramine production. However, our results regarding the tyramine reductions, as well as the results of some other authors concerning the reduction of this biogenic amine by using starter cultures [88,89], disagree with this statement. It seems, therefore, that more research is necessary to elucidate these discrepancies. It could be that other inhibitory mechanisms in addition to the pH reduction were involved in reducing the production of biogenic amines and that the strains that produce biogenic amines have a different sensitivity to these inhibitory mechanisms.

For additional information, the biogenic amine index (BAI) and the total vasoactive biogenic amines (TVBA) were calculated. Regarding the BAI, it was first developed and used by Mietz and Karmas [90], with the aim of assessing the freshness (bacterial quality) of tuna. The initial formula proposed by Mietz and Karmas does not take tyramine into account. In the present study, we used the formula for the BAI calculation proposed by Veciana-Nogués et al. [91], which does take into account this biogenic amine. Basically, a BAI quantifies the amines that come from microbial metabolism, and their evaluation is of great interest in foods in which any microbial growth is undesirable and indicates spoilage. In fermented foodstuffs (foods and beverages), there is a desirable and normal development of microorganisms during manufacturing. Therefore, these indices do not have an absolutely direct relationship with the microbiological quality of food. In the case of sausages, therefore, this index remains as an indicator of the degree of activity of the decarboxylating microorganisms in the product. The vasoactive amines (tyramine, histamine, tryptamine and 2-phenylethylamine) possess vasoactive and psychoactive properties and therefore indicate a food poisoning hazard. The use of starter cultures significantly reduced the BAI (19.66% in the EQU batch and 20.81% in the SAP batch) and the TVBA (12.12% and 16.23% in the EQU and SAP batches, respectively) in the final ripened sausages. This indicates that the use of these two starter cultures also improves the hygienic quality and safety of Galician chorizo sausage.

3.3. Effect on Lipolytic Changes during the Manufacturing Process

Lipolysis and fat oxidation are major sources of volatiles generated during the ripening of meat products. In order to investigate the effect of the addition of starter cultures on these chemical processes, we firstly analysed some parameters that indicate fat degradation. The results of these analyses are shown in Table 7. The acidity values that indicate the free fatty acid content increased significantly ($p < 0.001$) from 1.38, 1.32 and 1.47 mg KOH/g fat to 10.19, 10.05 and 13.39 mg KOH/g of fat in the CNT, EQU and SAP batches, respectively. At the end of the manufacturing process, the values were significantly ($p < 0.001$) higher in the SAP batch than in the CNT and EQU batches. This parameter therefore increased during the manufacture; specifically, 7.38, 7.61 and 9.10 times in the CNT, EQU and SAP batches, respectively. Increases of the acidity value reported in the literature in dry-fermented sausages are very variable. Similar increases than ours were observed by Salgado et al. [53] (10.42 times), but lower increases were reported by Lizaso et al. [64] (4.32 times), Franco et al. [52] (4 times) and Fernández-Fernández et al. [4] (1.7 times). The final values of fat acidity largely vary in the ripened sausages, as discussed by Franco et al. [52]. The final values in the present study are similar to those reported by Fernández-Fernández et al. [4] for the same sausage type, as well as to those observed by other authors in other ripened sausages [92,93]. These values indicate that this sausage undergoes during ripening a considerable lipolysis and that the use of the SAP starter enhances this process.

Peroxide values also increased significantly ($p < 0.001$), from 0.96, 1.10 and 1.18 to 6.02, 6.30 and 5.93 meq O_2/kg fat in the CNT, EQU and SAP batches, respectively. At the end of ripening, no significant ($p > 0.05$) differences were observed among the batches. Usually, an increase in this parameter is noticed during sausage ripening [4,24,29,53]. However, a decrease was observed in other cases [94]; in some others, after an initial increase, a decline was reported in the last stages of ripening [29,52]. In the present study, the initial values of this parameter were low and, despite the fact that the peroxides increased during the manufacturing, the final values were lower than most of those reported in the literature for other similar sausages [24,29,52,53,64,92,95]. The reasons for the low values of this parameter in the present study could be the high quality of the fat used in the manufacture, coming from pigs slaughtered 48–72 h before manufacturing and adequately stored under refrigeration, and also the fact that the mincing, mixing and stuffing were carried out under vacuum, thus avoiding air (and therefore oxygen) incorporation during these processes.

The TBA value is a measure of the secondary oxidation processes and quantifies the malondialdehyde, one of the most representative compounds of those coming from the hydroperoxide decomposition. In the present study, the TBA value significantly ($p < 0.001$) increased from 0.16–0.22 to 0.54, 0.74 and 0.77 mg malondialdehyde/kg. The use of starter cultures significantly ($p < 0.05$) increased the value of this parameter in the final product and no differences ($p > 0.05$) were observed between the two inoculated batches. The evolution of the TBA value during the sausage ripening is very variable. As in the present study, Franco et al. [52] observed a progressive increase during the ripening process. However, in other cases after an initial increase, the malondialdehyde decreases in the final steps of the ripening [24,53,61,94]. Usually the concentration of malondialdehyde at the end of the manufacture of this type of sausages is around or under 1 mg/kg [25,52,53,61,94,95]. However higher values were frequently reported in the literature [24,92,95]. Domínguez Fernández and Zumalacárregui Rodríguez [96] reported values of 2.21 mg malondialdehyde/kg in "Chorizo" sausage after 35 days of ripening and indicated that this concentration is insufficient for the sensorial perception of rancidity.

Table 6. Evolution of the biogenic amines (mg/kg of TS) along the manufacturing process of Galician chorizo made without and with starter cultures (means of three replicates in each sausage group).

Days of Ripening	0			2			5			9			14			21			30			SEM
	CNT	EQU	SAP	CNT	EQU	SAP	CNT	EQU	SAP	CNT	EQU	SAP	CNT	EQU	SAP	CNT	EQU	SAP	CNT	EQU	SAP	
Tryptamine	15.10 ^{a,1}	14.91 ^{a,1}	12.87 ^{a,2}	19.84 ^{b,1}	17.48 ^{b,2}	16.55 ^{b,3}	27.57 ^{c,1}	21.22 ^{c,2}	17.72 ^{c,3}	37.02 ^{d,1}	27.67 ^{d,2}	25.37 ^{d,3}	41.65 ^{e,1}	30.23 ^{e,2}	29.70 ^{e,2}	40.15 ^{f,1}	38.08 ^{f,2}	33.43 ^{f,3}	47.27 ^{g,1}	41.69 ^{g,3}	39.12 ^{g,3}	10.63
2-Phenyl-ethylamine	12.40 ^{a,1}	11.79 ^{a,1}	10.44 ^{a,2}	15.64 ^{b,1}	14.08 ^{b,2}	13.60 ^{b,2}	18.91 ^{c,1}	15.49 ^{c,2}	14.04 ^{c,3}	21.71 ^{d,1}	15.46 ^{c,2}	16.69 ^{d,3}	24.94 ^{e,1}	14.06 ^{b,2}	20.81 ^{e,3}	23.32 ^{f,1}	20.85 ^{d,2}	19.92 ^{f,3}	25.58 ^{g,1}	21.34 ^{d,2}	20.66 ^{g,3}	4.61
Putrescine	12.43 ^{a,1}	11.04 ^{a,2}	9.37 ^{a,3}	19.99 ^{b,1}	14.99 ^{b,2}	10.96 ^{b,3}	25.14 ^{c,1}	23.95 ^{c,2}	18.77 ^{c,3}	32.80 ^{d,1}	29.70 ^{d,2}	23.18 ^{d,3}	42.08 ^{e,1}	38.01 ^{e,2}	30.96 ^{e,3}	50.92 ^{f,1}	48.38 ^{f,2}	40.38 ^{f,3}	59.56 ^{g,1}	52.13 ^{g,2}	49.13 ^{g,3}	15.51
Cadaverine	10.84 ^{a,1}	10.38 ^{a,1}	8.21 ^{a,3}	16.22 ^{b,1}	13.75 ^{b,2}	9.73 ^{b,3}	21.96 ^{c,1}	16.57 ^{c,2}	13.03 ^{c,3}	27.49 ^{d,1}	20.51 ^{d,2}	17.09 ^{d,3}	30.41 ^{e,1}	22.41 ^{e,2}	21.16 ^{e,3}	39.92 ^{f,1}	25.41 ^{f,2}	23.34 ^{f,3}	44.17 ^{g,1}	24.28 ^{g,2}	28.15 ^{g,3}	9.47
Histamine	5.04 ^{a,1}	5.27 ^{a,1}	5.75 ^{a,1}	7.09 ^{b,1}	6.32 ^{b,2}	7.44 ^{b,1}	9.81 ^{c,1}	7.44 ^{c,2}	8.32 ^{c,3}	11.96 ^{d,1}	10.53 ^{d,2}	11.88 ^{d,1}	16.17 ^{e,1}	12.99 ^{e,2}	14.50 ^{e,3}	20.11 ^{f,1}	17.30 ^{f,2}	15.91 ^{f,3}	22.50 ^{g,1}	21.04 ^{g,2}	22.93 ^{g,1}	5.86
Tyramine	21.90 ^{a,1}	18.58 ^{a,2}	19.59 ^{a,3}	30.85 ^{b,1}	20.66 ^{b,2}	22.48 ^{b,3}	33.75 ^{c,1}	29.99 ^{c,2}	28.55 ^{c,3}	35.98 ^{d,1}	31.78 ^{d,2}	30.39 ^{d,3}	38.83 ^{e,1}	33.08 ^{e,2}	32.47 ^{e,2}	42.36 ^{f,1}	43.37 ^{f,1}	37.29 ^{f,2}	56.46 ^{g,1}	49.32 ^{g,2}	44.45 ^{g,3}	10.23
Spermidine	5.74 ^{a,1}	5.44 ^{a,1}	6.31 ^{a,2}	6.09 ^{a,1}	6.36 ^{b,1}	6.56 ^{a,1}	7.68 ^{b,1}	7.47 ^{c,1}	8.60 ^{b,2}	7.56 ^{b,1}	8.90 ^{d,2}	8.59 ^{b,2}	10.86 ^{c,1}	10.10 ^{e,12}	9.71 ^{c,2}	8.93 ^{d,1}	8.77 ^{d,1}	9.49 ^{c,1}	9.85 ^{d,1,2}	9.59 ^{e,1}	10.34 ^{c,2}	3.42
Spermine	20.04 ^{a,1}	19.65 ^{a,1}	20.13 ^{a,1}	23.58 ^{b,1}	19.48 ^{a,2}	21.37 ^{a,2}	18.88 ^{a1}	19.53 ^{a,2}	17.94 ^{b,3}	20.08 ^{a,1}	22.00 ^{b,2}	22.30 ^{a,2}	20.26 ^{a,1}	23.99 ^{b,2}	23.18 ^{a,2}	23.84 ^{b,1}	25.83 ^{c,2}	23.85 ^{b,1}	24.32 ^{b,1}	22.09 ^{b,2}	20.53 ^{a,2}	8.53
TBA	103.47 ^{a,1}	97.07 ^{a,2}	92.67 ^{a,3}	139.30 ^{b,1}	113.12 ^{b,2}	108.70 ^{b,3}	163.71 ^{c,1}	141.66 ^{c,2}	126.96 ^{c,3}	194.60 ^{d,1}	166.53 ^{d,2}	155.49 ^{d,3}	225.20 ^{e,1}	184.88 ^{e,2}	182.49 ^{e,2}	249.55 ^{f,1}	228.00 ^{f,2}	203.60 ^{f,3}	289.71 ^{g,1}	241.47 ^{g,2}	235.32 ^{g,2}	65.45
BAI	50.20 ^{a,1}	45.28 ^{a,2}	42.92 ^{a,3}	74.15 ^{b,1}	55.72 ^{b,2}	50.62 ^{b,3}	90.66 ^{c,1}	77.94 ^{c,2}	68.67 ^{c,3}	108.24 ^{d,1}	92.52 ^{d,2}	82.54 ^{d,3}	127.49 ^{e,1}	106.50 ^{e,2}	99.10 ^{e,3}	153.31 ^{f,1}	134.46 ^{f,2}	116.91 ^{f,3}	182.69 ^{g,1}	146.77 ^{g,2}	144.67 ^{g,2}	39.84
TVBA	54.44 ^{a,1}	50.55 ^{a,2}	48.66 ^{a,3}	73.42 ^{b,1}	58.54 ^{b,2}	60.08 ^{b,3}	90.04 ^{c,1}	74.15 ^{c,2}	68.63 ^{c,3}	106.67 ^{d,1}	85.43 ^{d,2}	84.33 ^{d,3}	121.59 ^{e,1}	90.36 ^{e,2}	97.48 ^{e,3}	125.94 ^{f,1}	119.60 ^{f,2}	106.55 ^{f,3}	151.80 ^{g,1}	133.39 ^{g,2}	127.16 ^{g,2}	30.29

TBA: Total biogenic amines; BAI: Biogenic amine index (sum of putrescine + cadaverine + histamine+ tyramine); TVBA: sum of the vasoactive amines (tyramine + histamine + tryptamine + 2 phenylethylamine). CNT: Non-inoculated control batches; EQU: Batches inoculated with *L. sakei* + *S. equorum*; SAP: Batches inoculated with *L. sakei* + *S. saprophyticus*. $^{a-g}$ Means in the same row and sausage group (CNT, EQU or SAP) not followed by a common letter differ significantly ($p < 0.05$) (differences associated to the ripening time). $^{1-3}$ Means in the same row and ripening time not followed by a common number differ significantly ($p < 0.05$) (differences associated to the use of starter cultures). SEM: standard error of the mean.

Table 7. Evolution of the fat parameters along the manufacturing process of Galician chorizo made without and with starter cultures (means of three replicates in each sausage group).

Days of Ripening	0			2			5			9			14			21			30			SEM
	CNT	EQU	SAP	CNT	EQU	SAP	CNT	EQU	SAP	CNT	EQU	SAP	CNT	EQU	SAP	CNT	EQU	SAP	CNT	EQU	SAP	
Acidity index #	1.38 ^{a,1}	1.32 ^{a,1}	1.47 ^{a,2}	2.55 ^{b,1}	3.12 ^{b,2}	2.93 ^{b,3}	3.61 ^{c,1}	3.87 ^{c,2}	3.65 ^{c,1}	4.49 ^{d,1}	4.65 ^{d,2}	4.70 ^{d,2}	6.09 ^{e,1}	6.84 ^{e,2}	6.11 ^{e,1}	7.82 ^{f,1}	8.94 ^{f,2}	10.28 ^{f,3}	10.19 ^{g,1}	10.05 ^{g,1}	13.39 ^{g,2}	0.26
Peroxide value *	0.96 ^{a,1}	1.10 ^{a,2}	1.18 ^{a,3}	1.77 ^{b,1}	2.13 ^{b,2}	1.86 ^{b,3}	3.18 ^{c,1}	3.27 ^{c,1}	2.43 ^{c,2}	3.99 ^{d,1}	4.17 ^{d,2}	3.24 ^{d,3}	4.05 ^{e,1}	4.56 ^{e,2}	4.17 ^{e,1}	4.86 ^{f,1}	5.07 ^{f,2}	5.22 ^{f,2}	6.02 ^{g,1}	6.30 ^{g,1}	5.93 ^{g,1}	0.05
TBA value §	0.20 ^{a,1,2}	0.22 ^{a,1}	0.16 ^{a,2}	0.19 ^{a,1}	0.17 ^{b,1}	0.29 ^{b,2}	0.18 ^{a,1}	0.21 ^{a,1}	0.28 ^{b,2}	0.34 ^{b,1}	0.21 ^{a,2}	0.24 ^{c,3}	0.46 ^{c,1}	0.32 ^{c,2}	0.34 ^{d,2}	0.48 ^{c,1}	0.37 ^{d,2}	0.60 ^{e,3}	0.54 ^{d,1}	0.74 ^{e,2}	0.77 ^{f,2}	0.02

Expressed as mg KOH/g fat; * Expressed as meq de O_2/kg fat; § Expressed as mg malondialdehyde/kg of sample; CNT: Non-inoculated control batches; EQU: Batches inoculated with *L. sakei* + *S. equorum*; SAP: Batches inoculated with *L. sakei* + *S. saprophyticus*. $^{a-g}$ Means in the same row and sausage group (CNT, EQU or SAP) not followed by a common letter differ significantly ($p < 0.05$) (differences associated to the ripening time). $^{1-3}$ Means in the same row and ripening time not followed by a common number differ significantly ($p < 0.05$) (differences associated to the use of starter cultures). SEM: standard error of the mean.

Free fatty acid release during ripening of meat products is an important phenomenon for the sensory characteristics of the final products since most of the volatiles come from fatty acid degradation, mainly via oxidation processes. The evolution of the free fatty acids (FFA) during the manufacture of the batches of sausage is shown in Table 8. No significant differences ($p > 0.05$) were observed in the total FFA content among the batches in the mix. Total FFA contents in the mixes before stuffing (253.62, 247.11 and 263.64 mg/100 g of fat for the CNT, EQU and SAP batches, respectively) are in agreement with those reported in the literature for the mix of other sausages [12,26,61,70], although higher initial values were reported at the beginning of the manufacture of some other sausages [14,29,67]. Regarding the FFA profile of the mixes in the present work, the oleic acid (C18:1) was the main fatty acid, followed, in decreasing order of abundance, by linoleic (C18:2), palmitic (C16), and stearic (C18), these four FFA accounting for 87.40%, 87.07% and 87.32% of the total FFA of the mix in the CNT, EQU and SAP batches, respectively. The FFA profile in the mixes slightly differed among the batches. In the CNT batch, the fifth, sixth, seventh and eighth most important FFAs were the palmitoleic (C16:1), linolenic (C18:3), myristic (C14) and arachidonic (C20:4) fatty acids, while these same places were occupied by the linolenic C18:3), palmitoleic (C16:1), myristic (C14) and docosadienoic (C22:2) FA in the EQU batch, and by the linolenic (C18:3), palmitoleic (C16:1), arachidonic (C20:4) and myristic (C14) FA in the SAP batch. This FFA acid profile is quite constant in sausages manufactured from pig fat. Regarding the main fatty acids, profiles only differ in the FFA that occupy the second and third place. In some cases, palmitic (C16) is more abundant than linoleic (C18:2) [26,64], and in some other, as in the present work, linoleic (C18:2) dominates over palmitic (C16) [12,23,70]. In any case, very small differences between the palmitic (C16) and linoleic (C18:2) fatty acid concentrations have been consistently reported.

The total FFA content increased to final values of 1814.41, 1805.65 and 2295.46 mg/100 g of fat, for the CNT, EQU and SAP batches, respectively. In agreement with the observations made in the acidity values of the fat (Table 7), the final total FFA content was significantly ($p < 0.001$) higher in the SAP than in the EQU and CNT batches, and no significant ($p > 0.05$) differences were observed between the CNT and EQU batches. The increases in FFA content (7.15, 7.30 and 8.70 times for the CNT, EQU and SAP batches, respectively) are in agreement with the increases reported by other authors (6–8 times [12], 6–7 times [29] and 6 times [26]). However, lower [61,67,70] and higher [12] increases have been reported in other works. Usually, the use of starter cultures increases the FFA content during the ripening process. However, in some cases lower FFA contents were reported when commercial starter cultures were added [14].

In the present study, during the ripening process the different free fatty acids increased at a different rate (e.g., eicosatrienoic acid (C20:3n3) increased 10.88, 21.34 and 25.13 times in the CNT, EQU and SAP batches, respectively, while lauric acid (C12) only increased 3.14, 3.45, and 5.36 times in the CNT, EQU and SAP batches, respectively). The final values of total FFA content in the present work are in agreement with the values observed in other studies [12,67]. However, as occurred for the mixes, lower [26,61,70] and higher [14,29] values than ours have been reported.

Table 8. Evolution of the free fatty acids (mg FFA/100g of fat) along the manufacturing process of Galician chorizo made without and with starter cultures (mean values of three batches in each sausage group).

Days of Ripening	0			2			5			9		
	CNT	EQU	SAP	CNT	EQU	SAP	CNT	EQU	SAP	CNT	EQU	SAP
C8	0.26 ^{a,1,2}	0.28 ^{a,1}	0.25 ^{a,2}	0.26 ^{a,1}	0.33 ^{b,2}	0.38 ^{b,3}	0.40 ^{b,1}	0.52 ^{c,2}	0.35 ^{c,3}	0.53 ^{c,1}	0.61 ^{d,2}	0.46 ^{d,3}
C10	0.45 ^{a,1}	0.25 ^{a,2}	0.34 ^{a,3}	0.46 ^{a,1}	0.39 ^{b,2}	0.45 ^{b,1}	0.52 ^{b,1}	0.61 ^{c,2}	0.46 ^{b,3}	0.64 ^{c,1}	0.72 ^{d,2}	0.60 ^{c,3}
C12	0.69 ^{a,1}	0.61 ^{a,2}	0.55 ^{a,3}	0.74 ^{b,1}	0.68 ^{b,2}	0.86 ^{b,3}	0.75 ^{b,1}	0.84 ^{c,2}	1.17 ^{c,3}	0.91 ^{c,1}	1.13 ^{d,2}	1.29 ^{d,3}
C14	4.59 ^{a,1}	3.58 ^{a,2}	3.22 ^{a,3}	4.51 ^{a,1}	5.55 ^{b,2}	5.54 ^{b,2}	5.11 ^{b,1}	6.48 ^{c,2}	6.95 ^{c,3}	6.26 ^{c,1}	8.24 ^{d,2}	8.76 ^{d,3}
C14:1	0.15 ^{a,1}	0.18 ^{a,2}	0.16 ^{a,1,2}	0.21 ^{b,1}	0.21 ^{b,1}	0.15 ^{a,2}	0.23 ^{b,c,1}	0.24 ^{c,1}	0.20 ^{b,2}	0.24 ^{c,1}	0.38 ^{d,2}	0.23 ^{c,1}
C15	0.40 ^{a,1}	0.39 ^{a,1}	0.41 ^{a,1}	0.42 ^{a,1}	0.44 ^{b,1}	0.48 ^{b,2}	0.45 ^{b,1}	0.72 ^{c,2}	0.53 ^{c,3}	0.52 ^{c,1}	0.77 ^{d,2}	0.65 ^{d,3}
C15:1	0.20 ^{a,1}	0.29 ^{a,2}	0.21 ^{a,1}	0.32 ^{b,1}	0.31 ^{a,1}	0.55 ^{b,2}	0.62 ^{c,1}	0.60 ^{b,1}	0.56 ^{b,2}	0.67 ^{d,1}	0.88 ^{c,2}	0.72 ^{c,3}
C16	53.63 ^{a,1}	41.28 ^{a,2}	50.01 ^{a,3}	72.78 ^{b,1}	93.76 ^{b,2}	82.77 ^{b,3}	91.58 ^{c,1}	110.43 ^{c,2}	123.56 ^{c,3}	105.66 ^{d,1}	134.64 ^{d,2}	154.04 ^{d,3}
C16:1	6.41 ^{a,1}	5.49 ^{a,2}	5.82 ^{a,3}	8.15 ^{b,1}	10.09 ^{b,2}	8.55 ^{b,3}	8.87 ^{c,1}	11.80 ^{c,2}	10.54 ^{c,3}	10.95 ^{d,1}	15.31 ^{d,2}	16.32 ^{d,3}
C17	0.65 ^{a,1}	0.64 ^{a,1}	0.71 ^{a,2}	1.06 ^{b,1}	1.33 ^{b,2}	1.06 ^{b,1}	1.14 ^{c,1}	1.67 ^{b,c,2}	1.32 ^{c,3}	1.41 ^{d,1}	2.01 ^{c,2}	1.44 ^{d,3}
C17:1	0.41 ^{a,1}	0.56 ^{a,2}	0.49 ^{a,3}	0.53 ^{b,1}	0.68 ^{b,2}	1.15 ^{b,3}	0.95 ^{c,1}	0.84 ^{c,2}	1.43 ^{c,3}	1.19 ^{d,1}	1.44 ^{d,2}	1.74 ^{d,3}
C18	30.98 ^{a,1}	25.45 ^{a,2}	30.10 ^{a,1}	51.51 ^{b,1}	74.90 ^{b,2}	61.16 ^{b,3}	57.53 ^{c,1}	94.73 ^{c,2}	65.80 ^{c,3}	74.64 ^{d,1}	103.16 ^{d,2}	89.27 ^{d,3}
C18:1n9	70.96 ^{a,1}	88.55 ^{a,2}	86.48 ^{a,2}	123.88 ^{b,1}	175.63 ^{b,2}	147.32 ^{b,3}	148.46 ^{c,1}	186.66 ^{c,2}	170.44 ^{c,3}	213.76 ^{d,1}	284.53 ^{d,2}	240.09 ^{d,3}
C18:2n6	66.11 ^{a,1}	59.90 ^{a,2}	63.63 ^{a,1}	110.65 ^{b,1}	156.17 ^{b,2}	121.71 ^{b,3}	138.36 ^{c,1}	165.10 ^{c,2}	163.70 ^{c,2}	165.58 ^{d,1}	210.54 ^{d,2}	184.86 ^{d,3}
C18:3n6	0.25 ^{a,1,2}	0.27 ^{a,1}	0.23 ^{a,2}	0.38 ^{b,1}	0.35 ^{b,2}	0.50 ^{b,3}	0.39 ^{b,1}	0.43 ^{c,2}	0.74 ^{c,3}	0.46 ^{c,1}	0.58 ^{d,2}	0.85 ^{d,3}
C18:3n3	6.08 ^{a,1}	7.71 ^{a,2}	7.83 ^{a,2}	7.05 ^{b,1}	15.51 ^{b,2}	11.04 ^{b,3}	8.77 ^{c,1}	17.44 ^{c,2}	12.83 ^{c,3}	9.01 ^{d,1}	22.32 ^{d,2}	12.81 ^{c,3}
C20	0.28 ^{a,1}	0.55 ^{a,2}	0.48 ^{a,3}	0.70 ^{b,1}	0.99 ^{a,b,2}	1.28 ^{b,3}	1.12 ^{c,1}	1.10 ^{b,c,1}	1.50 ^{c,1}	1.07 ^{d,1}	1.43 ^{b,c,2}	1.63 ^{d,3}
C20:1n9	1.38 ^{a,1}	1.01 ^{a,2}	1.35 ^{a,1}	2.37 ^{b,1}	3.31 ^{b,2}	3.39 ^{b,3}	2.55 ^{c,1}	5.06 ^{c,2}	4.60 ^{c,3}	4.04 ^{d,1}	7.74 ^{d,2}	6.92 ^{d,3}
C20:2	1.32 ^{a,1}	1.00 ^{a,2}	1.38 ^{a,3}	2.86 ^{b,1}	2.75 ^{b,2}	2.58 ^{b,3}	3.21 ^{c,1}	3.44 ^{c,2}	2.93 ^{c,3}	4.65 ^{d,1}	4.31 ^{d,2}	3.48 ^{d,3}
C20:3n6	0.25 ^{a,1}	0.34 ^{a,2}	0.33 ^{a,1}	0.80 ^{b,1}	0.65 ^{b,2}	0.53 ^{b,3}	1.02 ^{c,1}	1.06 ^{c,1,2}	1.13 ^{c,2}	1.67 ^{d,1}	2.07 ^{d,2}	1.26 ^{d,3}
C20:3n3	0.27 ^{a,1}	0.26 ^{a,1}	0.22 ^{a,3}	0.51 ^{b,1}	0.99 ^{b,2}	0.88 ^{b,3}	0.67 ^{c,1}	1.03 ^{c,2}	1.08 ^{c,3}	1.10 ^{d,1}	1.32 ^{d,2}	1.63 ^{d,3}
C20:4n6	3.30 ^{a,1}	2.41 ^{a,2}	3.60 ^{a,3}	6.62 ^{b,1}	8.24 ^{b,2}	6.70 ^{b,1}	8.51 ^{c,1}	12.85 ^{c,2}	11.69 ^{c,3}	13.00 ^{d,1}	14.73 ^{d,2}	16.60 ^{d,3}
C20:5n3	0.17 ^{a,1}	0.14 ^{a,1}	0.35 ^{a,3}	0.37 ^{b,1}	0.48 ^{b,2}	0.44 ^{b,2}	0.60 ^{c,1}	0.72 ^{c,2}	1.04 ^{c,3}	1.03 ^{d,1}	1.24 ^{d,2}	1.45 ^{d,3}
C22	0.33 ^{a,1}	0.48 ^{a,2}	0.46 ^{a,2}	0.66 ^{b,1}	0.82 ^{b,2}	0.77 ^{b,3}	1.33 ^{c,1}	0.93 ^{c,2}	1.36 ^{c,1}	1.88 ^{d,1}	1.87 ^{d,1}	1.32 ^{d,2}
C22:1n9	0.70 ^{a,1}	0.59 ^{a,2}	0.84 ^{a,3}	0.72 ^{b,1}	0.89 ^{b,2}	1.46 ^{b,3}	1.17 ^{c,1}	1.12 ^{c,1}	1.51 ^{c,3}	1.61 ^{d,1}	2.35 ^{d,2}	1.78 ^{d,3}
C22:2n6	1.55 ^{a,1}	2.80 ^{a,2}	2.10 ^{a,2}	4.16 ^{b,1}	4.56 ^{b,2}	4.60 ^{b,2}	9.22 ^{c,1}	5.40 ^{c,2}	8.13 ^{c,3}	11.90 ^{d,1}	9.35 ^{d,2}	9.32 ^{d,2}
C23	0.44 ^{a,1}	0.67 ^{a,2}	0.56 ^{a,3}	0.80 ^{b,1}	1.16 ^{b,2}	1.05 ^{b,3}	1.09 ^{c,1}	1.31 ^{c,2}	1.65 ^{c,3}	2.08 ^{d,1}	2.61 ^{d,2}	2.94 ^{d,3}
C24	1.44 ^{a,1}	1.42 ^{a,1}	1.55 ^{a,2}	1.98 ^{b,1}	1.58 ^{b,2}	2.53 ^{b,3}	2.40 ^{c,1}	3.18 ^{c,2}	4.38 ^{c,3}	3.68 ^{d,1}	3.55 ^{d,2}	5.41 ^{d,3}
Total FFA	253.62 ^{a,1}	247.11 ^{a,1}	263.64 ^{a,1}	403.55 ^{b,1}	562.72 ^{b,2}	469.87 ^{b,3}	497.00 ^{c,1}	636.29 ^{c,2}	601.61 ^{c,3}	640.18 ^{d,1}	839.84 ^{d,2}	767.85 ^{d,3}
SFA	94.13 ^{a,1}	75.60 ^{a,2}	88.63 ^{a,1,2}	135.20 ^{b,1}	181.91 ^{b,2}	158.32 ^{b,3}	163.42 ^{c,1}	222.52 ^{c,2}	209.04 ^{c,3}	199.28 ^{d,1}	260.73 ^{d,2}	267.80 ^{d,3}
UFA	159.49 ^{a,1}	171.51 ^{a,2}	175.01 ^{a,2}	268.35 ^{b,1}	380.80 ^{b,2}	312.18 ^{b,3}	333.58 ^{c,1}	413.77 ^{c,2}	392.57 ^{c,3}	440.90 ^{d,1}	579.11 ^{d,2}	500.05 ^{d,3}
MUFA	80.21 ^{a,1}	96.67 ^{a,2}	95.35 ^{a,2}	136.18 ^{b,1}	191.12 ^{b,2}	162.57 ^{b,3}	162.85 ^{c,1}	206.32 ^{c,2}	189.28 ^{c,3}	232.46 ^{d,1}	312.63 ^{d,2}	267.80 ^{d,3}
PUFA	79.29 ^{a,1}	74.83 ^{a,2}	79.66 ^{a,1}	132.63 ^{b,1}	189.81 ^{b,2}	149.15 ^{b,3}	170.74 ^{c,1}	207.46 ^{c,2}	203.50 ^{c,3}	208.73 ^{d,1}	266.73 ^{d,2}	237.72 ^{d,3}

Days of Ripening	14			21			30			SEM
	CNT	EQU	SAP	CNT	EQU	SAP	CNT	EQU	SAP	
C8	0.67 ^{d,1}	0.74 ^{e,2}	0.73 ^{e,2}	1.11 ^{e,1}	0.84 ^{f,2}	0.91 ^{f,3}	1.52 ^{f,1}	1.36 ^{g,2}	1.33 ^{g,3}	0.03
C10	0.87 ^{d,1}	1.05 ^{e,2}	1.07 ^{d,2}	1.21 ^{e,1}	1.44 ^{f,2}	1.92 ^{e,3}	1.37 ^{f,1}	1.93 ^{g,2}	2.49 ^{f,3}	0.05
C12	1.02 ^{d,1}	1.48 ^{e,2}	1.54 ^{e,3}	1.38 ^{e,1}	1.51 ^{f,2}	1.98 ^{f,3}	2.17 ^{f,1}	2.11 ^{g,2}	2.95 ^{g,3}	0.05
C14	8.15 ^{d,1}	12.03 ^{e,2}	12.35 ^{e,3}	12.09 ^{e,1}	14.16 ^{f,2}	18.22 ^{f,3}	17.89 ^{f,1}	18.69 ^{g,2}	26.14 ^{g,3}	0.49
C14:1	0.29 ^{d,1}	0.48 ^{e,2}	0.31 ^{d,1}	0.37 ^{e,1}	0.61 ^{f,2}	0.41 ^{e,3}	0.53 ^{f,1}	0.71 ^{g,2}	0.84 ^{f,3}	0.02
C15	0.58 ^{d,1}	1.09 ^{e,2}	1.05 ^{e,2}	0.90 ^{e,1}	1.34 ^{f,2}	1.58 ^{f,3}	1.19 ^{f,1}	1.75 ^{g,2}	1.98 ^{g,3}	0.04
C15:1	0.95 ^{e,1}	1.10 ^{d,2}	1.04 ^{d,3}	1.04 ^{f,1}	1.31 ^{e,2}	1.23 ^{e,3}	1.32 ^{g,1}	1.45 ^{f,2}	1.29 ^{f,3}	0.03
C16	133.40 ^{e,1}	184.20 ^{e,2}	183.27 ^{e,2}	157.80 ^{f,1}	274.36 ^{f,2}	323.97 ^{f,3}	345.90 ^{g,1}	318.31 ^{g,2}	401.05 ^{g,3}	8.36
C16:1	16.79 ^{e,1}	23.91 ^{e,2}	22.35 ^{e,3}	29.63 ^{f,1}	35.78 ^{f,2}	34.99 ^{f,3}	43.21 ^{g,1}	49.79 ^{g,2}	48.96 ^{g,3}	1.16
C17	2.28 ^{e,1}	2.65 ^{d,2}	3.43 ^{e,3}	2.54 ^{f,1}	2.99 ^{d,1}	4.56 ^{f,2}	4.71 ^{g,1}	4.33 ^{e,2}	6.61 ^{g,3}	0.13
C17:1	1.46 ^{e,1}	1.53 ^{e,2}	1.82 ^{e,3}	1.70 ^{f,1}	1.97 ^{f,2}	1.91 ^{f,3}	2.19 ^{g,1}	3.20 ^{g,2}	2.40 ^{g,3}	0.06
C18	85.67 ^{e,1}	150.59 ^{e,2}	158.91 ^{e,3}	121.79 ^{f,1}	158.60 ^{f,2}	287.21 ^{f,3}	204.19 ^{g,1}	202.07 ^{g,1}	324.32 ^{g,3}	6.40
C18:1n9	296.28 ^{e,1}	376.06 ^{e,2}	358.92 ^{e,3}	446.77 ^{f,1}	404.91 ^{f,2}	626.05 ^{f,3}	654.22 ^{g,1}	619.27 ^{g,2}	768.83 ^{g,3}	16.30
C18:2n6	235.53 ^{e,1}	276.81 ^{e,2}	249.35 ^{e,3}	288.59 ^{f,1}	295.00 ^{f,2}	475.26 ^{f,3}	364.82 ^{g,1}	369.26 ^{g,2}	515.13 ^{g,3}	9.87
C18:3n6	0.84 ^{d,1}	0.81 ^{e,1}	1.27 ^{e,3}	1.35 ^{e,1}	1.17 ^{f,2}	1.52 ^{f,3}	1.89 ^{g,1}	1.54 ^{g,2}	2.01 ^{g,3}	0.04
C18:3n3	16.06 ^{e,1}	26.92 ^{e,2}	20.07 ^{d,3}	25.23 ^{f,1}	28.82 ^{f,2}	26.65 ^{e,3}	29.30 ^{g,1}	34.54 ^{g,2}	28.67 ^{f,3}	0.71
C20	1.13 ^{c,1}	1.47 ^{b,c,2}	2.22 ^{e,3}	1.06 ^{d,1}	1.77 ^{c,d,2}	3.15 ^{f,3}	1.75 ^{g,1}	2.00 ^{d,2}	3.46 ^{g,3}	0.06
C20:1n9	6.30 ^{e,1}	8.88 ^{e,2}	10.21 ^{e,3}	8.51 ^{f,1}	9.36 ^{f,2}	19.09 ^{f,3}	12.72 ^{g,1}	13.13 ^{g,1}	17.65 ^{g,2}	0.40
C20:2	6.01 ^{e,1}	7.74 ^{e,2}	4.15 ^{e,3}	7.59 ^{f,1}	8.44 ^{f,2}	7.49 ^{f,1}	11.64 ^{g,1}	13.00 ^{g,2}	9.24 ^{g,3}	0.26
C20:3n6	1.87 ^{e,1}	2.43 ^{e,2}	1.39 ^{e,3}	2.39 ^{f,1}	4.05 ^{f,2}	2.29 ^{f,3}	3.50 ^{g,1}	4.83 ^{g,2}	3.21 ^{g,3}	0.10
C20:3n3	2.17 ^{e,1}	3.17 ^{e,2}	2.74 ^{e,3}	2.41 ^{f,1}	3.62 ^{f,2}	4.82 ^{f,3}	2.94 ^{g,1}	5.55 ^{g,2}	5.53 ^{g,2}	0.13
C20:4n6	23.14 ^{e,1}	25.93 ^{e,2}	27.97 ^{e,3}	25.42 ^{f,1}	33.82 ^{f,2}	53.27 ^{f,3}	38.03 ^{g,1}	47.86 ^{g,2}	51.19 ^{g,3}	1.26
C20:5n3	1.33 ^{e,1}	1.51 ^{e,2}	1.63 ^{e,3}	1.27 ^{f,1}	1.35 ^{f,2}	2.47 ^{f,3}	1.92 ^{g,1}	2.74 ^{g,2}	3.08 ^{g,3}	0.07
C22	1.03 ^{e,1}	2.14 ^{e,2}	2.58 ^{e,3}	1.65 ^{f,1}	2.48 ^{f,2}	3.71 ^{f,3}	2.09 ^{g,1}	4.57 ^{g,2}	4.17 ^{g,3}	0.10
C22:1n9	1.96 ^{e,1}	2.80 ^{e,2}	3.72 ^{e,3}	3.95 ^{f,1}	3.33 ^{f,2}	4.14 ^{f,3}	4.62 ^{g,1}	6.29 ^{g,2}	5.05 ^{g,3}	0.13
C22:2n6	15.27 ^{e,1}	14.25 ^{e,2}	14.20 ^{e,2}	16.89 ^{f,1}	20.73 ^{f,2}	21.33 ^{f,3}	37.31 ^{g,1}	43.34 ^{g,2}	34.23 ^{g,3}	0.93
C23	3.32 ^{e,1}	3.55 ^{e,2}	3.61 ^{e,3}	3.73 ^{f,1}	4.16 ^{f,2}	4.72 ^{f,3}	6.55 ^{g,1}	5.65 ^{g,2}	8.52 ^{g,3}	0.17
C24	4.59 ^{e,1}	6.88 ^{e,2}	6.75 ^{e,3}	10.95 ^{f,1}	15.15 ^{f,2}	8.16 ^{f,3}	14.95 ^{g,1}	26.40 ^{g,2}	15.13 ^{g,3}	0.52
Total FFA	868.97 ^{e,1}	1.142.20 ^{e,2}	1.098.57 ^{e,3}	1.179.31 ^{f,1}	1.333.06 ^{f,2}	1.943.04 ^{f,3}	1.814.41 ^{g,1}	1.805.65 ^{g,1}	2.295.46 ^{g,2}	46.72
SFA	242.70 ^{e,1}	367.89 ^{e,2}	382.48 ^{e,3}	316.20 ^{f,1}	478.81 ^{f,2}	660.11 ^{f,3}	604.26 ^{g,1}	589.16 ^{g,2}	798.15 ^{g,3}	16.05
UFA	626.27 ^{e,1}	774.31 ^{e,2}	721.14 ^{e,3}	863.11 ^{f,1}	854.24 ^{f,2}	1.282.93 ^{f,3}	1.210.15 ^{g,1}	1.216.49 ^{g,1}	1.497.32 ^{g,2}	30.90
MUFA	324.03 ^{e,1}	414.76 ^{e,2}	398.37 ^{e,3}	491.97 ^{f,1}	457.27 ^{f,2}	687.82 ^{f,3}	718.81 ^{g,1}	693.84 ^{g,2}	845.02 ^{g,3}	12.59
PUFA	302.23 ^{e,1}	359.56 ^{e,2}	322.77 ^{e,3}	371.57 ^{f,1}	396.99 ^{f,2}	592.63 ^{f,3}	492.08 ^{g,1}	522.66 ^{g,2}	652.29 ^{g,3}	13.18

SFA: sum of saturated fatty acids; UFA: sum of unsaturated fatty acids; MUFA: sum of monounsaturated fatty acids; PUFA: sum of polyunsaturated fatty acids. CNT: Non-inoculated control batches; EQU: Batches inoculated with *L. sakei* + *S. equorum*; SAP: Batches inoculated with *L. sakei* + *S. saprophyticus*. $^{a-g}$ Means in the same row and sausage group (CNT, EQU or SAP) not followed by a common letter differ significantly ($p < 0.05$) (differences associated to the ripening time). $^{1-3}$ Means in the same row and ripening time not followed by a common number differ significantly ($p < 0.05$) (differences associated to the use of starter cultures). SEM: standard error of the mean.

The FFA profile in the ripened sausages hardly varied with respect to the profile in the mixes and in the different batches. The main FFA was again oleic acid (18:1), followed by linoleic (C18:2), palmitic (C16) and stearic (C18) acid, these four fatty acids accounting for 86.48, 83.56 and 87.53% of the total FFA in the CNT, EQU and SAP batches, respectively. The other FFA, in descending order of quantitative importance, were palmitoleic (C16:1), arachidonic (C20:4), docosadienoic (C22:2), linolenic (C18:3) and myristic (C14). The FFA profile in the ripened sausages basically agreed with that reported by other authors in other dry-fermented sausages [12,23,26,64,70,92] and hardly varies between the different authors and works when pork and pig fat are used in the sausage manufacture. This profile notably varies, however, when other fats such as tallow [25] or hump fat [28] are used. The FFA acid profile in the ripened sausages basically reflects the fatty acid composition of the fat of the raw materials and the nature of the lipases acting during the ripening process. Monounsaturated fatty acids always dominate in the FFA fraction of the sausages made from pork and pig fat since monounsaturated fatty acids dominate in the triacylglycerols that are the main fraction in the fat used for the manufacture. Some authors [97] reported that the majority of the FFA derives from the triacylglycerols. In this same line, Muriel et al. [98] also indicated that, in ripened meat products, the lipolytic processes affecting neutral lipids have a higher incidence in the FFA fraction than those affecting the polar lipids.

4. Conclusions

The use of autochthonous starter cultures integrated by *Lactobacillus sakei* LS131 + *Staphylococcus equorum* SA25 (starter EQU), or by *L. sakei* LS131 + *Staphylococcus saprophyticus* SB12 (starter SAP) in the manufacture of Galician chorizo slightly but significantly reduced the pH values during the fermentation and improved the colour by increasing the percentage of transformation to nitrosyl-heme pigments as well as the a* and b* values in the final products. The two starters also significantly decreased the *Enterobacteriaceae* counts in the final product, but without completely eliminating this microbial group.

Both starter cultures significantly increased the α-amino acidic nitrogen and the total basic volatile nitrogen fractions during manufacturing, also increasing the free amino acid content. Moreover, the two cultures reduced the total biogenic amine content by 20%, also reducing in the same proportion the total vasoactive biogenic amine content. The SAP starter enhanced the lipolytic processes, increasing the content in free fatty acids without modifying the FFA profile.

Due to their performances, these two starter cultures seem to be suitable for increasing the quality and safety of the Galician chorizo sausage.

Author Contributions: Conceptualization, J.C.; methodology, J.C. and J.A.C.; formal analysis M.R.-G., S.F. and J.A.C.; data curation, M.R.-G. and J.A.C.; writing—original draft preparation J.C. All authors have read and agreed to the published version of the manuscript.

Funding: This research received no external specific funding.

Acknowledgments: This work was financially supported by the Xunta de Galicia (The Regional Government), Spain, through the CITACA Strategic Partnership ED431E 2018/07.

Conflicts of Interest: The authors declare no conflict of interest.

References

1. ANICE (Asociación Nacional de Industrias de la Carne de España). The Spanish Meat Industry: Dimension and Economic Data. [La Industria Cárnica Española: Dimensión y Datos Económicos.]. 2020. Available online: https://www.interempresas.net/Industria-Carnica/Articulos/301352-La-industria-carnica-espanola-dimension-y-datos-economicos.html (accessed on 19 September 2020).
2. Lorenzo, J.M.; Martínez, S.; Carballo, J. Microbiological and biochemical characteristics of Spanish fermented sausages. In *Beneficial Microbes in Fermented and Functional Foods*; Rai, V.R., Bai, J.A., Eds.; CRC Press: Boca Raton, FL, USA, 2015; pp. 55–72.

3. Fernández-Fernández, E.; Romero-Rodríguez, M.A.; Vázquez-Odériz, M.L. Physicochemical and sensory properties of Galician chorizo sausage preserved by refrigeration, freezing, oil-immersion, or vacuum-packing. *Meat Sci.* **2001**, *58*, 99–104. [CrossRef]
4. Fernández-Fernández, E.; Rozas-Barrero, J.; Romero-Rodríguez, M.A.; Vázquez-Odériz, M.L. Changes in the physicochemical properties and organoleptic quality of Galician chorizos during curing and after vacuum-packing. *Food Chem.* **1997**, *60*, 555–558. [CrossRef]
5. Fernández-Fernández, E.; Vázquez-Odériz, M.L.; Romero-Rodríguez, M.A. Colour changes during manufacture of Galician chorizo sausage. *Zeitschrift für Lebensmitteluntersuchung und Forschung A* **1998**, *207*, 18–21. [CrossRef]
6. Vázquez, R.; Prieto, B.; Carballo, J.; Franco, I. Estudio del contenido en macro y microelementos minerales en embutidos tradicionales gallegos. *Alimentaria* **2001**, *327*, 31–36.
7. Dolazo, F.; Steinhof, U.; Pfeiffer, S.; Ring, C. Microbiological status of the Spanish fermented sausage "Chorizo gallego". *Fleischwirtschaft* **1998**, *78*, 1089–1092.
8. Fonseca, S.; Cachaldora, A.; Gómez, M.; Franco, I.; Carballo, J. Monitoring the bacterial population dynamics during the ripening of Galician chorizo, a traditional dry-fermented Spanish sausage. *Food Microbiol.* **2013**, *33*, 77–84. [CrossRef] [PubMed]
9. Ordóñez, J.A.; Hierro, E.M.; Bruna, J.M.; de la Hoz, L. Changes in the components of dry-fermented sausages during ripening. *Crit. Rev. Food Sci. Nutr.* **1999**, *39*, 329–367. [CrossRef] [PubMed]
10. Hierro, E.; de la Hoz, L.; Ordóñez, J.A. Contribution of microbial and meat endogenous enzymes to the lipolysis of dry fermented sausages. *J. Agric. Food Chem.* **1997**, *45*, 2989–2995. [CrossRef]
11. Hierro, E.; de la Hoz, L.; Ordóñez, J.A. Contribution of the microbial and meat endogenous enzymes to the free amino acid and amine contents of dry fermented sausages. *J. Agric. Food Chem.* **1999**, *47*, 1156–1161. [CrossRef] [PubMed]
12. Casaburi, A.; Di Monaco, R.; Cavella, S.; Toldrá, F.; Ercolini, D.; Villani, F. Proteolytic and lipolytic starter cultures and their effect on traditional fer-mented sausages ripening and sensory traits. *Food Microbiol.* **2008**, *25*, 335–347. [CrossRef]
13. Ciuciu Simion, A.M.; Vizireanu, C.; Alexe, P.; Franco, I.; Carballo, J. Effect of the use of selected starter cultures on some quality, safety and sensorial properties of Dacia sausage, a traditional Romanian dry-sausage variety. *Food Control* **2014**, *35*, 123–131. [CrossRef]
14. Domínguez, R.; Munekata, P.E.; Agregán, R.; Lorenzo, J.M. Effect of commercial starter cultures on free amino acid, biogenic amines and free fatty acid contents in dry-cured foal sausage. *LWT Food Sci. Technol.* **2016**, *71*, 47–53. [CrossRef]
15. Essid, I.; Hassouna, M. Effect of inoculation of selected *Staphylococcus xylosus* and *Lactobacillus plantarum* strains on biochemical, microbiological and textural characteristics of a Tunisian dry fermented sausage. *Food Control* **2013**, *32*, 707–714. [CrossRef]
16. Leroy, F.; Verluyten, J.; De Vuyst, L. Functional meat starter cultures for improved sausage fermentation. *Int. J. Food Microbiol.* **2006**, *106*, 270–285. [CrossRef] [PubMed]
17. Samelis, J.; Metaxopoulos, J.; Vlassi, M.; Pappa, A. Stability and safety of traditional Greek salami—A microbiological ecology study. *Int. J. Food Microbiol.* **1998**, *44*, 69–82. [CrossRef]
18. Frece, J.; Markov, K. Autochthonous starter cultures. In *Fermented Meat Products. Health Aspects*; Zdolec, N., Ed.; CRC Press: Boca Raton, FL, USA, 2017; pp. 270–293.
19. García Fontán, M.C.; Lorenzo, J.M.; Martínez, S.; Franco, I.; Carballo, J. Microbiological characteristics of Botillo, a Spanish traditional pork sausage. *LWT Food Sci. Technol.* **2007**, *40*, 1610–1622. [CrossRef]
20. García Fontán, M.C.; Lorenzo, J.M.; Parada, A.; Franco, I.; Carballo, J. Microbiological characteristics of "androlla", a Spanish traditional pork sausage. *Food Microbiol.* **2007**, *24*, 52–58. [CrossRef]
21. Cachaldora, A.; Fonseca, S.; Franco, I.; Carballo, J. Technological and safety characterization of Staphylococcaceae isolated from Spanish traditional dry-cured sausages. *Food Microbiol.* **2013**, *33*, 61–68. [CrossRef]
22. Fonseca, S.; Ouoba, L.I.I.; Franco, I.; Carballo, J. Use of molecular methods to characterize the bacterial community and to monitor different native starter cultures throughout the ripening of Galician chorizo. *Food Microbiol.* **2013**, *34*, 215–226. [CrossRef]

23. Casaburi, A.; Aristoy, M.C.; Cavella, S.; Di Monaco, R.; Ercolini, D.; Toldrá, F.; Villani, F. Biochemical and sensory characteristics of traditional fermented sausages of Vallo di Diano (Southern Italy) as affected by the use of starter cultures. *Meat Sci.* **2007**, *76*, 295–307. [CrossRef]
24. Pasini, F.; Soglia, F.; Petracci, M.; Caboni, M.F.; Marziali, S.; Montanari, C.; Gardini, F.; Grazia, L.; Tabanelli, G. Effect of fermentation with different lactic acid bacteria stater cultures on biogenic amine content and ripening patterns in dry fermented sausages. *Nutrients* **2018**, *10*, 1497. [CrossRef] [PubMed]
25. Wang, D.; Zhao, L.; Su, R.; Jin, Y. Effects of different starter culture combinations on microbial counts and physico-chemical properties in dry fermented mutton sausages. *Food Sci. Nutr.* **2019**, *7*, 1957–1968. [CrossRef] [PubMed]
26. Xiao, Y.; Liu, Y.; Chen, C.; Xie, T.; Li, P. Effect of *Lactobaillus plantarum* and *Staphylococcus xylosus* on flavour development and bacterial communities in Chinese dry fermented sausages. *Food Res. Int.* **2020**, *135*, 109247. [CrossRef] [PubMed]
27. El Adab, S.; Essid, I.; Hassouna, M. Microbial, biochemical and textural characteristics of a Tunisian dry fermented poultry meat sausage inoculated with selected starter cultures. *J. Food Saf.* **2015**, *35*, 75–85. [CrossRef]
28. Mejri, L.; Ziadi, A.; El Adab, S.; Boulares, M.; Essid, I.; Hassouna, M. Effect of commercial starter cultures on physicochemical, microbiological and textural characteristics of a traditional dry fermented sausage reformulated with camel meat and hump fat. *J. Food Meas. Charact.* **2017**, *11*, 758–767. [CrossRef]
29. Visessanguan, W.; Benjakul, S.; Smitinont, T.; Kittikun, C.; Thepkasikul, P.; Panya, A. Changes in microbiological, biochemical and physico-chemical properties of Nham inoculated with different inoculum levels of *Lactobacillus curvatus*. *LWT Food Sci. Technol.* **2006**, *39*, 814–826. [CrossRef]
30. Cachaldora, A. Study of the Technological Aptitude of Microbial Strains Isolated from Traditional Galician Sausages. Ph.D. Thesis, University of Vigo, Ourense, Spain, 2011. (In Spanish).
31. ISO 1442. *International Standards Meat and Meat Products-Determination of Moisture Content*; International Organization for Standardization: Geneva, Switzerland, 1997.
32. ISO 1443. *International Standards Meat and Meat Products-Determination of Total Fat Content*; International Organization for Standardization: Geneva, Switzerland, 1973.
33. ISO 937. *International Standards Meat and Meat Products-Determination of Nitrogen Content*; International Organization for Standardization: Geneva, Switzerland, 1978.
34. ISO 936. *International Standards Meat and Meat Products-Determination of Ash Content*; International Organization for Standardization: Geneva, Switzerland, 1998.
35. ISO 1841-1. *International Standards Meat and Meat Products-Determination of Chloride Content—Part 1: Volhard Method*; International Organization for Standardization: Geneva, Switzerland, 1996.
36. Zaika, L.L.; Zell, T.E.; Smith, J.L.; Palumbo, S.A.; Kissinger, J.C. The role of nitrite and nitrate in Lebanon bologna, a fermented sausage. *J. Food Sci.* **1976**, *41*, 1457–1460. [CrossRef]
37. CIE (Commission Internationale de l'Eclairage). *International Commission on Illumination, Recommendations on Uniform Color Spaces, Color Difference Equations, Psychometric Color Terms*; Supplement No. 15 to CIE Publication No. 15 (E-1.3.1) 1971/(TO-1.3); Bureau Central of the CIE: Paris, France, 1978.
38. Johnson, M.J. Isolation and properties of pure yeast polypeptidase. *J. Biol. Chem.* **1941**, *7*, 575–586.
39. Moore, S.; Stein, W.M.H. Photometric ninhydrin method for use in the chromatography of amino acids. *J. Biol. Chem.* **1948**, *176*, 367–388.
40. Pearson, D. Application of chemical methods for the assessment of beef quality. II. Methods related to protein breakdown. *J. Sci. Food Agric.* **1968**, *19*, 366–369. [CrossRef]
41. De Ketelaere, A.; Demeyer, D.; Vandekerckhove, P.; Vervaeke, I. Stoichiometry of carbohydrate fermentation during dry sausage ripening. *J. Food Sci.* **1974**, *39*, 297–300. [CrossRef]
42. Alonso, M.L.; Álvarez, A.I.; Zapico, J. Rapid analysis of free amino acids in infant foods. *J. Liq. Chromatogr.* **1994**, *17*, 4019–4030. [CrossRef]
43. Eerola, S.; Hinnkanen, R.; Lindfors, E.; Hirvi, T.K. Liquid chromatographic determination of biogenic amines in dry sausages. *J. AOAC Int.* **1993**, *76*, 575–577. [CrossRef] [PubMed]
44. Lorenzo, J.M.; Martínez, S.; Franco, I.; Carballo, J. Biogenic amine content duringthe manufacture of dry-cured lacón, a Spanish traditional meat product: Effect of some additives. *Meat Sci.* **2007**, *77*, 287–293. [CrossRef] [PubMed]

45. Folch, J.; Lees, M.; Stanley, G.H.S. A simple method for isolation and purification of total lipids from animal tissues. *J. Biol. Chem.* **1957**, *226*, 497–509. [PubMed]
46. Spanish Government Presidency. *Orden Del 31 De Enero De 1977 Por La Que Se Establecen Los Métodos Oficiales De Análisis De Aceites y Grasas, Cereales y Derivados, Productos Lácteos y Productos Derivados De La Uva, B.O.E. no. 167 (14/07/1977)*; Spanish Government Presidency: Madrid, Spain, 1977; pp. 15800–16708.
47. Tarladgis, B.G.; Watts, B.M.; Younathan, M.T.; Dugan, L.R. A distillation method for the quantitative determination of malonaldehyde in rancid foods. *J. Am. Oil Chem. Soc.* **1960**, *37*, 44–48. [CrossRef]
48. Kaluzny, M.; Duncan, L.A.; Merritt, M.V.; Epps, D.E. Rapid separation of lipid classes in high yield and purity using bonded phase columns. *J. Lipid Res.* **1985**, *26*, 135–140.
49. Shehata, A.J.; de Man, J.M.; Alexander, J.C. A simple and rapid method for the preparation of methyl esters of fats in milligram amounts for gas chromatography. *Can. Inst. Food Sci. Technol. J.* **1970**, *3*, 85–89. [CrossRef]
50. Méndez-Cid, F.J.; Franco, I.; Martínez, S.; Carballo, J. Lipid characteristics of dry-cured "Tocino" during the manufacturing process. Effects of salting intensity and ripening temperature. *J. Food Compos. Anal.* **2016**, *52*, 33–43. [CrossRef]
51. Coppola, R.; Iorizzo, M.; Saotta, R.; Sorrentino, E.; Grazia, L. Characterization of micrococci and staphylococci isolatedfrom *Soppressata molisana*, a southern Italy fermented sausage. *Food Microbiol.* **1997**, *14*, 47–53. [CrossRef]
52. Franco, I.; Prieto, B.; Cruz, J.M.; López, M.; Carballo, J. Study of the biochemical changes during the processing of Androlla, a Spanish dry-cured pork sausage. *Food Chem.* **2002**, *78*, 339–345. [CrossRef]
53. Salgado, A.; García Fontán, M.C.; Franco, I.; López, M.; Carballo, J. Biochemical changes during the ripening of *Chorizo de cebolla*, a Spanish traditional sausage. Effect of the system of manufacture (homemade or industrial). *Food Chem.* **2005**, *92*, 413–424. [CrossRef]
54. Drosinos, E.H.; Paramithiotis, S.; Kolovos, G.; Tsikouras, I.; Metaxopoulos, I. Phenotypic and technological diversity of lactic acid bacteria and staphylococci isolated from traditionally fermented sausages in Southern Greece. *Food Microbiol.* **2007**, *24*, 260–270. [CrossRef] [PubMed]
55. Molly, K.; Demeyer, D.; Johansson, G.; Raemaekers, M.; Ghistelinck, M.; Geenen, I. The importance of meat enzymes in ripening and flavour generation in dry fermented sausages. First results of a European project. *Food Chem.* **1997**, *59*, 539–545. [CrossRef]
56. Ferrer, J.; Arboix, P. The "salchichón de Vich" (Vich sausage). II—Evolution of chemical parameters during the curing process and valoration of his organoleptic quality. In Proceedings of the 32nd European Meeting Meat Research Workers, Ghent, Belgium, 24–29 August 1986; pp. 279–281.
57. Ling, P.P.; Ruzhitsky, V.N.; Kapanidis, A.N.; Lee, T.-C. Measuring the colour of food. *Chem. Technol.* **1996**, *11*, 46–52.
58. Gómez, R.; Álvarez-Orti, M.; Pardo, J.E. Influence of the paprika type on redness loss in red line meat products. *Meat Sci.* **2008**, *80*, 823–828. [CrossRef]
59. Salgado, A.; García Fontán, M.C.; Franco, I.; López, M.; Carballo, J. Effect of the type of manufacture (homemade or industrial) on the biochemical characteristics of *Chorizo de cebolla* (a Spanish traditional sausage). *Food Control* **2006**, *17*, 213–221. [CrossRef]
60. Olivares, A.; Navarro, J.L.; Salvador, A.; Flores, M. Sensory acceptability of slow fermented sausages based on fat content and ripening time. *Meat Sci.* **2010**, *86*, 251–257. [CrossRef]
61. Lorenzo, J.M.; Franco, D. Fat effect on physico-chemical, microbial and textural changes through the manufactured of dry-cured foal sausage. Lipolysis, proteolysis and sensory properties. *Meat Sci.* **2012**, *92*, 704–714. [CrossRef]
62. Gómez, M.; Lorenzo, J.M. Effect of fat level on physicochemical, volatile compounds and sensory characteristics of dry-ripened "chorizo" from Celta pig breed. *Meat Sci.* **2013**, *95*, 658–666. [CrossRef]
63. Baka, A.M.; Papavergou, E.J.; Pragalaki, T.; Bloukas, J.G.; Kotzekidou, P. Effect of selected autochthonous starter cultures on processing and quality characteristics of Greek fermented sausages. *LWT Food Sci. Technol.* **2011**, *44*, 54–61. [CrossRef]
64. Lizaso, G.; Chasco, J.; Beriain, M.J. Microbiological and biochemical changes during ripening of Salchichón, a Spanish dry cured sausage. *Food Microbiol.* **1999**, *16*, 219–228. [CrossRef]
65. Bruna, J.M.; Ordóñez, J.A.; Fernández, M.; Herranz, B.; de la Hoz, L. Microbial and physico-chemical changes during the ripening of dry-fermented sausages superficially inoculated with or having added an intracellular cell-free extract of Penicillium aurantiogriseum. *Meat Sci.* **2001**, *59*, 87–96. [CrossRef]

66. Roseiro, L.C.; Santos, C.; Sol, M.; Borges, M.J.; Anjos, M.; Gonçalves, H.; Carvalho, A.S. Proteolysis in Painho de Potalegre dry fermented sausage in relation to ripening time and salt content. *Meat Sci.* **2008**, *79*, 784–794. [CrossRef] [PubMed]
67. Martín-Sánchez, A.M.; Chaves-López, C.; Sendra, E.; Sayas, E.; Fernández-López, J.; Pérez-Álvarez, J.A. Lipolysis, proteolysis and sensory characteristics of a Spanish fermented dry-cured meat product (salchichón) with oregano essential oils as surface mold inhibitor. *Meat Sci.* **2011**, *89*, 35–44. [CrossRef]
68. Ikonic, P.; Tasic, T.; Petrovic, L.; Skaljac, S.; Jokanovic, M.; Mandic, A.; Ikonic, B. Proteolysis and biogenic amine formation during the ripening of *Petrovská Klobása*, traditional dry-fermented sausage from Northern Serbia. *Food Control* **2013**, *30*, 69–75. [CrossRef]
69. El Abad, S.; Hassouna, M. Proteolysis, lipolysis and sensory characteristics of of a Tunisian dry fermented poultry meat sausage with oregano and thyme essential oil. *J. Food Saf.* **2016**, *36*, 19–32.
70. Bolumar, T.; Nieto, P.; Flores, J. Acidity, proteolysis and lipolysis changes in rapid-cured fermented sausage dried at different temperatures. *Food Sci. Technol. Int.* **2001**, *7*, 269–276. [CrossRef]
71. Du, S.; Cheng, H.; Ma, J.-K.; Li, Z.-J.; Wang, C.-H.; Wang, Y.-L. Effect of starter culture on microbiological, physicochemical and nutrition quality of Xiangxi sausage. *J. Food Sci. Technol.* **2019**, *56*, 811–823. [CrossRef]
72. Roseiro, L.C.; Gomes, A.; Gonçalves, H.; Sol, M.; Cercas, R.; Santos, C. Effect of processing on proteolysis and biogenic amines formation in a Portuguese traditional dry-fermented ripened sausage "*Chouriço Grosso de Estremoz e Borba PGI*". *Meat Sci.* **2010**, *84*, 172–179. [CrossRef]
73. Tang, K.X.; Shi, T.; Gänzle, M. Effect of starter cultures on taste-active amino acids and survival of pathogenic *Escherichia coli* in dry-fermented beef sausages. *Eur. Food Res. Technol.* **2018**, *244*, 2203–2212. [CrossRef]
74. Serio, A.; Laika, J.; Maggio, F.; Sacchetti, G.; D'Alessandro, F.; Rossi, C.; Martuscelli, M.; Chaves-López, C.; Paparella, A. Casing contribution to proteolytic changes and biogenic amines content in the production of an artisanal naturally fermented dry sausage. *Foods* **2020**, *9*, 1286. [CrossRef]
75. Zhu, S.W. Study on taste and taste substances of Jinhua ham. *Food Sci. (Chin.)* **1993**, *159*, 8–11.
76. Belitz, H.-D.; Grosch, W.; Schieberle, P. Amino acids, peptides, proteins. In *Food Chemistry*, 4th ed.; Belitz, H.-D., Grosch, W., Schieberle, P., Eds.; Springer: Berlin, Germany, 2009; pp. 8–92.
77. Hernández-Jover, T.; Izquierdo-Pulido, M.; Veciana-Nogués, M.T.; Mariné-Font, A.; Vidal-Carou, C. Effect of starter cultures on biogenic amine formation during fermented sausage production. *J. Food Prot.* **1997**, *60*, 825–830. [CrossRef]
78. González-Fernández, C.; Santos, E.M.; Jaime, I.; Rovira, J. Influence of starter cultures and sugar concentrations on biogenic amine contents in chorizo dry sausage. *Food Microbiol.* **2003**, *20*, 275–284. [CrossRef]
79. Suzzi, G.; Gardini, F. Biogenic amines in dry fermented sausages: A review. *Int. J. Food Microbiol.* **2003**, *88*, 41–54. [CrossRef]
80. Komprda, T.; Smelá, D.; Pechová, P.; Kalhotka, L.; Stencl, J.; Klejdus, B. Effect of starter culture, spice mix and storage time and temperature on biogenic amine content of dry fermented sausages. *Meat Sci.* **2004**, *67*, 607–616. [CrossRef]
81. Lorenzo, J.M.; Franco, D.; Carballo, J. Biogenic amines in fermented meat products. In *Fermented Meat Products Health Aspects*; Zdolec, N., Ed.; CRC Press: Boca Raton, FL, USA, 2017; pp. 450–473.
82. Hernandez-Jover, T.; Izquierdo-Pulido, M.; Veciana-Nogués, M.T.; Mariné-Font, A.; Vidal-Carou, M.C. Biogenic amine andpolyamine contents in meat and meat products. *J. Agric. Food Chem.* **1997**, *45*, 2098–2102. [CrossRef]
83. Bover-Cid, S.; Hugas, M.; Izquierdo-Pulido, M.; Vidal-Carou, M.C. Amino-acid decarboxylase activity of bacteria isolated from fermented pork sausages. *Int. J. Food Microbiol.* **2001**, *66*, 185–189. [CrossRef]
84. Bover-Cid, S.; Miguélez-Arrizado, M.J.; Latorre Moratalla, L.L.; Vidal Carou, M.C. Freezing of meat raw materials affects tyramine and diamine accumulation in spontaneously fermented sausages. *Meat Sci.* **2006**, *72*, 62–68. [CrossRef]
85. Van Ba, H.; Seo, H.-W.; Kim, J.-H.; Cho, S.-H.; Kim, Y.-S.; Ham, J.-S.; Park, B.-Y.; Kim, H.-W.; Kim, T.-B.; Seong, P.-N. The effects of starter culture types on the technological quality, lipid oxidation and biogenic amines in fermented sausages. *LWT Food Sci. Technol.* **2016**, *74*, 191–198. [CrossRef]
86. Hu, Y.; Xia, W.; Liu, X. Changes in biogenic amines in fermented silver carp sausages inoculated with mixed starter cultures. *Food Chem.* **2007**, *104*, 188–195. [CrossRef]

87. Anderegg, J.; Fischer, M.; Dürig, J.; Die, A.; Lacroix, C.; Meile, L. Detection of biogenic amines and tyramine-producing bacteria in fermented sausages from Switzerland. *J. Food Prot.* **2020**, *83*, 1512–1519. [CrossRef]
88. Maijala, R.; Eerola, S.; Lievonen, S.; Hill, P.; Hirvi, T. Formation of biogenic amines during ripening of dry sausages as affect by starter cultures and thawing time of raw materials. *J. Food Sci.* **1995**, *69*, 1187–1190. [CrossRef]
89. Bover-Cid, S.; Izquierdo-Pulido, M.; Vidal-Carou, M.C. Mixed starter cultures to control biogenic amine production in dry fermented sausages. *J. Food Prot.* **2000**, *63*, 1556–1562. [CrossRef]
90. Mietz, J.L.; Karmas, E. Chemical quality index of canned tuna as determined by HPLC. *J. Food Sci.* **1977**, *42*, 155–158. [CrossRef]
91. Veciana-Nogués, M.T.; Mariné-Font, A.; Vidal-Carou, M.C. Biogenic amines as hygienic quality indicators of tuna. Relationships with microbial counts, ATP-related compounds, volatile amines, and organoleptic changes. *J. Agric. Food Chem.* **1997**, *45*, 2036–2041. [CrossRef]
92. Beriain, M.J.; Peña, M.P.; Bello, J. A study of the chemical components which characterize Spanish saucisson. *Food Chem.* **1993**, *48*, 31–37. [CrossRef]
93. Safa, H.; Gatellier, P.; Lebert, A.; Picgirard, L.; Mirade, P.-S. Effect of combined salt and animal fat reductions on physicochemical and biochemical changes durin the manufacture of dry-fermented sausages. *Food Bioprocess Technol.* **2015**, *8*, 2109–2122. [CrossRef]
94. Nagy, A.; Mihályi, V.; Incze, K. Ripening and storage of Hungarian salami. Chemical and organoleptic changes. *Fleischwirtschaft* **1989**, *69*, 587–588.
95. Lorenzo, J.M.; Michinel, M.; López, M.; Carballo, J. Bio-chemical characteristics of two Spanish traditional dry-cured sau-sage varieties: Androlla and Botillo. *J. Food Compos. Anal.* **2000**, *13*, 809–817. [CrossRef]
96. Domínguez Fernández, M.C.; Zumalacárregui Rodríguez, J.M. Lipolytic and oxidative changes in "Chorizo" during ripening. *Meat Sci.* **1991**, *29*, 99–107. [CrossRef]
97. Marco, A.; Navarro, J.L.; Flores, M. The influence of nitrite and nitrate onmicrobial, chemical and sensory parameters of slow dry fermented sausage. *Meat Sci.* **2006**, *73*, 660–673. [CrossRef] [PubMed]
98. Muriel, E.; Andrés, A.I.; Petron, M.J.; Antequera, T.; Ruiz, J. Lipolytic and oxidative changes in Iberian dry-cured loin. *Meat Sci.* **2007**, *75*, 315–323. [CrossRef] [PubMed]

foods

MDPI

Article

Dry-Cured Meat Products According to the Smoking Regime: Process Optimization to Control Polycyclic Aromatic Hydrocarbons

Maria João Fraqueza [1], Marta Laranjo [2], Susana Alves [1], Maria Helena Fernandes [1], Ana Cristina Agulheiro-Santos [2,3], Maria José Fernandes [1], Maria Eduarda Potes [2,4] and Miguel Elias [2,3,*]

1 CIISA-Centro de Investigação Interdisciplinar em Sanidade Animal, Faculdade de Medicina Veterinária, Universidade de Lisboa, Avenida da Universidade Técnica, 1300-477 Lisboa, Portugal; mjoaofraqueza@fmv.ulisboa.pt (M.J.F.); susanaalves@fmv.ulisboa.pt (S.A.); helenafernandes@fmv.ulisboa.pt (M.H.F.); mjfernandes@fmv.ulisboa.pt (M.J.F.)

2 MED-Mediterranean Institute for Agriculture, Environment and Development, IIFA-Instituto de Investigação e Formação Avançada, Universidade de Évora, Pólo da Mitra, Ap. 94, 7006-554 Évora, Portugal; mlaranjo@uevora.pt (M.L.); acsantos@uevora.pt (A.C.A.-S.); mep@uevora.pt (M.E.P.)

3 Departamento de Fitotecnia, Escola de Ciências e Tecnologia, Universidade de Évora, Pólo da Mitra, Ap. 94, 7006-554 Évora, Portugal

4 Departamento de Medicina Veterinária, Escola de Ciências e Tecnologia, Universidade de Évora, Pólo da Mitra, Ap. 94, 7006-554 Évora, Portugal

* Correspondence: elias@uevora.pt

Received: 11 December 2019; Accepted: 11 January 2020; Published: 15 January 2020

check for updates

Abstract: The manufacturing of dry-cured meat products usually includes a smoking step. Polycyclic aromatic hydrocarbons (PAHs) are potentially carcinogenic chemical compounds that may result from smoking. The aim of the present study was to optimize the smoking regime of traditional dry-cured meat products in order to minimize the presence of PAHs. Dry-cured sausages were submitted to different smoking regimes: (A) no smoking; (B) 20 h effective smoking; (C) 60 h effective smoking; (D) effective smoking until reaching 38%–40% weight losses. Three independent batches were produced per smoking regime, and three samples per batch were analyzed. Microbiological, physicochemical, and sensory analyses were performed. The total PAHs content was generally low and did not differ significantly in meat products submitted to the four different smoking regimes. The PAH4 and benzo(α)pyrene levels were below the established legal limits in all analyzed dry-cured sausages. Nevertheless, non-smoked sausages always showed lower PAHs values for all PAHs groups.

Keywords: cured meat products; smoking; chemical hazards; polycyclic aromatic hydrocarbons (PAHs); food safety; food quality

---

1. Introduction

Dry-cured meat products result from ancestral know-how passed onto the new generations. The meat is salted and mixed with condiments and additives, particularly nitrate or nitrite. This meat batter is afterwards fermented, dried and smoked according to traditional processes. The main objectives of this technology are to give the meat a different appearance and distinct flavors and textures and to extend its shelf-life. These meat products are considered to comply with the 'history of safe use' concept of European Food Safety Authority (EFSA)'s safety assessment guidance, due to the large evidence of safe production and consumption by a genetically diverse human population collected over the years [1].

However, in 2015, the International Agency for Research on Cancer (IARC) communicated an opinion from a working group of experts based on a systematized analysis of thousands of research articles concluding that the relationship between meat products and colorectal cancer was unquestionable [2]. Case–control and cohort studies demonstrated the association between meat products consumption and colorectal cancer. In fact, a significant dose–response relationship was specifically established with an increased risk of 17% when approximately 100 g of processed meat were consumed per day. Based on this strong evidence, the IARC working group classified processed meat as "carcinogenic to humans" (Group 1) [3].

The two main mechanisms involved in the increased risk of cancer by consumption of processed meat are related to the presence of N-nitroso compounds (NOCs) and polycyclic aromatic hydrocarbons (PAHs).

PAHs are ubiquitous environmental contaminants [4]; however, several authors have reported human exposure to PAHs to occur mainly through food [5–7]. Different sources of food contamination include the production of PAHs during the thermal processing of foods, such as grilling and smoking, contamination from food-packaging materials, and direct deposition of PAHs from the atmosphere [8].

PAHs can have severe harmful effects on human health. They have carcinogenic, mutagenic, and teratogenic properties. PAHs are formed throughout the smoking process and can be deposited on the surface of smoked meat products or barbecued meat [3,9]. Traditional cold and hot smoking are carried out in small manufacturing units by burning wood or wood chips. Under cold smoking, environment temperatures usually do not exceed 20 °C, while in hot smoking, temperatures are around 80 °C [10,11]. To prevent excessive accumulation of PAHs in smoked processed meats, it is recommended to avoid high pyrolysis temperatures and the generation of direct smoking. In modern smoking chambers, the smoke generator is located aside the smoking chamber. Moreover, this separation avoids fat dripping over the fire and minimizes the formation of toxic compounds [12]. It should be emphasized that the main products of wood pyrolysis are phenols, carbonyls, and organic acids, which are also responsible for the flavor, color, and antimicrobial properties of smoke [13].

The aim of this study was to submit a dry-cured meat product to different levels of smoking in order to minimize the presence of PAHs. Furthermore, we aimed to establish good smoking practices in order to avoid and control this hazard.

2. Materials and Methods

2.1. Cured Products Processing and Sampling

"Paio", a traditional Portuguese dry-cured sausage (DCS), was manufactured in a local factory using commercial hybrid Iberian × Duroc pig breed meat.

Trimmings (80/20) were mechanically cut into cubes of 35 to 45 mm and mixed with red pepper (*Capsicum annuum* L.) paste (6% w/w), white wine (1% v/v), garlic (*Allium sativum* L.) paste (1% w/w), and powder laurel (*Laurus nobilis* L.) (0.006% w/w). Polyphosphates (0.15% w/w) were added to the commercial mixture AGLO P (MANE IBERICA S.A., Barcelona, Spain), while nitrates (0.003% w/w) and nitrites (0.003% w/w) were added in the form of the commercial additive NITROS 5/5 (Formulab, Moreira, Portugal). Since red pepper and garlic pastes have in their original composition approximately 17% of salt, supplementary addition of salt to the mixture was done to obtain a final concentration of 4% in the end products.

The meat batter was stored under controlled conditions at 5 °C for 48 h for ripening purposes. Afterwards, the batter was stuffed into pork natural casings of 50–55 mm.

From the initial batter, four experimental groups were submitted to different smoking regimes: (A) no smoking; (B) 20 h effective smoking (3 days of smoking); (C) 60 h effective smoking (8.5 days of smoking); (D) effective smoking until reaching 38%–40% weight losses.

The general curing procedure occurred in two steps. All DCS, except group (D), were dried in a cure chamber under controlled conditions at a temperature of 5 °C and relative humidity of 80%–85%

until reaching 38%–40% weight losses, throughout the curing period or after a smoking step. Smoking was generated by burning oak (*Quercus ilex* L.) wood. Smoked DCS were indirectly exposed to smoke during approximately 7 h/day.

Three independent batches of "Paio" were produced in different working days. Three samples were collected for each of the smoking groups.

2.2. Microbiological Analyses

Microbiological analyses were performed following established procedures: mesophiles following ISO 4833-1 [14]; lactic acid bacteria (LAB) according to ISO 15214 [15], under anaerobiosis; staphylococci as described by Laranjo et al. [16]; enterococci as described by Talon et al. [17]; enterobacteria according to ISO 21528-2 [18]. All microbiological analyses were performed in triplicate, and the results were expressed in log colony-forming units (cfu)/g.

The detection of *Listeria monocytogenes* was not performed, because all studied products had water activity (a_W) values below 0.92.

2.3. Physicochemical Analyses

2.3.1. Determination of pH, a_W, and Chlorides

DCS casings were removed, and pH values were measured with a Crison 507 (Crison, Barcelona, Spain) pH-meter following the procedures described in ISO 2917 [19]. Water activity (a_W) was determined at 25 °C with a hygrometer (Hygroskop Rotronic DT, Zurich, Switzerland) equipped with a WA-40 probe. Salt content was confirmed through the determination of chlorides according to the Volhard method, as described in ISO 1841-1 [20].

2.3.2. Color

Color was measured on cross sections of DCS with a Konica Minolta CR-400 colorimeter (Konica Minolta Inc., Tokyo, Japan) in five replicates per sample. The chromatic coordinates L* a* b* were determined using the CIELab System. All measurements were performed using the standard illuminant D65.

2.3.3. Texture Profile Analysis

Texture profile analysis (TPA) was performed using a Stable Micro System TA-Hdi (Stable Micro Systems, Godalming, England) following the procedures described before [21,22] and adapted by Laranjo et al. [23] using a cylindrical flat-ended plunger (with an area of 1 cm^2). The tests were carried out at room temperature (20 °C ± 1 °C). The samples were cut into 1 cm-thick slices, with a diameter of approximately 2.5 cm, which were compressed twice in two consecutive cycles of 50% compression with 5 s intervals between cycles, while the plunger was actioned at a constant speed of 1 mm s^{-1}. Force–time curves were used to calculate the following parameters: hardness, adhesiveness, springiness, cohesiveness, resilience, and chewiness. Five replicates per sample were used.

2.4. Sensory Analysis

Sensory analysis was performed in a special room with the requirements described in ISO 8589 [24]. Sample preparation was done 30 min before each session. The samples of each batch and condition were cut into 3 mm-thick slices. Six slices codified with a three-digit number were randomly disposed in white dishes. The trained group included five men and five women (40–60 years old) selected according to ISO 8586-1 [25]. The group was asked to evaluate the products according to the following attributes, using a quantitative descriptive analysis (QDA®) with a scale ranging from 0 to 100 corresponding to "no perception" or "maximum perception", respectively: color intensity, aroma intensity, flavor intensity, salt perception, hardness, fibrousness, succulence, off colors, off aromas, off flavors, overall appreciation. The evaluation of hardness was the exception, being communicated to the group that

50% of the scale would correspond to the optimum value. Each panelist tasted six sausages per session. Crackers and mineral water were provided, for the panelists to rinse their mouths between evaluations.

2.5. Polycyclic Aromatic Hydrocarbons Determination

Polycyclic aromatic hydrocarbons were determined by GC–MS using deuterated labelled PAHs, according to Alves et al. [26] and a procedure adapted from Mottier, Parisod, and Turesky [8]. The deuterium-labelled internal standard PAHs mix (LGC standards, Middlesex, UK) contained 16 PAHs (chemical purity >98%; isotope purity >99%): naphthalene-d8, acenaphthylene-d8, acenaphthene-d10, fluorene-d10, anthracene-d10, phenanthrene-d10, fluoranthene-d10, pyrene-d10, chrysene-d12, benzo[a]anthracene-d12, benzo[b]fluoranthene-d12, benzo[k]fluoranthene-d12, benzo[a]pyrene-d12, benzo[g,h,i]perylene-d12, indeno [1,2,3-c,d]-pyrene-d12, and dibenzo[a,h]anthracene-d14. Briefly, PAHs were saponified with 2M potassium hydroxide in ethanol/distilled water (9:1, v/v) under reflux for 5 h. Solid-phase extraction (SPE) was used for the purification of the PAHs, using two SPE column phases in a vacuum manifold (Varian, Palo Alto, CA, USA). First, the extract was applied into an aminopropyl SPE column (Supelco, Bellefonte, PA, USA) previously conditioned with cyclohexane and, afterwards, it was applied into a C18 SPE column (Supelco, Bellefonte, PA, USA). The PAHs were evaporated in a rotary evaporator to about 0.5 mL of volume and transferred into an amber GC vial. An internal standard (400 μL of a solution 20 μg/kg of the deuterated mixture containing 16 PAHs) was added before saponification. PAHs were analyzed by gas chromatography coupled to mass spectrometry using a GC–MS QP2010-Plus (Shimadzu, Kyoto, Japan) equipped with a SPB-5 chromatographic column (30 m × 0.25 mm × 0.25 μm film thickness, Supelco, Bellefonte, PA, USA). The chromatographic conditions were as follow: injector temperature, 250 °C; injection mode, splitless; column flow, 1.18 mL/min; carrier gas, helium; column oven temperature program, the initial temperature of 80 °C (maintained 0.5 min) was increased to 230 °C at 8 °C/min, then it was increased to 300 °C at 5 °C/min (maintained for 6 min). The mass spectrometer conditions were as follows: ion source temperature, 200 °C; interface temperature, 280 °C; ionization energy, 70 eV. The analysis was performed by selected ion monitoring (SIM). Each PAH was quantified using external calibration curves, which were constructed using commercial standard solutions containing unlabeled PAHs mixtures (Sigma-Aldrich, St. Louis, MO, USA), namely, naphthalene (NAP), acenaphthylene (ACY), acenaphthene (ACE), fluorine (FLR), phenanthrene (PHE), anthracene (ANT), fluoranthene (FLT), pyrene (PYR), chrysene (CHR), benzo[a]anthracene (BaH), benzo[b]fluoranthene (BbF), benzo[k]fluoranthene (BkF), benzo[a]pyrene (BaP), indeno[1,2,3-c,d]pyrene (IcP), dibenzo[a,h]anthracene (DhA), and benzo[g,h,i]perylene (BgP) at 13 concentrations, ranging from 0 to 30 μg/kg of sausage, and deuterated PAHs mixtures at a concentration set to 20 μg/kg. The calibration standards were injected before and after the samples, and both data sets were used to build the calibration curves.

2.6. Statistical Analysis

The Shapiro–Wilk test was used in order to evaluate whether the data followed a normal distribution. When not normally distributed ($p < 0.05$), some PAHs data were transformed before further analysis. For ACY, ACE, FLR, ANT, FLT, PYR, BaA, total PAHs, heavy, light, and PAH8, the data were log-transformed, whereas for NAT and PHE, the inversed and square root transformations were used, respectively. After statistical analysis, means and standard error of the means were back-transformed.

All data were analyzed with ANOVA using Statistica^{TM} v.12.0 software (1984–2014) from Statsoft (StatSoft Inc., Tulsa, OK, USA). Significant differences ($p < 0.05$) were identified using Tukey's honest significant difference (HSD) test.

3. Results and Discussion

3.1. Microbial Characterization

Table 1 summarizes the results of the microbial analysis performed in meat products submitted to different smoking regimes. Total mesophilic bacteria counts reflect the higher counts obtained for LAB, as expected in dry-cured products, where these microorganisms are responsible for a slight fermentation during a long period [27,28]. The LAB counts reached 8.0 to 8.8 log cfu/g, which is in accordance with the values reported in dry-cured sausages by several authors [29–31]. *Staphylococci* counts were between 3.6 and 4.9 log cfu/g, with significantly lower counts in non-smoked control sausages. This fact reflects the higher LAB counts (8.8 log cfu/g) that outcompete *Staphylococci* in these non-smoked products, due to a higher acid production (lower pH values) and other factors related to the technological process [32–34]. Moreover, the dominant species in Portuguese dry-cured sausages is *Staphylococcus xylosus* [35], which is less tolerant to a decrease in pH, and the level of acidification is a key component of sausage fermentation and of microbiota modulation [33,36]. Furthermore, *Enterococci* were present among other LAB at a concentration between 3.15 and 4.21 log cfu/g, in agreement with the number of *Enterococci* present in other Mediterranean dry-cured sausages [29]. Enterobacteria counts were 2.1–2.7 log cfu/g without significant differences between the different smoking regimes. According to the UK guidelines for ready-to-eat foods, these values are borderline (2–4 log cfu/g) [37]. These results are similar to those reported before regarding dry-fermented sausages from Portugal and other Mediterranean countries [29,35]. However, these values are higher than those reported by others for Portuguese dry-fermented sausages [30,31], denoting the need to improve both the quality of the raw materials and the hygiene procedures.

Table 1. Effect of smoking on microbiological and physiochemical parameters of dry-cured sausages.

Microbiological and Physicochemical Parameters	Smoking Regimes				p-Value
	A	B	C	D	
Mesophiles	8.01 ^{a} ± 0.22	7.27 ^{b} ± 0.09	7.34 ^{ab} ± 0.12	7.43 ^{ab} ± 0.27	$p = 0.034$ *
Lactic acid bacteria (LAB)	8.80 ^{a} ± 0.13	8.12 ^{b} ± 0.05	8.02 ^{b} ± 0.09	8.01 ^{b} ± 0.11	$p = 0.000$ ***
Staphylococci	3.59 ^{b} ± 0.29	4.57 ^{ab} ± 0.24	4.89 ^{a} ± 0.17	4.26 ^{ab} ± 0.34	$p = 0.011$ *
Enterococci	3.15 ± 0.18	4.21 ± 0.30	4.04 ± 0.39	3.89 ± 0.33	$p = 0.094$
Enterobacteria	2.25 ± 0.31	2.39 ± 0.62	2.71 ± 0.34	2.06 ± 0.36	$p = 0.738$
pH	4.89 ± 0.04	4.93 ± 0.05	4.95 ± 0.05	4.96 ± 0.04	$p = 0.629$
a_W	0.852 ± 0.011	0.844 ± 0.009	0.833 ± 0.010	0.840 ± 0.010	$p = 0.619$
salt content (%)	4.37 ± 0.46	4.77 ± 0.15	4.55 ± 0.35	4.56 ± 0.27	$p = 0.836$

(A) no smoking; (B) 20 h effective smoking; (C) 60 h effective smoking; (D) effective smoking until reaching 38%–40% weight losses. Microbial counts are expressed in log colony-forming units (cfu)/g (mean ± SEM); a_W: water activity. Within the same row, different letters (^{a} and ^{b}) represent significantly different arithmetic means (Tukey's honest significant difference (HSD) test). Significance: * $p < 0.05$, *** $p < 0.001$.

3.2. Physicochemical Characterization and Sensory Analysis

The pH and a_W mean values were generally low, namely, 4.89–4.96 and 0.833–0.852, respectively, as it has been found in similar long dry-cured meat products [38,39]. Nevertheless, in non-smoked meat products, which had the highest level of LAB, the mean pH value was the lowest. The standardization of meat products weight losses was reflected by the similar a_W values obtained for all smoking regimes. The salt content of the studied dry-cured products ranged from 3.48% to 5.26%. The commercial formulation of the dry-cured sausages under analysis was not changed.

Table 2 shows the results obtained for color, textural parameters, and sensory attributes. A brighter (L*) and yellower (b*) color was observed for meat products submitted to longer smoking regimes (Table 2). The red (a*) color of the products was not affected by the smoking regime, since it is related to the nitrosomyoglobin pigment formed in cured meat products due to the addition of nitrite to the formula. Fat color was affected by the smoking regime, being more yellow and darker in products submitted to 60 h and continuous smoking regimes. Only slight changes in color were noticed, since color was measured in interior sliced sausages, and the deposition of smoking compounds occurs primarily on the surface of a product, with penetration increasing over time [40]. Therefore, the a* and b* values were slightly higher in sausages with longer smoking regimes, as has been reported by others for "chorizo" and Frankfurter-type sausages [40,41].

Table 2. Effect of smoking on color, textural parameters, and sensory attributes of dry-cured sausages.

Color, Textural Parameters, and Sensory Attributes	Smoking Regimes				*p*-Value
	A	B	C	D	
L*	38.48 ^{bc} ± 0.39	36.97 ^{c} ± 0.51	39.75 ^{ab} ± 0.44	40.36 ^{a} ± 0.47	$p = 0.000$ ***
a*	17.68 ± 0.39	16.81 ± 0.47	18.10 ± 0.37	18.05 ± 0.43	$p = 0.107$
b*	10.53 ^{ab} ± 0.49	9.19 ^{b} ± 0.46	10.42 ^{ab} ± 0.47	11.30 ^{a} ± 0.57	$p = 0.029$ *
Hardness (N)	58.82 ± 2.18	66.87 ± 1.94	64.70 ± 2.72	59.28 ± 3.08	$p = 0.060$
Adhesiveness (N × s)	−2.47 ^{b} ± 0.25	−3.28 ^{a} ± 0.24	−2.42 ^{bc} ± 0.22	−1.63 ^{c} ± 0.16	$p = 0.000$ ***
Cohesiveness	0.60 ^{ab} ± 0.01	0.58 ^{b} ± 0.01	0.59 ^{ab} ± 0.01	0.60 ^{a} ± 0.01	$p = 0.020$ *
Springiness	0.87 ± 0.01	0.87 ± 0.01	0.86 ± 0.01	0.91 ± 0.03	$p = 0.293$
Resilience	0.15 ^{ab} ± 0.00	0.13 ^{b} ± 0.00	0.14 ^{ab} ± 0.00	0.16 ^{a} ± 0.00	$p = 0.001$ **
Chewiness (N × mm)	30.97 ± 1.29	33.66 ± 1.23	33.03 ± 1.77	32.55 ± 2.09	$p = 0.689$
Color intensity	69 ^{b} ± 2	76 ^{a} ± 1	66 ^{b} ± 1	70 ^{b} ± 1	$p = 0.000$ ***
Aroma intensity	70 ^{ab} ± 2	73 ^{a} ± 3	65 ^{b} ± 2	65 ^{ab} ± 2	$p = 0.017$ *
Flavor intensity	68 ± 2	72 ± 1	67 ± 1	70 ± 2	$p = 0.117$
Hardness	52 ^{b} ± 1	51 ^{b} ± 1	53 ^{ab} ± 1	56 ^{a} ± 1	$p = 0.019$ *
Fibrousness	24 ± 3	20 ± 4	34 ± 3	35 ± 4	$p = 0.019$ *
Succulence	65 ^{ab} ± 2	68 ^{a} ± 2	63 ^{ab} ± 2	57 ^{b} ± 2	$p = 0.010$ *
Off colors	0 ± 0	0 ± 0	0 ± 0	0 ± 0	$p = 0.470$
Off aromas	2 ± 1	1 ± 0	1 ± 1	1 ± 0	$p = 0.233$
Off flavors	3 ± 1	1 ± 1	5 ± 1	3 ± 1	$p = 0.074$
Salt perception	51 ^{c} ± 1	60 ^{a} ± 2	54 ^{bc} ± 1	58 ^{ab} ± 1	$p = 0.000$ ***
Overall appreciation	66 ^{ab} ± 14	69 ^{a} ± 9	62 ^{b} ± 13	61 ^{b} ± 11	$p = 0.019$ *

(A) no smoking; (B) 20 h effective smoking; (C) 60 h effective smoking; (D) effective smoking until reaching 38%–40% weight losses. L*: light, a*: red, b*: yellow. Values are represented as mean ± SEM. Within the same row, different letters (^{a,b} and ^{c}) represent significantly different arithmetic means (Tukey's HSD test). Significance: * $p < 0.05$, ** $p < 0.01$, *** $p < 0.001$.

Regarding the texture profile analysis, significantly different adhesiveness, cohesiveness, and resilience values were observed according to the smoking regime (Table 2). It was noticed that lower adhesiveness and higher cohesiveness and resilience values were found in products with an effective smoking until reaching 38%–40% weight losses. Elias and co-workers [42] obtained similar texture values, although the products analyzed in the present study were harder, probably due to the lower a_W and pH values.

Concerning the sensory analysis (Table 2), the panel observed significant differences in color intensity, aroma intensity, hardness, fibrousness, succulence, salt perception, and overall appreciation. The meat products subjected to the longer smoking periods were considered to be harder by the panelists, which agrees with the opinion expressed by the sensory panel of the work described by Carrapiso et al. [43]. Salt perception was higher for sausages from the smoking regime B. Sausages subjected to the smoking regimes C and D were evaluated by the panelists to possess intense smoking flavor and aroma, which could influence salt perception. Regarding the non-smoked sausages, the panelists reported the samples to be raw and exhibit a characteristic casing flavor, as negative attributes. The overall appreciation was higher for meat products smoked for 20 h.

3.3. Polycyclic Aromatic Hydrocarbons

The total PAHs and light PAHs (<4 rings) mean concentration values (Figure 1) were generally low, when compared to those reported in other studies [26], and did not differ significantly in meat products submitted to the four different smoking regimes. Nevertheless, non-smoked sausages always showed lower PAHs values for all PAHs groups. Multiple factors could explain the PAHs content and profile in dry-cured meat products. The obtained results could be related to the fact that the deposition of PAHs occurs mainly on the surface of products, without massive diffusion to the inside [44]. *Paio* is a traditional large-caliber dry-cured meat product approximately 20–30 cm long. When compared to small-caliber meat products of a similar size, the mean content of PAHs (μg/kg product) could be lower, as mentioned before [26]. Moreover, all dry-cured meat products (smoked and non-smoked) were dried until they achieved the same 38%–40% weight losses and similar water activity values, thus minimizing any variability in the content of PAHs related to the drying/curing process.

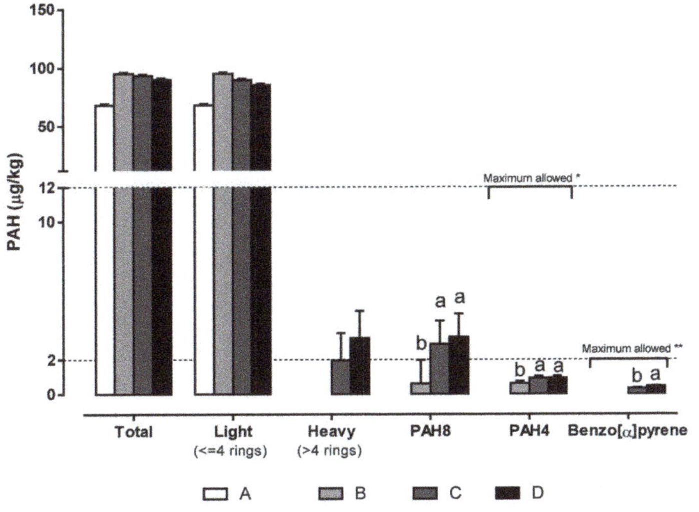

Figure 1. Polycyclic aromatic hydrocarbons (PAHs) (μg/kg wet weight) present in dry-cured sausages under different smoking regimes. Different letters for each group of PAHs represent significantly different means. * Maximum value of PAH4 allowed: 12 μg/kg; ** Maximum value of Benzo[α]pyrene allowed: 2 μg/kg. (**A**) no smoking; (**B**) 20 h effective smoking; (**C**) 60 h effective smoking; (**D**) effective smoking until reaching 38%–40% weight losses.

Heavy PAHs were only detected in the products subjected to longer smoking periods (C and D). Eight PAHs (PAH8, benzo[α]pyrene, benz[α]anthracene, benzo[α]fluoranthene, benzo[k]fluoranthene, benzo[ghi]perylene, chrysene, dibenz[a,h]anthracene, and indeno[1,2,3-437 cd]pyrene) were present in all smoked meat products, with significantly higher values in those subjected to longer smoking periods (C and D). This same profile was observed for the four PAHs PAH4, benzo(α)pyrene, benz(α)anthracene, benzo(β)fluoranthene, and chrysene. Benzo(α)pyrene was detected only in products subjected to the smoking regimes C and D, with a significantly higher concentration in those obtained with the longer regime D. Nevertheless, the levels of all these hazardous PAHs were below the maximum limits established by the Commission Regulation (EC) No. 1881/2006 [45] in its consolidated version of 19 March 2018, namely, 12 μg/kg wet weight for PAH4 and 2 μg/kg wet weight for benzo(α)pyrene.

Table 3 shows the content of individual PAHs under different smoking regimes. The most abundant PAHs were naphthalene and phenanthrene. The contents of individual PAHs was generally similar to those previously determined for distinct Portuguese and Serbian fermented sausages [26]. A different screening of Portuguese traditional smoked meat products reported significantly higher

PAHs values [46]. The observed difference is due to the distinct ways of expressing the results: our results were expressed in μg/kg wet weight according to the units patent in the legislation [45], while the values reported by Santos et al. [46] were expressed in μg/kg dry weight.

Table 3. Individual PAHs (μg/kg wet weight) determined in dry-cured sausages under different smoking regimes.

PAHs	Smoking Regimes				SEM	*p*-Value
	A	B	C	D		
naphthalene (NAP)	38.53	46.04	45.91	42.23	6.988	0.765
acenaphthylene (ACY)	2.77	4.73	6.90	6.47	1.418	0.208
acenaphthene (ACE)	1.37 ^{ab}	1.97 ^{a}	1.28 ^{ab}	1.12 ^{b}	1.194	0.013 *
fluorene (FLR)	3.68	5.67	6.41	3.87	1.260	0.199
phenanthrene (PHE)	10.57	24.97	14.67	17.25	6.616	0.104
anthracene (ANT)	0.53	1.56	2.07	2.11	1.536	0.050
fluoranthene (FLT)	1.79	2.49	2.44	3.01	1.164	0.140
pyrene (PYR)	1.82	2.40	1.95	2.81	1.169	0.219
benzo[a]anthracene (BaA)	n.d.	1.22 ^{a}	0.39 ^{b}	0.21 ^{b}	0.179	0.002 **
chrysene (CHR)	n.d.	0.55	0.54	0.57	0.049	0.947
benzo[α]pyrene (BaP)	n.d.	n.d.	0.32 ^{b}	0.44 ^{a}	0.028	0.015 *
benzo[β] fluoranthene (BbF)	n.d.	n.d.	n.d.	n.d.	n.d.	-
benzo[k]fluoranthene (BkF)	n.d.	n.d.	n.d.	n.d.	n.d.	-
indeno[1,2,3-c,d]pyrene (IcP)	n.d.	n.d.	1.78	0.90	0.791	0.289
dibenzo[a,h]anthracene (DhA)	n.d.	n.d.	3.37	0.56	3.464	0.611
benzo[g,h,i]perylene (BgP)	n.d.	n.d.	0.93^{b}	2.86 ^{a}	0.438	0.009 **

(A) no smoking; (B) 20 h effective smoking; (C) 60 h effective smoking; (D) effective smoking until reaching 38%–40% weight losses. n.d.: not detected; the levels of individual PAHs are mean values. Within the same row, different letters (^{a} and ^{b}) represent significantly different arithmetic means (Tukey's HSD test). Significance: * $p < 0.05$, ** $p < 0.01$.

The processing of traditional dry-cured sausages of different types and in different countries usually includes a smoking step. It is essential to monitor the PAHs content in the final meat products, optimizing the smoking step to a maximum of 20 h cold smoking, evaluating raw materials contamination, and using other technological strategies to control and reduce the presence of potentially carcinogenic chemicals in dry-cured meat products [47].

Nevertheless, the presence of PAHs in dry-cured meat products is due not only to the PAHs released and deposited on the surface of sausages during the smoking process but also to spices and/or aromatic herbs used for seasoning [48,49].

4. Conclusions

Chemical safety of dry-cured meat products can be assured through the optimization of the smoking regime. In the present study, we have demonstrated that a reduced smoking step allows the control of PAHs levels in dry-cured meat sausages, while maintaining the products' sensory characteristics. A good manufacturing process and an adequate selection of raw materials should be considered in order to avoid contamination.

Future perspectives include the study of low-salt dry-cured meat products in association with the control of the smoking step, in order to guarantee the stabilization of dry-cured meat products without increasing the content of PAHs.

Author Contributions: Conceptualization, M.J.F. (Maria João Fraqueza) and M.L.; methodology, M.J.F. (Maria João Fraqueza), M.L., S.A., A.C.A.-S., M.E.P.; formal analysis, M.H.F. and M.J.F. (Maria José Fernandes); investigation, M.J.F. (Maria João Fraqueza), M.L., S.A., A.C.A.-S., M.E.P.; writing—original draft preparation, M.J.F. (Maria João Fraqueza) and M.L.; writing—review and editing, M.J.F. (Maria João Fraqueza), M.L., S.A., and M.E.P.; supervision, M.J.F. (Maria João Fraqueza) and M.E.; funding acquisition, M.J.F. (Maria João Fraqueza) and M.E. All authors have read and agreed to the published version of the manuscript.

Funding: This work was supported by the projects PDR2020-1.0.1-FEADER-031373 and PDR2020-1.0.1-FEADER-031359, funded by national funds through Fundação para a Ciência e a Tecnologia (FCT)/MCTES and co-funded through the European Regional Development Fund (ERDF), and by the projects UIDB/05183/2020 (MED) and UIDP/00276/2020 (CIISA) financed by national funds through FCT.

Acknowledgments: Maria João Fraqueza is a member of the network Red CYTED HEALTHY MEAT (119RT0568).

Conflicts of Interest: The authors declare no conflict of interest.

References

1. EFSA. Opinion of the Scientific Panel on Dietetic Products, Nutrition and Allergies on a request from the Commission related to the Tolerable Upper Intake Level of Sodium. *EFSA J.* **2005**, *2*, 59. [CrossRef]
2. IARC. *Consumption of Red Meat and Processed Meat*; World Health Organization (WHO): Lyon, France, 2015.
3. IARC. *Red Meat and Processed Meat*; World Health Organisation-WHO: Lyon, France, 2018; Volume 114.
4. Abdel-Shafy, H.I.; Mansour, M.S.M. A review on polycyclic aromatic hydrocarbons: Source, environmental impact, effect on human health and remediation. *Egypt. J. Pet.* **2016**, *25*, 107–123. [CrossRef]
5. Roseiro, L.C.; Gomes, A.; Santos, C. Influence of processing in the prevalence of polycyclic aromatic hydrocarbons in a Portuguese traditional meat product. *Food Chem. Toxicol.* **2011**, *49*, 1340–1345. [CrossRef] [PubMed]
6. Guillen, M.D.; Sopelana, P.; Partearroyo, M.A. Food as a source of polycyclic aromatic carcinogens. *Rev. Environ. Health* **1997**, *12*, 133–146. [CrossRef] [PubMed]
7. Falcó, G.; Domingo, J.L.; Llobet, J.M.; Teixido, A.; Casas, C.; Muller, L. Polycyclic aromatic hydrocarbons in foods: Human exposure through the diet in Catalonia, Spain. *J. Food Prot.* **2003**, *66*, 2325–2331. [CrossRef] [PubMed]
8. Mottier, P.; Parisod, V.; Turesky, R.J. Quantitative Determination of Polycyclic Aromatic Hydrocarbons in Barbecued Meat Sausages by Gas Chromatography Coupled to Mass Spectrometry. *J. Agric. Food. Chem.* **2000**, *48*, 1160–1166. [CrossRef]
9. Yebra-Pimentel, I.; Fernández-González, R.; Martínez-Carballo, E.; Simal-Gándara, J. A Critical Review about the Health Risk Assessment of PAHs and Their Metabolites in Foods. *Crit. Rev. Food Sci. Nutr.* **2015**, *55*, 1383–1405. [CrossRef]
10. Ledesma, E.; Rendueles, M.; Díaz, M. Contamination of meat products during smoking by polycyclic aromatic hydrocarbons: Processes and prevention. *Food Control* **2016**, *60*, 64–87. [CrossRef]
11. Ahmad, J.I. SMOKED FOODS Applications of Smoking. In *Encyclopedia of Food Sciences and Nutrition*, 2nd ed.; Caballero, B., Ed.; Academic Press: Oxford, UK, 2003; pp. 5309–5316. [CrossRef]
12. Palma, S. Hidrocarbonetos Aromáticos Policíclicos em Produtos Cárneos Fumados. *Segurança E Qual. Aliment.* **2008**, *5*, 41–43.
13. Lingbeck, J.M.; Cordero, P.; O'Bryan, C.A.; Johnson, M.G.; Ricke, S.C.; Crandall, P.G. Functionality of liquid smoke as an all-natural antimicrobial in food preservation. *Meat Sci.* **2014**, *97*, 197–206. [CrossRef]
14. ISO. *Microbiology of the Food Chain—Horizontal Method for the Enumeration of Microorganisms. Part 1: Colony Count at 30 Degrees C by the Pour Plate Technique*; International Organization for Standardization: Geneva, Switzerland, 2013; Volume ISO 4833-1.
15. ISO. *Microbiology of Food and Animal Feeding Stuffs—Horizontal Method for the Enumeration of Mesophilic Lactic Acid Bacteria—Colony-Count Technique at 30 Degrees C*; International Organization for Standardization: Geneva, Switzerland, 1998; Volume ISO 15214.
16. Laranjo, M.; Gomes, A.; Agulheiro-Santos, A.C.; Potes, M.E.; Cabrita, M.J.; Garcia, R.; Rocha, J.M.; Roseiro, L.C.; Fernandes, M.J.; Fraqueza, M.J.; et al. Impact of salt reduction on biogenic amines, fatty acids, microbiota, texture and sensory profile in traditional blood dry-cured sausages. *Food Chem.* **2017**, *218*, 129–136. [CrossRef] [PubMed]
17. Talon, R.; Lebert, I.; Lebert, A.; Leroy, S.; Garriga, M.; Aymerich, T.; Drosinos, E.H.; Zanardi, E.; Ianieri, A.; Fraqueza, M.J.; et al. Traditional dry fermented sausages produced in small-scale processing units in Mediterranean countries and Slovakia. 1: Microbial ecosystems of processing environments. *Meat Sci.* **2007**, *77*, 570–579. [CrossRef] [PubMed]

18. ISO. *Microbiology of the Food Chain-Horizontal Method for the Detection and Enumeration of Enterobacteriaceae-Part 2: Colony-Count Technique*; International Organization for Standardization: Geneva, Switzerland, 2017; Volume ISO 21528-2.
19. ISO. *Meat and Meat Products—Measurement of pH—Reference Method*; International Organization for Standardization: Geneva, Switzerland, 1999; Volume 2917.
20. ISO. *Meat and Meat Products-Determination of Chloride Content*; International Organization for Standardization: Geneva, Switzerland, 1996; Volume ISO 1841.
21. Caine, W.R.; Aalhus, J.L.; Best, D.R.; Dugan, M.E.R.; Jeremiah, L.E. Relationship of texture profile analysis and Warner-Bratzler shear force with sensory characteristics of beef rib steaks. *Meat Sci.* **2003**, *64*, 333–339. [CrossRef]
22. Honikel, K.O. Reference methods supported by OECD and their use in Mediterranean meat products. *Food Chem.* **1997**, *59*, 573–582. [CrossRef]
23. Laranjo, M.; Agulheiro-Santos, A.C.; Potes, M.E.; Cabrita, M.J.; Garcia, R.; Fraqueza, M.J.; Elias, M. Effects of genotype, salt content and calibre on quality of traditional dry-fermented sausages. *Food Control* **2015**, *56*, 119–127. [CrossRef]
24. ISO. *Sensory Analysis-General Guidance for the Design of Test Rooms*; International Organization for Standardization: Geneva, Switzerland, 2012; Volume ISO 8589.
25. ISO. *Sensory Analysis-General Guidance for the Selection, Training and Monitoring of Assessors-Part 1: Selected Assessors*; International Organization for Standardization: Geneva, Switzerland, 1993; Volume ISO 8586-1.
26. Alves, S.P.; Alfaia, C.M.; Škrbić, B.D.; Živančev, J.R.; Fernandes, M.J.; Bessa, R.J.B.; Fraqueza, M.J. Screening chemical hazards of dry fermented sausages from distinct origins: Biogenic amines, polycyclic aromatic hydrocarbons and heavy elements. *J. Food Compos. Anal.* **2017**, *59*, 124–131. [CrossRef]
27. Demeyer, D.; Raemaekers, M.; Rizzo, A.; Holck, A.; De Smedt, A.; ten Brink, B.; Hagen, B.; Montel, C.; Zanardi, E.; Murbrekk, E.; et al. Control of bioflavour and safety in fermented sausages: First results of a European project. *Food Res. Int.* **2000**, *33*, 171–180. [CrossRef]
28. Garriga, M.; Aymerich, T. The Microbiology of Fermentation and Ripening. In *Handbook of Fermented Meat and Poultry*, 2nd ed.; Toldrá, F., Ed.; John Wiley & Sons, Ltd.: Chichester, UK, 2015; pp. 107–115.
29. Talon, R.; Lebert, I.; Leroy, S.; Garriga, M.; Aymerich, T.; Drosinos, E.H.; Zanardi, E.; Ianieri, A.; Fraqueza, M.J.; Patarata, L.; et al. Microbial Ecosystem of Traditional Dry Fermented Sausages in Mediterranean Countries and Slovakia. In *Mediterranean Ecosystems: Dynamics, Management and Conservation*; Williams, G.S., Ed.; Mediterranean Ecosystems: Dynamics, Management and Conservation: New York, NY, USA, 2012; pp. 115–128.
30. Elias, M.; Carrascosa, A.V. Characterisation of the Paio do Alentejo, a traditional Portuguese Iberian sausage, in respect to its safety. *Food Control* **2010**, *21*, 97–102. [CrossRef]
31. Alves, S.P.; Fernandes, M.J.; Fernandes, M.H.; Bessa, R.J.B.; Laranjo, M.A.; Santos, A.C.; Elias, M.; Fraqueza, M.J. Quality and Acceptability of Dry Fermented Sausages Prepared with Low Value Pork Raw Material. *J. Food Process. Preserv.* **2017**, *41*, e12823. [CrossRef]
32. Stavropoulou, D.A.; De Maere, H.; Berardo, A.; Janssens, B.; Filippou, P.; De Vuyst, L.; De Smet, S.; Leroy, F. Species Pervasiveness within the Group of Coagulase-Negative Staphylococci Associated with Meat Fermentation Is Modulated by pH. *Front. Microbiol.* **2018**, *9*, 2232. [CrossRef]
33. Stavropoulou, D.A.; De Maere, H.; Berardo, A.; Janssens, B.; Filippou, P.; De Vuyst, L.; De Smet, S.; Leroy, F. Pervasiveness of Staphylococcus carnosus over Staphylococcus xylosus is affected by the level of acidification within a conventional meat starter culture set-up. *Int. J. Food Microbiol.* **2018**, *274*, 60–66. [CrossRef] [PubMed]
34. Fraqueza, M.J. Antibiotic resistance of lactic acid bacteria isolated from dry-fermented sausages. *Int. J. Food Microbiol.* **2015**, *212*, 76–88. [CrossRef] [PubMed]
35. Fraqueza, M.J.; Rocha, J.M.; Laranjo, M.; Potes, M.E.; Fialho, A.; Fernandes, M.J.; Fernandes, M.H.; Barreto, A.; Semedo-Lemsaddek, T.; Elias, M. What is the main processing factor influencing *Staphylococcus* species diversity in different manufacturing units? *J. Food Sci.* **2019**, *84*, 2932–2943. [CrossRef] [PubMed]
36. Fraqueza, M.J.; Patarata, L. Fermented meat products-from the technology to the quality control. In *Fermented Food Products*; Sankaranarayanan, A., Amaresan, N., Dhanasekaran, D., Eds.; CRC Press-Taylor & Francis Group: Boca Raton, FL, USA, 2020.
37. HPA. *Guidelines for Assessing the Microbiological Safety of Ready-to-Eat Foods*; Health Protection Agency, Ed.; Health Protection Agency (HPA): London, UK, 2009.

38. Puolanne, E.; Petäjä-Kanninen, E. Principles of Meat Fermentation. In *Handbook of Fermented Meat and Poultry*, 2nd ed.; Toldrá, F., Ed.; John Wiley & Sons, Ltd.: Chichester, UK, 2015; pp. 13–17.
39. Laranjo, M.; Gomes, A.; Agulheiro-Santos, A.C.; Potes, M.E.; Cabrita, M.J.; Garcia, R.; Rocha, J.M.; Roseiro, L.C.; Fernandes, M.J.; Fernandes, M.H.; et al. Characterisation of "Catalão" and "Salsichão" Portuguese traditional sausages with salt reduction. *Meat Sci.* **2016**, *116*, 34–42. [CrossRef]
40. Ledesma, E.; Rendueles, M.; Díaz, M. Benzo(a)pyrene penetration on a smoked meat product during smoking time. *Food Addit. Contam. Part A* **2014**, *31*, 1688–1698. [CrossRef]
41. Pöhlmann, M.; Hitzel, A.; Schwägele, F.; Speer, K.; Jira, W. Contents of polycyclic aromatic hydrocarbons (PAH) and phenolic substances in Frankfurter-type sausages depending on smoking conditions using glow smoke. *Meat Sci.* **2012**, *90*, 176–184. [CrossRef]
42. Elias, M.; Potes, M.E.; Roseiro, L.C.; Santos, C.; Gomes, A.; Agulheiro-Santos, A.C. The Effect of Starter Cultures on the Portuguese Traditional Sausage "Paio do Alentejo" in Terms of Its Sensory and Textural Characteristics and Polycyclic Aromatic Hydrocarbons Profile. *J. Food Res.* **2014**, *3*, 45–56.
43. Carrapiso, A.I.; Martín-Cabello, L.; Torrado-Serrano, C.; Martín, L. Sensory Characteristics and Consumer Preference of Smoked Dry-Cured Iberian Salchichon. *Int. J. Food Prop.* **2015**, *18*, 1964–1972. [CrossRef]
44. Gomes, A.; Santos, C.; Almeida, J.; Elias, M.; Roseiro, L.C. Effect of fat content, casing type and smoking procedures on PAHs contents of Portuguese traditional dry fermented sausages. *Food Chem. Toxicol.* **2013**, *58*, 369–374. [CrossRef]
45. EC. Commission Regulation (EC) No 1881/2006 of 19 December 2006 setting maximum levels for certain contaminants in foodstuffs (Text with EEA relevance). *Off. J. Eur. Union* **2006**, *364*, 5–24.
46. Santos, C.; Gomes, A.; Roseiro, L.C. Polycyclic aromatic hydrocarbons incidence in Portuguese traditional smoked meat products. *Food Chem. Toxicol.* **2011**, *49*, 2343–2347. [CrossRef] [PubMed]
47. Fraqueza, M.J.; Borges, A.; Patarata, L. Strategies to Reduce the Formation of Carcinogenic Chemicals in Dry Cured Meat Products. In *Food Control and Biosecurity*, 1st ed.; Holban, A.M., Grumezescu, A.M., Eds.; Academic Press: Cambridge, MA, USA, 2018; pp. 295–342.
48. Rozentale, I.; Yan Lun, A.; Zacs, D.; Bartkevics, V. The occurrence of polycyclic aromatic hydrocarbons in dried herbs and spices. *Food Control* **2018**, *83*, 45–53. [CrossRef]
49. Singh, L.; Varshney, J.G.; Agarwal, T. Polycyclic aromatic hydrocarbons' formation and occurrence in processed food. *Food Chem.* **2016**, *199*, 768–781. [CrossRef] [PubMed]

foods

MDPI

Article

Fat Inclusion Level, NaCl Content and LAB Starter Cultures in the Manufacturing of Italian-Type Ostrich Salami: Weight Loss and Nutritional Traits †

Marco Cullere [1], **Enrico Novelli** [2] and **Antonella Dalle Zotte** [1,*]

1 Department of Animal Medicine, Production and Health, University of Padova, Agripolis, Viale dell'Università, 16, 35020 Legnaro, Italy; marco.cullere@unipd.it
2 Department of Comparative Biomedicine and Nutrition, University of Padova, Agripolis, Viale dell'Università, 16, 35020 Legnaro, Italy; enrico.novelli@unipd.it
* Correspondence: antonella.dallezotte@unipd.it
† This paper is a part of Marco Cullere's Ph.D. Thesis.

Received: 13 March 2020; Accepted: 7 April 2020; Published: 10 April 2020

Abstract: The experiment studied the effect of two different fat inclusion levels (30% and 40%), NaCl contents (2.4 and 2.6%) and starter cultures (lactic acid bacteria (LAB) 6: *L. curvatus/S. xylosus*; LAB 8: *L. sakei/S. xylosus*) on the weight loss and nutritional composition of Italian-type ostrich salami. With this purpose, 8 batches of 9 salami each ($n = 72$) were prepared. Salami were ripened for 20 weeks: weight loss was monitored throughout the experiment, while salami nutritional composition was evaluated at 10 and 20 weeks of ripening. The lowest fat and highest salt inclusion levels provided the highest cumulative weight loss throughout the trial. At 10 weeks of ripening, salami with 40% fat were the richest in moisture and fat, whereas the leanest ones had the highest protein, ash and cholesterol contents. LAB 6 provided salami with the highest moisture and protein, while LAB 8 increased fat and cholesterol contents. At 20 weeks of ripening the proximate composition of ostrich salami was solely affected by fat inclusion level, with similar findings to those observed at 10 weeks. Overall, fat inclusion level had a great impact on the weight loss and nutritional composition of Italian-style ostrich salami. Reducing the NaCl inclusion from 2.6% to 2.4%, the weight loss of ostrich salami was retarded by approximately 1 week, without affecting the nutritional composition of the final product. Results of the study suggested that it is feasible to produce salami with lower fat and salt contents, while ensuring satisfactory product quality.

Keywords: Italian-type salami; ostrich meat; sodium reduction; fat reduction; starter cultures; meat processing

1. Introduction

Italian-type salami are intended to be slow ripened sausages, rarely smoked, with a pH not below 5 and generally between 5.3 and 6.2. They have been produced for centuries, starting from Roman times, and traditionally they are made out of pork meat and fat in variable ratios, salt, and eventually sugar and nitrate/nitrite [1]. As no starter cultures are used in the manufacturing of artisanal-made salami, the fermentation process is driven by autochthonous microflora and this originates a vast regional diversity which is typical of artisanal-made salami. Nowadays, starter cultures are being increasingly used in salami manufacturing as they ensure product safety and acceptable quality, together with reducing the ripening time [2]. Lactic acid bacteria (LAB) mainly ferment sugars into lactic acid, being thus responsible for the acidification of the product, but they generally lack the main aroma production pathways [3]. By contrast, coagulase negative cocci (CNC) such as *Staphylococcus* degrade free amino acids and inhibit the oxidation of unsaturated free fatty acids, ultimately contributing to the color and flavor formation in the salami. Among LAB species, *Lactobacillus* (*L.*) *sakei* is often the dominant one in traditional salami, followed by *L. curvatus* and *L. plantarum*. Considering CNC, *Staphylococcus xylosus* is reported to be the most common microbial species in Italian-type salami [4]. Overall, many studies have demonstrated that the aforementioned microbial species are suitable and adapted to the meat environment and ripening process with specificities linked to different ripening conditions, ingredients and meat species [5–8].

Salt (NaCl) is a key ingredient for salami manufacturing because it affects the final taste of the product, being a flavor enhancer, as well as the texture, together with ensuring microbiological stability mainly through water activity reduction [9]. For these reasons, in order to provide a satisfactory fermentation process and quality of the final product, its level should always be >2% [10]. Despite raw meat generally having a low NaCl content, meat products can provide 20–30% of NaCl dietary intake [11]. In salami manufacturing, fat can account for 40–50% of the final product and it has a pivotal role as it affects flavor formation, texture and color, thus guaranteeing satisfactory quality attributes.

In recent years, however, consumer and health organization concerns regarding products containing significant quantities of fat and salt have notably increased. The reduction of total fat intake, and more specifically the reduction of saturated fat intake, have been identified as a pivotal part in the prevention of unhealthy weight gain in adults and in the reduction of noncommunicable diseases [12]. In fact, obesity is associated with an increased risk of numerous cancers, coronary heart disease and strokes [13]. A reduction in salt consumption is also a key factor in lowering the risk of heart disease, strokes and hypertension [14]. The recommendations by the joint Food and Agriculture Organization of the United Nations (FAO)/World Health Organization (WHO) expert consultation on Fats and fatty acids in human nutrition point out the concerns regarding these pathological conditions, both in industrialized and middle-income countries. In an attempt to meet the nutrition and health needs, the new approaches to develop healthy food products, including fermented sausages, have been steadily growing [15–17].

On the basis of the aforementioned considerations, this research work aimed to study the effects of two different fat and salt inclusion levels and two different LAB starter cultures on the quality of an Italian-type slow-fermented salami manufactured with ostrich meat and evaluated at two different ripening times. Ostrich meat was chosen because it is a healthy alternative to pork meat thanks to its low fat, mainly polyunsaturated fatty acids, content [18]. Moreover, the meat of this species could replace pork meat in countries where ostriches are farmed on a large scale, thus possibly being used by the meat industry to develop new products that, potentially, could provide an added value to a local resource. This paper, which is the first part of a wide research project, focused on the weight loss and nutritional traits (proximate composition and cholesterol content) of the studied product.

2. Materials and Methods

2.1. Animal, Diet and Meat

The present work was a collaboration between the Department of Animal Medicine, Production and Health-MAPS of the University of Padova and the private farm "Azienda Struzzo 2000" (Volto, RO, Italy). For the experiment, a 90 kg male ostrich (*Struthio camelus*) was used. The ostrich was reared in a 100 m^2 outdoor paddock of the aforementioned private farm where it was fed with a crumbled mix of 60% alfalfa (*Medicago sativa*), 20% of maize and 20% of fresh carrots provided by the horticultural production of the farm. The chemical composition of the ostrich diet is shown in Table 1. The ostrich was transported to the slaughterhouse where, according to the animal welfare dispositions [19], it was electrically stunned and bled. The carcass was then plucked, skinned, eviscerated and passed through a pre-cooling tunnel (+2 °C). Afterwards, it was cut in two halves (thighs) which were cooled for 24 h at +4 °C. The carcass was then dissected and, after bones, tendons and cartilage elimination, the meat was used for salami preparation.

Table 1. Chemical composition (g/kg as is basis) of the ostrich diet.

Analyzed Composition	g/kg
Water	126
Crude protein	147
Crude fat	17.8
Ash	60.7
Crude fiber	143
Neutral detergent fiber	231
Acid detergent fiber	155
Acid detergent lignin	22.7
Acid Insoluble Ash	4.74
Starch	213
Gross Energy, MJ/kg	8.90
Ca	10.5
P	2.50
Fe	0.12

2.2. *Experimental Design and Salami Preparation*

The present experiment was planned as a 2 × 2 × 2 design: 8 batches of 9 ostrich salami each (n = a total of 72 salami of approximately 650 g/each, fresh weight) using 2 different levels of fat (30% and 40%), NaCl (2.4% and 2.6%) and two different LAB starter cultures (LAB 6 and LAB 8) were produced. For the salami preparation, 30.5 kg of fresh ostrich meat and 16.5 kg of fresh pork back-fat were used. Meat and fat were ground separately by means of a professional meat grinder, equipped with a plate with 6 and 7 mm diameter holes, respectively. Subsequently, following the experimental design, ground meat and fat were divided into two batches according to two different fat inclusion percentages (30 and 40%) and then mixed. Subsequently, each batch was separated in two equal units and added with 2.4 or 2.6% of NaCl, followed by a spices mix (0.78% black pepper, 0.009% cinnamon, 0.009% cloves, 0.009% nutmeg) and 3.55% red wine. Each unit was finally split into two equal parts which were inoculated (1.6 g/kg) and mixed with two different LAB starter cultures: LAB 6 (*Lactobacillus curvatus/Staphylococcus xylosus* + dextrose) or LAB 8 (*Lactobacillus sakei/Staphylococcus xylosus* + dextrose). Batches were then individually mixed and put in a refrigerated chamber at +4 °C for 12 h. The following day all the batters were stuffed into natural casings (salami diameter ranged 6–8 cm), labeled and put in a dedicated chamber using controlled ripening conditions. All the salami were nitrite/nitrate free. During drying and ripening processes surface mold growth was not prevented, which is coherent with artisanal-type salami. Despite this, only a limited growth of white molds was observed along the process.

2.3. *Drying and Ripening*

Initially, salami were dried for a 5-day long period with RH (Relative Humidity) ranging from 65 to 85% and T (Temperature) starting from 19 °C and decreasing of 1 °C/day. When T was 14 °C, ripening phase started: RH was maintained between 70 and 80% and T decreased of 1 °C/day until 12 °C, afterwards it remained constant. Ripening of the first group of salami (4 salami/treatment) was stopped when the first of them lost up to 35% of its initial weight, at 10 weeks, while the second group of salami (5 salami/treatment) was ripened for further 10 weeks.

2.4. *Weight Loss Determination and Chemical Analyses*

Once a week, for 10 weeks, and once every two weeks, for the further 10 weeks, salami were individually weighed using a commercial weight scale and cumulative weight loss for each treatment was subsequently calculated. After collection, at 10 and 20 weeks of ripening, salami were transported to the MAPS Department, individually freed from casings, frozen in liquid nitrogen, homogenized using a Retsch Grindomix GM 200 (15 s at 10,000 rpm) and then analyzed. At 10 and 20 weeks of ripening, proximate composition was determined (dry matter: method 950.46; ash: method 923.03; crude fat:

method 960.39), with protein content calculated by difference [20]. Cholesterol content of the salami was also analyzed at 10 and 20 weeks of ripening, following the method described by Casiraghi et al. [21].

2.5. Chemical Composition of the Diet

Analyses of the ostrich diet was carried out in duplicate following the Association of Official Analytical Chemists (AOAC) [22] methods to determine the concentrations of dry matter (method: 934.01), crude protein (method: 2001.11), crude fiber (method: 978.10), ash (method: 967.05) and starch (amyloglucosidase-α-amylase method: 996.11). Crude fat was determined after acid-hydrolysis [23]. Gross energy content of the diet was determined with an adiabatic bomb calorimeter [24]. Neutral detergent fiber (NDF, without sodium sulfite), acid detergent fiber (ADF), acid detergent lignin (ADL) and acid-insoluble ash (AIA) were analyzed according to Mertens [25], AOAC (procedure 973.187) [22] and Van Soest et al. [26], respectively, using the sequential procedure and the filter bag system (Ankom Technology, New York, NY, USA). Mineral analyses of the diet (Ca, P, Fe) was performed by ICP-OES (Spectro Ciros Vision EOP) after microwave digestion AOAC (procedure 999.10) [22].

2.6. Statistical Analysis

Data were analyzed using SAS 9.1 statistical analysis software for Windows [27] General Linear Model (GLM) procedures. A three-way ANOVA, which was stratified by ripening time (2.5 and 5 months), tested fat, salt and LAB starter cultures as fixed effects on cumulative weight loss, proximate composition and cholesterol content of artisanal-made Italian-type ostrich salami. The statistical analysis considered also interactions (Fat × Salt; Fat × LAB; Salt × LAB; Fat × Salt × LAB). When no significant interactions were found, only main effects were considered. Least square means were obtained using the Bonferroni test. p-values were considered significant when < 0.05.

3. Results

3.1. Weight Loss of Ostrich Salami

Weight loss of ostrich salami was significantly affected by fat (F) and salt (S) inclusion levels, from the beginning until the end of the ripening time (20 weeks), but not by LAB starter culture (L) (Table 2). After the drying period, salami manufactured with 30% fat already showed a more intense weight loss compared to those with 40% fat content ($p < 0.0001$).

Table 2. Effect of two levels of fat, salt, and lactic acid bacteria (LAB) starter cultures and their interaction on cumulative weight loss (% of the initial weight) of ostrich salami ripened for 20 weeks [1].

	FAT (F)		SALT (S)		LAB (L)		Significance					RSD [2]
	30	40	2.4	2.6	6	8	F	S	L	F × S	F × L	
Drying	13.2	10.1	11.3	11.9	11.5	11.7	<0.0001	<0.01	ns	<0.05	ns	0.65
Week 2	22.2	16.5	19.1	20.0	19.4	19.7	<0.0001	<0.001	ns	<0.05	ns	0.86
Week 3	25.6	18.5	21.5	22.6	21.9	22.2	<0.0001	<0.001	ns	<0.01	ns	0.92
Week 4	28.1	20.3	23.6	24.8	24.1	24.3	<0.0001	<0.001	ns	<0.05	<0.05	0.98
Week 5	30.1	21.6	25.2	26.5	25.7	26.0	<0.0001	<0.0001	ns	<0.05	<0.05	0.98
Week 6	31.9	23.0	26.8	28.1	27.3	27.6	<0.0001	<0.05	ns	ns	ns	1.95
Week 7	33.5	24.4	28.3	29.6	28.9	29.0	<0.0001	<0.0001	ns	<0.05	<0.05	1.04
Week 8	34.8	25.4	29.4	30.8	30.0	30.2	<0.0001	<0.0001	ns	<0.05	<0.05	1.04
Week 9	36.1	26.5	30.6	32.0	31.2	31.4	<0.0001	<0.0001	ns	<0.05	<0.05	1.06
Week 10	37.4	27.5	31.7	33.2	32.3	32.5	<0.0001	<0.0001	ns	<0.05	ns	1.06
Week 12	38.2	28.3	32.5	33.9	33.1	33.4	<0.0001	<0.0001	ns	<0.01	<0.05	0.95
Week 14	39.8	29.8	34.0	35.6	34.6	35.0	<0.0001	<0.0001	ns	<0.05	<0.05	0.97
Week 16	41.0	30.9	35.2	36.7	35.8	36.2	<0.0001	<0.0001	ns	<0.05	<0.05	0.97
Week 18	42.1	32.0	36.3	37.8	36.8	37.3	<0.0001	<0.0001	ns	<0.01	ns	0.99
Week 20	43.1	32.9	37.2	38.8	37.8	38.2	<0.0001	<0.0001	ns	<0.05	ns	1.01

ns = not significant; [1] Up to 10 weeks of ripening the sample size was of 9 salami/treatment, while from 10 to 20 weeks of ripening it was 5 salami/treatment; [2] Residual Standard Deviation.

This difference tended to increase along the trial (Figure 1) and resulted in a 10% discrepancy in cumulative weight loss, in the two fat groups, at the end of the ripening phase (43.1 vs. 32.9% for F30 and F40 salami, respectively).

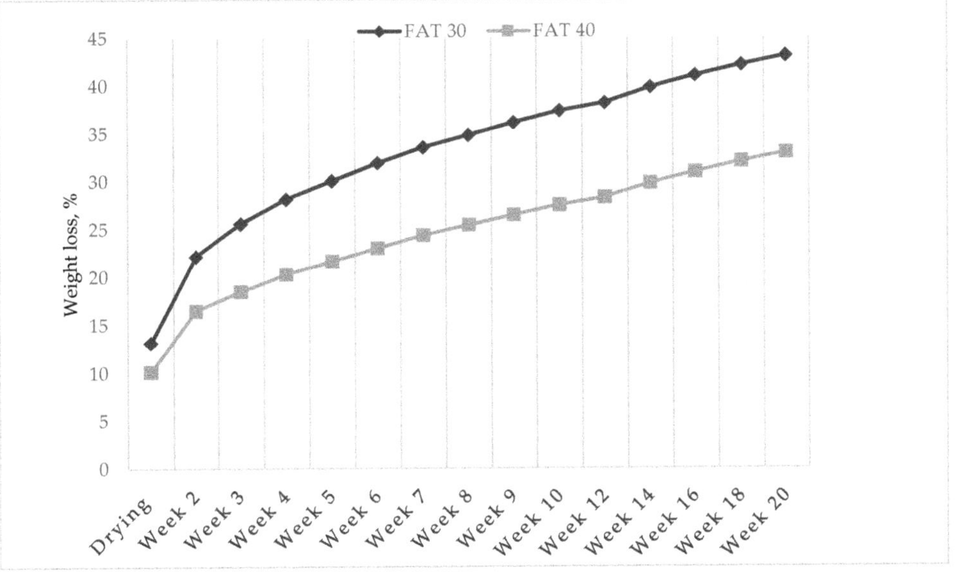

Figure 1. Effect of fat level on cumulative weight loss (% of the initial weight) of ostrich salami ripened for 20 weeks.

A higher NaCl content favored the weight loss of salami: in fact, it was higher in the 2.6% group compared to the 2.4% one ($p < 0.05$). In this case, the divergence between the two groups of salami increased from the drying phase until the fifth week of ripening, but after that it remained constant until the end of the trial (Figure 2).

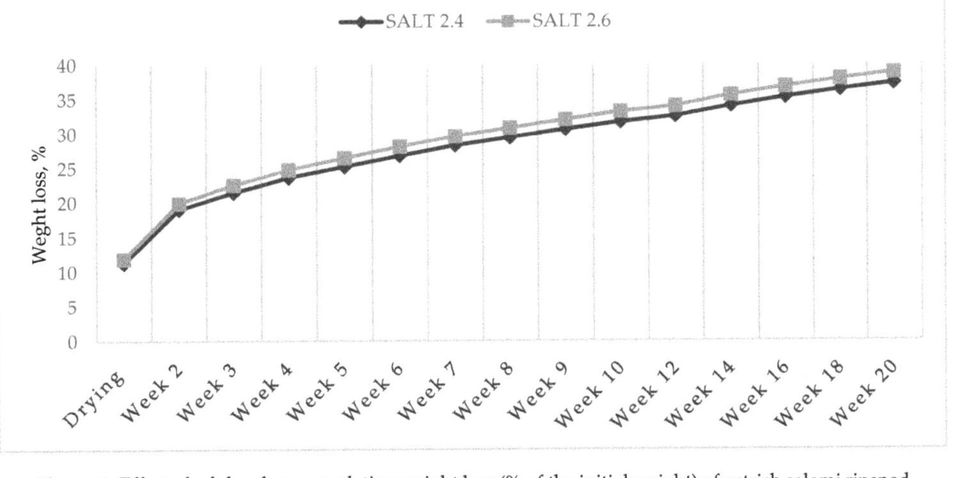

Figure 2. Effect of salt level on cumulative weight loss (% of the initial weight) of ostrich salami ripened for 20 weeks.

The F × S interaction showed an effect on the cumulative weight loss of ostrich salami: it was observed that within the group of salami with 30% fat, those manufactured with 2.4% and 2.6% salt inclusions exhibited the same cumulative weight loss, whereas at 40% fat inclusion level the salami with a higher salt content showed a more intense weight loss compared to those belonging to the 2.4% NaCl inclusion level (Figure 3).

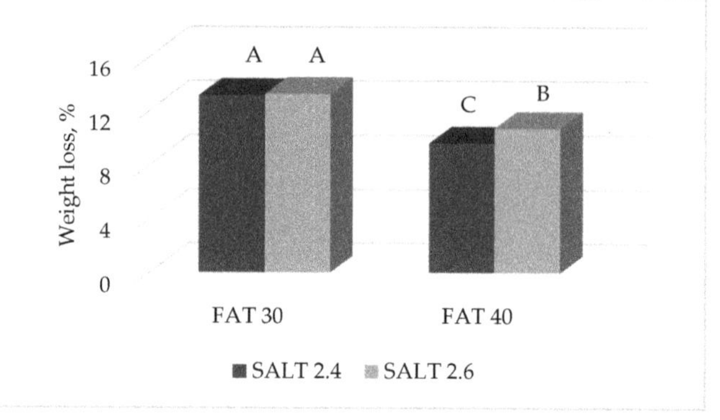

Figure 3. Effect of the interaction FAT × SALT on the cumulative weight loss (% of the initial weight) of ostrich salami at the end of the drying phase. Histograms reporting different A, B, C letters significantly differ for $p < 0.05$.

Results of the present study highlighted also that different LAB starter cultures did not affect cumulative weight loss of ostrich salami over a 20 weeks ripening period. However, the F × L interaction highlighted a statistically significant difference for the cumulative weight loss up to 16 weeks of ripening (Figure 4; $p < 0.05$); despite this, no differences were observed between the two starter cultures within the FAT group.

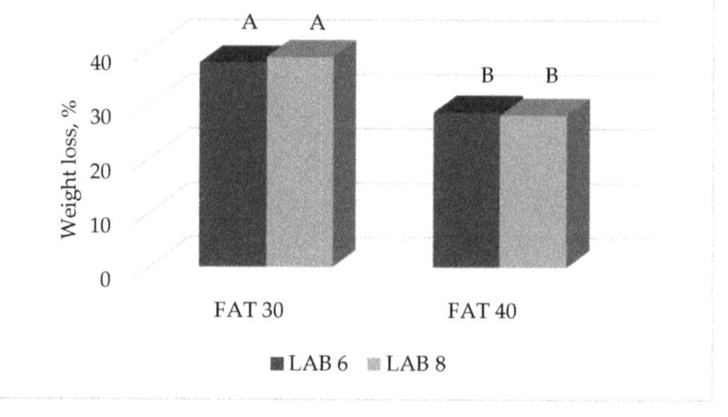

Figure 4. Effect of the interaction FAT × LAB on the cumulative weight loss (% of the initial weight) of ostrich salami ripened for 16 weeks. Histograms reporting different A, B letters significantly differ for $p < 0.05$.

3.2. *Proximate Composition and Cholesterol Content of Ostrich Salami*

The results presented in Table 3 show that fat level affected moisture, protein, fat, ash and cholesterol contents of Italian-style ostrich salami, analyzed at 10 weeks of ripening ($p < 0.0001$). As expected, higher presence of pork back-fat directly increased the fat content of the salami, which was higher in the 40% than the 30% groups (35.6 vs. 32.8 g/100 g, respectively). Interestingly, the leanest salami exhibited the highest cholesterol content (93.4 vs. 83.1 mg cholesterol/100 g meat, for 30 and 40% fat inclusion levels, respectively). Salami containing 30% pork back-fat had also lower moisture (33.2 vs. 36.4 g/100 g, respectively), higher protein (25.8 vs. 20.8 g/100 g, respectively) and ash (4.91 vs. 4.28 g/100 g, respectively) contents, compared to those manufactured with 40% back-fat.

Table 3. Effect of two levels of fat, salt, and two LAB starters cultures and their interaction on proximate composition (g/100 g) and cholesterol content (mg/100 g) of ostrich salami ripened for 10 weeks.

	FAT (F)		SALT (S)		LAB (L)		Significance					RSD [1]
	30	40	2.4	2.6	6	8	F	S	L	F × S	F × L	
Salami	4	4	4	4	4	4						
Moisture	33.2	36.4	34.9	34.6	33.8	35.7	<0.0001	ns	<0.0001	<0.0001	<0.0001	0.36
Protein	25.8	20.8	23.2	23.4	23.1	23.5	<0.0001	ns	<0.05	<0.05	ns	0.55
Fat	32.8	35.6	34.4	34.0	35.3	33.1	<0.0001	ns	<0.001	<0.05	<0.01	1.37
Ash	4.91	4.28	4.34	4.86	4.57	4.63	<0.0001	<0.0001	ns	<0.05	<0.01	0.07
Cholesterol [2]	93.4	83.1	88.6	87.9	90.8	85.7	<0.0001	ns	<0.001	ns	ns	3.83

ns = not significant; [1] Residual Standard Deviation; [2] Cholesterol content of the batter was 58.2 mg/100 g.

Different salt percentages affected the ash content of ostrich salami, being the highest in the 2.6% salt group (4.86 vs. 4.34 g/100 g, respectively; $p < 0.0001$), whereas the other variables remained unaffected.

In the present trial, the use of different starter cultures had a considerable effect on the proximate composition of ostrich salami: when inoculated with LAB 8 salami showed lower fat (33.1 vs. 35.3 g/100 g, respectively) and cholesterol (85.7 vs. 90.8 mg/100 g, respectively) contents compared to those inoculated with LAB 6. By contrast, the presence of *L. curvatus* and *S. xylosus* lowered protein (23.1 vs. 23.5 g/100 g, respectively) and moisture (33.8 vs. 35.7 g/100 g, respectively) contents compared to salami inoculated with *L. sakei* and *S. xylosus*.

A significant F × S interaction was observed for moisture ($p < 0.0001$), protein ($p < 0.05$), fat ($p < 0.05$) and ash ($p < 0.05$) contents of ostrich salami ripened for 10 weeks. When 2.4% salt was added, the moisture content of salami manufactured with 40% fat (Figure 5a) was higher compared to salami prepared with 2.6% salt (37.4 vs. 35.4%, for 2.4 and 2.6% Salt groups, respectively). However, when the fat content of salami was 30%, the situation was exactly the opposite with 2.4% Salt group showing a lower moisture content than the 2.6% Salt group (32.5 vs. 33.9%, for 2.4 and 2.6% Salt groups, respectively).

Protein content was the highest in 30% fat salami, with no differences between the two levels of Salt within the same Fat group (Figure 5b). Considering the 40% fat inclusion level, salami manufactured with 2.6% salt had a higher fat content compared to those having 2.4% salt; this difference, however, was not significant considering salami having 30% fat, as 2.4 and 2.6% salt levels showed similar fat values (Figure 5c). Ash content was higher in the low fat compared to high fat salami and, within each fat group, a higher presence of salt increased the ash content compared to salami manufactured with the lowest salt level (Figure 5d).

A significant F × L interaction was observed for moisture ($p < 0.0001$), fat ($p < 0.01$) and ash ($p < 0.01$) contents of ostrich salami ripened for 10 weeks (Figure 6a–c). Specifically, salami inoculated with LAB 6 starter culture showed a lower moisture content compared to those inoculated with LAB 8, both at 30 and 40% fat inclusion levels. Salami of the 40% Fat group inoculated with LAB 6 had the highest fat amount, with other combinations not differing among each other. Ash content was different and greater in high fat salami compared to low fat ones, with different starter cultures showing similar results within the same fat inclusion level ($p < 0.01$).

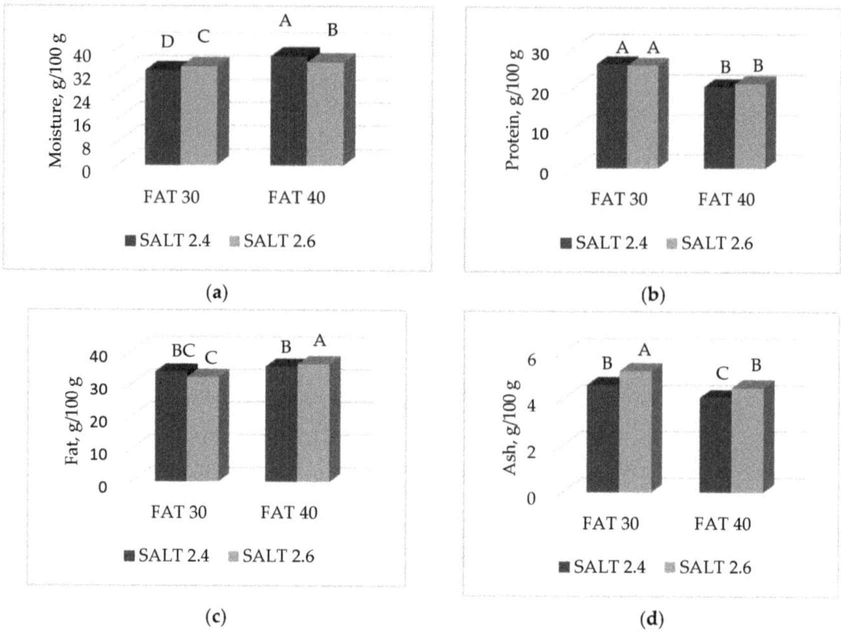

Figure 5. Effect of the interaction FAT × SALT on: (**a**) moisture ($p < 0.0001$); (**b**) protein ($p < 0.05$); (**c**) fat ($p < 0.05$); (**d**) ash ($p < 0.05$). Histograms reporting different A, B, C, D letters significantly differ.

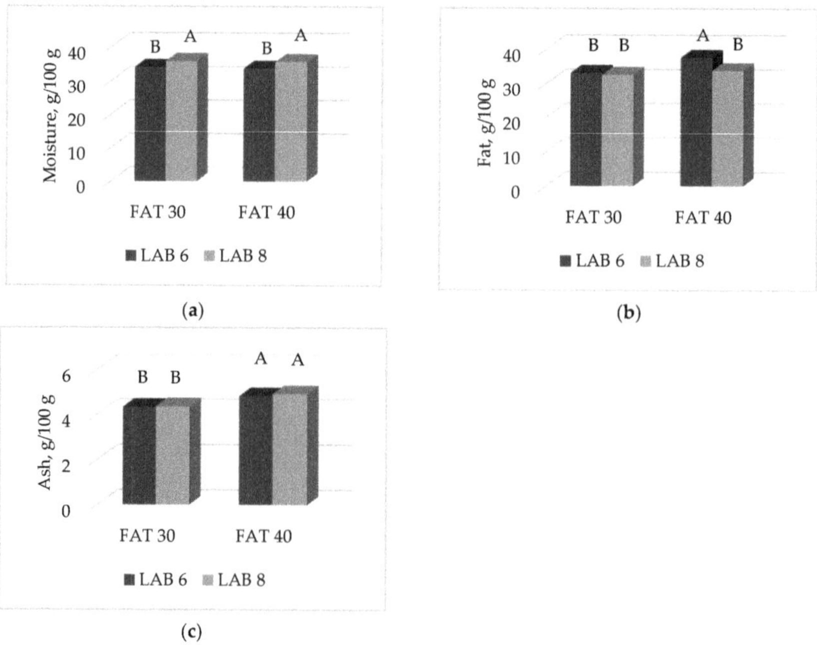

Figure 6. Effect of the interaction FAT × LAB on: (**a**) moisture ($p < 0.0001$); (**b**) fat ($p < 0.01$); (**c**) ash ($p < 0.01$). Histograms reporting different A, B letters significantly differ.

At 20 weeks of ripening, only fat content significantly affected the proximate composition of ostrich salami, whereas Salt and LAB starter cultures were ineffective (Table 4). Moisture content was higher in salami manufactured with 40% fat than those made with 30% fat ($p < 0.0001$); leaner salami had the highest protein (27.5 vs. 21.5 g/100 g for 30 and 40% fat inclusion levels, respectively), ash (5.54 vs. 4.79 g/100 g for 30 and 40% fat inclusion levels, respectively) and cholesterol (104.3 vs. 86 mg/100 g for 30 and 40% fat inclusion levels, respectively) contents. Conversely, the amount of fat was similar in salami manufactured either with 30 or 40% pork back-fat.

Table 4. Effect of two levels of fat, salt, and two LAB starters cultures and their interaction on proximate composition (g/100 g) and cholesterol content (mg/100 g) of ostrich salami ripened for 20 weeks.

	FAT (F)		SALT (S)		LAB (L)		Significance				RSD [1]
	30	40	2.4	2.6	6	8	F	S	L	F × L	
Salami	5	5	5	5	5	5					
Moisture	26.4	31.1	29.0	28.5	28.0	29.5	<0.0001	ns	ns	ns	2.61
Protein	27.5	21.5	24.3	24.7	24.2	24.8	<0.0001	ns	ns	ns	1.18
Fat	36.5	36.5	36.4	36.6	37.1	35.9	ns	ns	ns	ns	3.58
Ash	5.55	4.79	5.07	5.28	5.22	5.13	<0.0001	ns	ns	ns	0.42
Cholesterol	104.3	86.0	95.7	94.7	94.5	95.9	<0.0001	ns	ns	<0.05	6.54

ns = not significant; [1] Residual Standard Deviation.

The only interaction observed at 20 weeks of ripening was F × L on the cholesterol content (Figure 7): independently to the starter culture, salami manufactured with 30% fat had a higher cholesterol content than those with 40% fat. However, the histogram suggested a numerical difference between LAB 6 and LAB 8 within the same fat level: at 30% fat, LAB 6 had a lower cholesterol content than LAB 8 salami (101.0 vs. 107.6 mg/100 g for LAB 6 and LAB 8, respectively), whereas at 40% fat the opposite was observed (88.0 vs. 84.1 mg/100 g for LAB 6 and LAB 8, respectively).

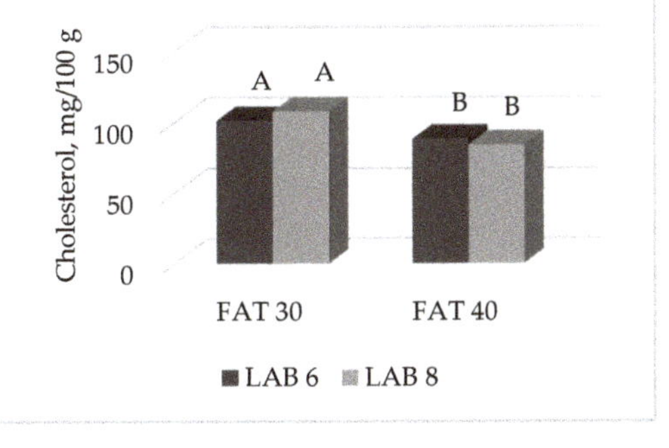

Figure 7. Effect of the interaction FAT × LAB on the cholesterol content (mg/100 g product) of ostrich salami ripened for 20 weeks. Histograms reporting different A, B letters significantly differ for $p < 0.05$.

4. Discussion

Lean meat contains more water than fat tissue, which explains why salami manufactured with a higher proportion of ostrich meat lost more weight compared to the fattest group of salami. Furthermore, fat has a crucial role for rheological and textural properties of meat products, especially when the product is manipulated thoroughly, as is the case of salami, as it binds water to form stable emulsions [28]. As a consequence, a lower fat content leads to a lower emulsion stability which causes a higher moisture loss [29]; the latter also contributes to explaining the findings of the present experiment. Similarly, Muguerza et al. [30] evidenced that weight losses of fermented sausages made of pork meat were significantly affected by fat level and that the lower the fat level the higher the weight loss of sausages. Moreover, at 28 days of ripening, the average weight loss of sausages manufactured with 30% pork back-fat was 38.5% which is much higher than the values observed in our study (28.1% of cumulative weight loss for salami belonging to the 30% fat inclusion level, at 4 weeks of ripening). As weight losses are known to depend on many processing factors such as temperature, relative humidity and air movement [31,32], this difference is not surprising as processing parameters differed in the two experiments.

The amount of water and its state in meat can change depending on numerous factors related to the tissue itself and to how the product is handled [33]. Salt (NaCl) is responsible for the solubilization of myofibrillar proteins in meat, which stimulates the proteins to increase hydration and water-binding capacity resulting in an improved texture, tenderness and juiciness [10]. Despite this, results of the present experiment highlighted that a higher NaCl content favored weight loss of ostrich salami. This finding could be explained by the fact that salt leads to the loss of free water too, which is also responsible for the decrease in water activity [34]. The different pattern in the observed salami weight loss as a result of the F × S interaction, indicated that, at 30% fat level, the 2.4% salt was enough to extract the free water from the salami, whereas at 40% fat only a higher salt concentration could reduce water availability as a higher stability of the emulsion fat-water was reached. Conversely, it is reasonable to expect that the ionic strength in the salami with 40% fat and 2.6% salt could reach values close to the threshold of the so called salting out (ionic strength >1 M), where water holding capacity starts to decrease thus favoring water loss [35].

The fact that different LAB starter cultures did not affect cumulative weight loss of ostrich salami was in line with the study by Kenneally et al. [36] on salami made of beef/pork meat inoculated with different microbial starters. Sometimes, in our study the interaction F × L on salami weight loss showed significant results (Table 2), but observing the histogram showing this interaction at 16 weeks of ripening (Figure 4) shows that different LAB starter cultures did not produce significant variations in terms of weight loss within the same fat inclusion level. For this reason, it was hypothesized that the significance of the interaction was mostly attributable to the different fat content rather than different LAB starter cultures. Trying to understand the reason for such interaction, authors have also considered the pH value of salami made with different starter cultures (Novelli et al., personal communication), a physical trait known to be strictly connected with water holding capacity. However, at 10 weeks of ripening salami belonging to LAB 6 and LAB 8 groups had an identical pH value (5.46), which confirmed the findings of this first part of the study: starter cultures were not a pivotal factor in affecting salami weight loss. Also, even if the pH value of the final product (20 weeks of ripening) slightly differed in salami manufactured with different starter cultures (5.47 for LAB 6 and 5.52 for LAB 8), this was not enough to determine diverse weight losses too.

Initial different fat contents changed the proportion of nutrients as well as the moisture content of ostrich salami, thus affecting the proximate composition of the product throughout the ripening process. Such findings could be explained by the fact that a lower fat content in ostrich salami determined a higher weight (water) loss during ripening, thus causing a higher concentration of nutrients compared to salami manufactured with 40% fat. This was the cause for the similar fat content observed at 20 weeks of ripening in the salami belonging to the two Fat groups. The clear different weight loss along the trial in salami belonging to 30% and 40% Fat groups, was also partially responsible for the highest cholesterol content observed in the 30% Fat salami. Moreover, as cholesterol is naturally present in the sarcolemma of muscle cells [37], a higher inclusion of lean meat from ostrich in our 30% Fat salami also explains the higher cholesterol content found in this group.

In the present study the salt content had a negligible effect on the proximate composition of ostrich salami and the different ash contents observed in the two groups was a direct consequence of the highest presence of salt. This finding was supported by results presented in a study evaluating the effect of NaCl partial substitution on proximate composition of Italian salami [11], where salami manufactured with 2.7% NaCl had the same proximate composition of those prepared with low NaCl formulations. Despite salt content of salami can affect the growth of starter cultures [38], results of the present experiment seemed to indicate that this was not a key factor neither for weight loss nor for the nutritional composition of ostrich salami as no interactions were observed between salt level and type of starter culture used.

Tissue lipase are primarily responsible for lipolysis during the fermentation process, but it has now been accepted that staphylococci can act as lipolytic bacteria, thus playing an important role in aroma formation [39]. Similarly, in the first stages of ripening, proteolysis in meat is due to endogenous enzymes such as calpains and cathepsins which break sarcoplasmic and myofibrillar proteins, while it is in the last stages of ripening that microbial enzymes play a predominant role in the secondary hydrolysis of oligopeptides and small peptides [40]. However, an appropriate choice of a combination of strains in the formulation of a starter culture is fundamental for successful fermentation and ripening processes, as different strains and microbial species are known to act differently according to different meat types, technological characteristics of the fermentation and ripening parameters [8,38,41,42]. In the current research, the presence of *L. curvatus* and *S. xylosus* (LAB 6) led to salami characterized by lower protein and moisture contents compared to those inoculated with *L. sakei* and *S. xylosus* (LAB 8). This could be the consequence of a different intensity in the degradation of the meat proteins, possibly as a combination of endogenous and microbial enzymes; in fact, the water activity of the two groups of salami, also analyzed at 10 weeks of ripening, differed, as well as their content of nonprotein nitrogen (Novelli et al., personal communication). This hypothesis found confirmation in the results of a study on the proteolytic activity of LAB [43], showing that different enzyme combinations from *L. sakei* and *L. curvatus* provided different results: the first exhibited exopeptidase activity and the second modifying the peptide profile. In a study considering the effect of the starter culture on proteolytic changes during processing of fermented sausages [44], it was observed that sausages inoculated with *L. sakei* or *L. carnosus* differed in terms of some individual amino acids. However, different from the processing stage, which was the main factor affecting proteolytic process, the type of starter culture was not a factor affecting total free amino acids concentration. In the present study, as a result of their metabolic activity, different LAB starter cultures generated salami with different nutritional composition until 10 weeks of ripening, whereas at 20 weeks the concentration of salt was probably too high for their survival [45], thus their activity might have been reduced or completely hampered. In fact, it was demonstrated that growth and survival of staphylococci, but especially of LAB, is markedly affected by NaCl concentration [6].

From results of the present study it emerged that ostrich salami manufactured with 30% pork back-fat and inoculated with different starter cultures showed the same fat content, whereas at 40% fat inclusion level, the combination *Lactobacillus curvatus*/*Staphylococcus xylosus* provided salami with a higher fat content compared to those containing *Lactobacillus sakei*/*Staphylococcus xylosus*. This finding suggested that the significance of the main effect of the starter culture on the fat content of ostrich salami was particularly evident when the fat level of the salami was 40%, whereas at 30% inclusion percentage only a tendency was observed. As no specific studies evaluating the effect of different fat contents on the growth of starter cultures for the production of fermented sausages have been conducted until now, further research on this topic is required; this is because the choice of the appropriate starter mix for each specific fermented meat product is of fundamental importance to ensure a satisfactory product quality [46].

5. Conclusions

Fat inclusion level had a great impact on the weight loss and nutritional composition of Italian-style ostrich salami, independent of the ripening phase. A lower fat content consistently shortened ripening time, thus being a positive aspect in terms of productivity, and it determined a higher nutrients concentration compared to high fat salami, with the only drawback of a higher cholesterol content compared to high fat salami. Reducing the NaCl inclusion from 2.6 to 2.4% retarded the weight loss of ostrich salami by approximately 1 week, without affecting the proximate composition and cholesterol content of the final product. At 10 weeks of ripening, *L. sakei* provided salami with a healthier nutritional composition compared to salami inoculated with *L. curvatus*. The metabolic activity of tested LAB starter cultures seemed to be affected by the fat inclusion level, even if further investigations to clarify this point is necessary. Understanding the latter could help to ensure a high quality product.

Author Contributions: Conceptualization, A.D.Z., E.N.; methodology, A.D.Z. and E.N.; formal analysis, M.C., E.N.; investigation, A.D.Z., E.N., and M.C.; data curation, A.D.Z. and M.C.; writing—original draft preparation, M.C. and A.D.Z.; writing—review and editing, M.C., A.D.Z. and E.N.; supervision and project administration, A.D.Z.; funding acquisition, A.D.Z. All authors have read and agreed to the published version of the manuscript.

Funding: Research funded by Padova University research funds (Progetto di Ricerca di Ateneo 2011) code: CPDA117509/11, and the GOP-1.3.1-11/B-2011-0045.

Acknowledgments: Authors thank Barbara Contiero for her technical support with the statistical analysis. Special thanks to Terenzio Boscolo for providing the ostrich meat and all the necessary equipment for salami preparation and ripening.

Conflicts of Interest: The authors declare no conflict of interest.

References

1. Comi, G.; Urso, R.; Iacumin, L.; Rantsiou, K.; Cataneo, P.; Cantoni, C.; Cocolin, L. Characterisation of naturally fermented sausages produced in the North east of Italy. *Meat Sci.* **2005**, *69*, 381–392. [CrossRef] [PubMed]
2. Rantsiou, K.; Drosino, E.H.; Gialitaki, M.; Urso, R.; Krommer, J.; Gasparik-Reichardt, J.; Tóth, S.; Metaxopoulos, I.; Comi, G.; Cocolin, L. Molecular characterization of *Lactobacillus* species isolated from naturally fermented sausages produced in Greece, Hungary and Italy. *Food Microbiol.* **2005**, *22*, 19–28. [CrossRef]
3. Talon, R.; Leroy, S.; Lebert, I. Microbial ecosystems of traditional fermented meat products: The importance of indigenous starters. *Meat Sci.* **2007**, *77*, 55–62. [CrossRef]
4. Reckem, E.V.; Geeraerts, W.; Charmpi, C.; Van der Veken, D.; De Vuyst, L.; Leroy, F. Exploring the link between the geographical origin of European fermented foods and the diversity of their bacterial communities: The case of fermented meats. *Front. Microbiol.* **2019**, *10*, 2302. [CrossRef] [PubMed]
5. Dicks, L.M.T.; Mellet, F.D.; Hoffman, L.C. Use of bacteriocin-producing starter cultures of *Lactobacillus plantarum* and *Lactobacillus curvatus* in production of ostrich salami. *Meat Sci.* **2004**, *66*, 703–708. [CrossRef]
6. Olesen, P.T.; Meyer, A.S.; Stahnke, L.H. Generation of flavor compounds in fermented sausages-the influence of curing ingredients, *Staphylococcus* starter culture and ripening time. *Meat Sci.* **2004**, *66*, 675–687. [CrossRef]
7. Leroy, F.; Verluyten, J.; De Vuyst, L. Functional meat starter cultures for improved sausage fermentation. *Int. J. Food Microbiol.* **2006**, *106*, 270–285. [CrossRef] [PubMed]
8. Todorov, S.D.; Koep, K.S.C.; Van Reenen, C.A.; Hoffman, L.C.; Slinde, E.; Dicks, L.M.T. Production of salami from beef, horse, mutton, Blesbok (*Damaliscus dorcas phillipsi*) and Springbok (*Antidorcas marsupialis*) with bacteriocinogenic strains of *Lactobacillus* plantarum and *Lactobacillus curvatus*. *Meat Sci.* **2007**, *77*, 405–412. [CrossRef]
9. Hoppu, U.; Hopia, A.; Ponhjanheimo, T.; Rotola-Pukkila, M.; Mäkinen, S.; Pihlanto, A.; Sandell, M. Efefct of salt reduction on consumer acceptance and senssory quality of food. *Foods* **2017**, *6*, 103. [CrossRef]
10. Ruusunen, M.; Puolanne, E. Reducing sodium intake from meat products. *Meat Sci.* **2005**, *70*, 531–541. [CrossRef]
11. Zanardi, E.; Ghidini, S.; Conter, M.; Ianieri, A. Mineral composition of Italian salami and effect of NaCl partial replacement on compositional, physico-chemical and sensory parameters. *Meat Sci.* **2010**, *86*, 742–747. [CrossRef]
12. World Health Organization (WHO). Healthy Diet. Available online: https://www.who.int/news-room/fact-sheets/detail/healthy-diet (accessed on 23 October 2018).
13. Hooper, L.; Abdelhamid, A.; Bunn, D.; Brown, T.; Summerbell, C.D.; Skeaff, C.M. Effects of total fat intake on body weight. *Cochrane Database Syst. Rev.* **2015**, *8*, CD011834. [CrossRef]
14. World Health Organization (WHO). *Regional Office for Europe Using Dietary Intake Modelling to Achieve Population Salt Reduction: A Guide to Developing a Country-Specific Salt Reduction Model*; World Health Organization (WHO): Geneva, Switzerland, 2018.

15. Flores, M.; Olivares, A.; Corral, S. Healthy trends affect the quality of traditional meat products in Mediterranean area. *Acta Agr. Slov.* **2013**, *4*, 183–188.
16. Leroy, F.; Geyzen, A.; Janssens, M.; De Vuyst, L.; Scholliers, P. Meat fermentation at the crossroads of innovation and tradition: A historical outlook. *Trends Food Sci. Tech.* **2013**, *31*, 130–137. [CrossRef]
17. Olmedilla-Alonso, B.; Jiménez-Colmenero, F.; Sánchez-Muniz, F. Development and assessment of healthy properties of meat and meat products designed as functional foods. *Meat Sci.* **2013**, *95*, 919–930. [CrossRef]
18. Cullere, M.; Hoffman, L.C.; Dalle Zotte, A. First evaluation of unfermented and fermented rooibos (*Aspalathus linearis*) in preventing lipid oxidation in meat products. *Meat Sci.* **2013**, *95*, 72–77. [CrossRef]
19. COUNCIL REGULATION (EC). No 1099/2009 of 24 September 2009 on the protection of animals at the time of killing. *Off. J. Eur. Union* **2009**, *303*, 30.
20. AOAC. Official methods of analysis of AOAC International. In *Association of Official Analytical Chemists*, 15th ed.; AOAC: Washington, DC, USA, 1995.
21. Casiraghi, E.; Lucisano, M.; Pompei, C.; Dellea, C. Cholesterol determination in butter by high performance chromatography. *Milchwissenschaft* **1994**, *49*, 194–196.
22. AOAC. Official methods of analysis of AOAC International. In *Association of Official Analytical Chemists*, 17th ed.; AOAC: Gaithersburg, MD, USA, 2000.
23. EC 1998. Commission Directive 98/64/EC of 3 September 1998 establishing Community methods of analysis for the determination of amino acids, crude oils and fats, and olaquindox in feeding stuffs and amending Directive 71/393/EEC. *Off. J. Eur. Union L257* **1998**, *257*, 0014–0028.
24. ISO-International Organization for Standardization 1998. *Animal Feeding Stuffs, Animal Products and Faeces or Urine. Determination of Gross Calorific Value—Bomb Calorimetric Method*; Reference number 9831, prepared by Technical Committee ISO/TC 34, Agricultural food products, Subcommittee SC 10, Animal feeding stuffs; ISO: Geneva, Switzerland, 1998.
25. Mertens, D.R. Gravimetric determination of amylase-treated neutral detergent fibre in feeds with refluxing beakers or crucibles: Collaborative study. *J. AOAC Int.* **2002**, *85*, 1217–1240. [PubMed]
26. Van Soest, P.J.; Robertson, J.B.; Lewis, B.A. Methods for dietary fiber, neutral detergent fiber, and nonstarch polysaccharides in relation to animal nutrition. *J. Dairy Sci.* **1991**, *74*, 3583–3597. [CrossRef]
27. SAS. *Statistical Analysis Software for Windows*; Version 9.2; SAS Institute Inc.: Cary, NC, USA, 2008.
28. Claus, J.R.; Hunt, M.C.; Kastner, C.L.; Kropf, D.H. Low-fat, high-added water Bologna: Effects of massaging, preblending, and time of addition of water and fat on physical and sensory characteristics. *J. Food Sci.* **1990**, *55*, 338–341. [CrossRef]
29. Crehan, C.M.; Hughes, E.; Troy, D.J.; Buckley, D.J. Effects of fat level and maltodextrin on the functional properties of frankfurters formulated with 5, 12 and 30% fat. *Meat Sci.* **2000**, *55*, 463–469. [CrossRef]
30. Muguerza, E.; Fista, G.; Ansorena, D.; Astiasaran, I.; Bloukas, J.G. Effect of fat level and partial replacement of pork backfat with olive oil on processing and quality characteristics of fermented sausages. *Meat Sci.* **2002**, *61*, 397–404. [CrossRef]
31. Baldini, P.; Cantoni, E.M.; Colla, F.; Diaferia, C.; Gabba, L.; Spotti, E.; Marchelli, R.; Dossena, A.; Virgili, E.; Sforza, S.; et al. Dry sausages ripening: Influence of thermohygrometric conditions on microbiological, chemical and physico-chemical characteristics. *Food Res. Int.* **2000**, *33*, 161–170. [CrossRef]
32. Olivares, A.; Navarro, J.L.; Salvador, A.; Flores, M. Sensory acceptability of slow fermented sausages based on fat content and ripening time. *Meat Sci.* **2010**, *86*, 251–257. [CrossRef]
33. Huff-Lonergan, E.; Lonergan, S.M. Mechanisms of water-holding capacity of meat: The role of postmortem biochemical and structural changes. *Meat Sci.* **2005**, *71*, 194–204. [CrossRef]
34. Guàrdia, M.D.; Guerrero, L.; Gelabert, J.; Gou, P.; Arnau, J. Consumer attitude towards sodium reduction in meat products and acceptability of fermented sausage with reduced sodium content. *Meat Sci.* **2006**, *73*, 484–490. [CrossRef]
35. Warner, R.D. The Eating Quality of Meat—IV Water-Holding Capacity and Juiciness. In *Lawrie's Meat Science*, 8th ed.; Toldrá, F., Ed.; Series in Food Science, Technology and Nutrition; Woodhead Publishing: Sawston, Cambridge, UK, 2017; pp. 419–453.
36. Kenneally, P.M.; Schwarz, G.; Fransen, N.G.; Arendt, E.K. Lipolytic starter culture effects on production of free fatty acids in fermented sausages. *J. Food Sci.* **1998**, *63*, 538–543. [CrossRef]
37. Chizzolini, R.; Zanardi, E.; Dorigoni, V.; Ghidini, S. Calorific value and cholesterol content of normal and low-fat meat and meat products. *Trends Food Sci. Tech.* **1999**, *10*, 119–128. [CrossRef]

38. Charmpi, C.; Van der Veken, D.; Van Reckem, E.; De Vuyst, L.; Leroy, F. Raw meat quality and salt levels affect the bacterial species diversity and community dynamics during the fermentation of pork mince. *Food Microbiol.* **2020**, *89*, 103434. [CrossRef] [PubMed]
39. Bedia, M.; Méndez, L.; Bañón, S. Evaluation of different starter cultures (Staphylococci plus Lactic Acid Bacteria) in semi-ripened Salami stuffed in swine gut. *Meat Sci.* **2011**, *87*, 381–386. [CrossRef] [PubMed]
40. Casaburi, A.; Di Monaco, R.; Cavella, S.; Toldrá, F.; Ercolini, D.; Villani, F. Proteolytic and lipolytic starter cultures and their effect on traditional fermented sausages ripening and sensory traits. *Food Microbiol.* **2008**, *25*, 335–347. [CrossRef] [PubMed]
41. Baka, A.M.; Papavergou, E.J.; Pragalaki, T.; Bloukas, J.G.; Kotzekidou, P. Effect of selected autochthonous starter cultures on processing and quality characteristics of Greek fermented sausages. *LWT-Food Sci. Technol.* **2011**, *44*, 54–61. [CrossRef]
42. Casquete, R.; Benito, M.J.; Martín, A.; Ruiz-Moyano, S.; Hernández, A.; Córdoba, M.G. Effect of autochthonous starter cultures in the production of "salchichón", a traditional Iberian dry-fermented sausage, with different ripening processes. *LWT-Food Sci. Technol.* **2011**, *44*, 1562–1571. [CrossRef]
43. Fadda, S.; Sanz, Y.; Vignolo, G.; Aristoy, M.C.; Oliver, G.; Toldrá, F. Hydrolysis of pork muscle sarcoplasmic proteins by *Lactobacillus curvatus* and *Lactobacillus sake*. *Appl. Environ. Microb.* **1999**, *65*, 578–584. [CrossRef]
44. Candogan, K.; Wardlaw, F.B.; Acton, J.C. Effect of starter culture on proteolytic changes during processing of fermented beef sausages. *Food Chem.* **2009**, *116*, 731–739. [CrossRef]
45. Ordóñez, J.A.; Hierro, E.M.; Bruna, J.M.; de la Hoz, L. Changes in the components of dry-fermented sausages during ripening. *Crit. Rev. Food Sci.* **1999**, *39*, 329–367. [CrossRef]
46. Leroy, F.; De Vuyst, L. Lactic acid bacteria as functional starter cultures for the food fermentation industry. *Trends Food Sci. Tech.* **2004**, *15*, 67–78. [CrossRef]

foods

MDPI

Article

Mango Peel Pectin by Microwave-Assisted Extraction and Its Use as Fat Replacement in Dried Chinese Sausage

Malaiporn Wongkaew [1,2,3], Sarana Rose Sommano [3,4,5,*], Tibet Tangpao [3], Pornchai Rachtanapun [5,6] and Kittisak Jantanasakulwong [5,6]

1. Major of Biotechnology, Graduate School, Chiang Mai University, Chiang Mai 50200, Thailand; malaiporn@rmutl.ac.th
2. Programme of Food Production and Innovation, Faculty of Integrated of Science and Technology, Rajamangala University of Technology Lanna, Chiang Mai 50300, Thailand
3. Plant Bioactive Compound Laboratory, Faculty of Agriculture, Chiang Mai University, Chiang Mai 50200, Thailand; gr.zhk.88@gmail.com
4. Department of Plant and Soil Sciences, Faculty of Agriculture, Chiang Mai University, Chiang Mai 50200, Thailand
5. Cluster of Agro Bio-Circular-Green Industry (Agro BCG), Chiang Mai University, Chiang Mai 50200, Thailand; pornchai.r@cmu.ac.th (P.R.); antanasakulwong.k@gmail.com (K.J.)
6. School of Agro-Industry, Faculty of Agro-Industry, Chiang Mai University, Chiang Mai 50200, Thailand

* Correspondence: sarana.s@cmu.ac.th

Received: 10 March 2020; Accepted: 3 April 2020; Published: 7 April 2020

check for updates

Abstract: In this research, low-fat dried Chinese sausage was formulated with mango peel pectin (MPP; 0%, 5%, 10%, and 15% (*w/w*)) extracted by microwave assisted extraction (MAE). The extractable yield of pectin attained from peel of Nam Dok Mai variety was achieved at 13.85% using 700-watt power. The extracted MPP were of high equivalent weight (1485.78 mg/mol), degree esterification (77.19%) and methoxyl content (19.33%) with a structure of greater porosity as compared to that of the conventional method. Spectrum scans by Fourier transform infrared spectrophotometer (FT-IR) indicated that the extracted MPP gave similar wave number profiles as the commercial pectin. Quality attributes of the Chinese sausages were assessed and compared with the control formula (CTRL). At higher concentrations of MPP, the intensity of redness and yellowness in sausage increased. The texture profile of the sausage illustrated that only the hardness value was comparable with the CTRL, while springiness, cohesiveness, gumminess and chewiness were statistically lower ($p < 0.05$). Furthermore, the sensory evaluation by experienced panellists ($n = 12$) indicated that 5% MPP similarly represented overall acceptability with the CTRL. Consequently, MPP can be effectively incorporated in the formula at low level to replace fat in Chinese sausage, allowing colour improvement and production of a healthier option.

Keywords: dried Chinese sausage; fat replacement; mango peel pectin; microwave-assisted extraction technique

1. Introduction

Fear of noncommunicable diseases (NCDs) has influenced the awareness of naturally functional ingredients in the human diet [1]. Additionally, this trend has motivated concerns of animal fat consumption, which encourages the novel formulation of products with reduced fat content [2]. Processed meats are usually products of high fat content, providing that fat could significantly improve texture, flavour, mouthfeel and perceived juiciness [3,4]. However, adding fat replacer into the product could adversely affect texture and their original sensory properties [5–7]. Moreover, excessive decrease in fat content can considerably alter the structural characteristics of meat products [8]. Chinese or Cantonese-style sausage, also called Kunchiang in Thai, is one of the preserved meat products of East Asian culinary heritage. The main ingredients are meats (pork or chicken) mixed with a high content of pork fat [9]. Attempts have been made to partly decrease or absolutely remove fat from Chinese sausage [10,11]. One option is by integrating functional ingredients, such as rice starch, gum and pectin, to replace the sum amount of lipid ingredient [12].

Dietary fibre is a carbohydrate polymer with more than 10 monomeric units, making it difficult to be hydrolysed by endogenous enzymes in the human small intestine [13,14]. The fibre can be classified into two groups videlicet, insoluble (cellulose and hemicellulose) and soluble (pectin, galactomannan, inulin and gum), depending on its solubility in aqueous solution [15]. Additionally, pectin is generally required as a food ingredient for the functional food industry [16]. Extractable pectin is utilised as a food additive that is promoted in the processes of gelling, stabilising and thickening [17]. Méndez-Zamora et al. [18] claimed that fat could be replaced with pectin and inulin in frankfurter sausages to produce healthy and functional products. The supplementation could also maintain the physical properties of meat product [19,20].

Mango peel is a potential source of dietary fibre with 5%–11% pectin depending on the extraction methods and also of fruit varieties [21–23]. Moreover, it comprises considerable various classes of polyphenols, carotenoids and vitamins with excellent antioxidative and functional properties [24,25], thus making this byproduct a promising target for commercial valorisation [26,27].

To recover pectin from plant resources, microwave-assisted extraction (MAE) is more effective for the extraction of high-quality pectin, compared with conventional heating techniques [28–31]. Such a technique has been adopted with pectin-rich biomasses such as banana peels [32], mango peels [22,33,34], pumpkin [35], and orange peels [36]. For Thai 'Sampee' mango variety, Sommano et al. [22,34] reported the improved recovering yield of mango peel pectin (MPP) by moderate microwave radiation and the process could also preserve bound phenolic content and antioxidant scavenging activities. Chaiwarit et al. [37] reported that MPP from 'Nam Dok Mai' variety could be a potential biopolymer for film formulation in drug delivery systems or edible film for food packaging. There is however, no research conducted on the functionality of MPP as a food additive in particular as fat replacer. With this rationale, the objectives of the present study were first to quantify the effect of MAE on functional properties of MPP of Nam Dok Mai variety and later to evaluate its potential to reduce fat content in Chinese sausage.

2. Materials and Methods

2.1. Preparation of Mango Peel Powder

Peel was removed from fully ripe mangoes Nam Dok Mai variety ($L^* = 50.90 \pm 4.34$, $a^* = 4.82 \pm 2.35$, $b^* = 16.59 \pm 3.09$; peel thickness = 138.76 ± 10.55 mm and percentage of peel to fruit weight = 5.31 ± 0.38%). The peels were cut into small pieces, washed with tap water, blanched with hot water at 95 °C for 10 min, drained and left cooled at room temperature, prior to drying at 60 ± 1 °C until the moisture content of 4%–6% was reached [38]. The dried peel was ground to a fine powder in a high-speed food processor and passed through a sieve, resulting in a final mass of particles smaller than 0.6 mm in diameter [39,40].

2.2. Extraction of Mango Peel Pectin Using Microwave-Assisted Technique

Twenty grams of mango peel powder was suspended in 600 mL of diluted acidic solution (distilled H_2O adjusted to pH 1.5 with 2 M HCl) and soaked for 20 min at room temperature. The slurry was heated in a microwave oven (ME711K-XST, Thai Samsung electronics Co., Ltd., Bangkok, Thailand) with an output power of optimal condition (700 watts for 3 min) followed by recooling to room temperature [22]. The solution was filtered and pressed manually using a nylon cloth. The filtrates were centrifuged at 5000× *g* for 20 min to eliminate any remaining coarse particles. Pectin was precipitated from this clear supernatant by adding the same volumes of ethanol (95%); mixed and stored in a refrigerator at 4 °C for 30 min. The separation was achieved by vacuum filtration. The obtained pectin was dried in a hot air-oven at 40 °C until constant weight [41]. The yield (%) of pectin was calculated from the following equation [40];

$$\text{Yield}\ (\%) = \left(\frac{M_0}{M}\right) \times 100 \quad (1)$$

where M_0 (g) = the weight of dried pectin and M (g) = the weight of dried mango peel powder.

2.3. Scanning Electron Microscope

Pectin powder was attached onto a specimen stub with a double-sided tape and sputter coated with gold [22,42]. The images were viewed at magnifications of × 100 and × 500 using SEM (JEOL JSM-5910, Japan Electron Optics Laboratory Co., Ltd., Tokyo, Japan) with an accelerating voltage of 10 kV.

2.4. Fourier Transform Infrared Spectrophotometer (FT-IR)

FT-IR analysis was implemented using an infrared spectrometer (Nicolet, USA) equipped with an MCT Detector (mercury cadmium telluride). Each sample was scanned by placing the sample side down on the ATR diamond crystal and applying the pressure tower. The spectrum was verified in the transparent mode from 900 to 4000 cm^{-1}, with a resolution of 4.0 cm^{-1} [22]. Each IR spectrum was improved for optical effects with the ATR correction algorithm (OMNIC software).

2.5. Mango Peel Pectin Characterisations

The equivalent weight (Eq.W) was determined by the method of Ranganna [43]. Briefly, 0.5 g of dried pectin was dissolved in 100 mL of distilled water at 25 °C and stirred for 2 h until completely dissolved. One gram of sodium chloride was added and titrated with 0.1 M of sodium hydroxide (NaOH) using 5 drops of phenol red as an indicator. Eq.W was calculated using the following equation:

$$\text{Eq.W} = \frac{1000 \times \text{pectin powder (g)}}{\text{NaOH concentration (N)} \times \text{NaOH volume (mL)}} \quad (2)$$

Methoxyl content (Mox) and degree of esterification (DE), the methods suggested in Ranganna [44] and Pinheiro et al. [45], were followed. Dried pectin (0.2 g) was stirred in CO_2-free distilled water (20 mL) until fully dissolved. One gram of NaCl was added to the solution, prior to titrating with 0.1 N NaOH in the presence of phenolphthalein. The volume was recorded as the initial titre (V_1). Then, 0.1 N NaOH solution (10 mL) was added to a neutralised polygalacturonic acid sample after the determination of the free carboxyl groups. The solution was mixed thoroughly until the colour of the solution became purple. A few drops of the indicator (0.25 N HCl) were added, and the mixture was titrated with 0.1 N NaOH until the colour turned from yellow to pink. The volume was noted as V_2. The Mox and DE were then calculated using the following equations;

$$\text{Mox} = \frac{(N)(V_2)(E)}{1000\ (S)} \quad (3)$$

$$\text{DE} = \frac{V_2 \times 100}{V_1 + V_2} \quad (4)$$

where S = mass of dried pectin (g); N = NaOH concentration (N); V1 = volume of NaOH used (mL); V2 = volume of NaOH used (mL) and E = equivalent weight of methoxyl = 31

The water holding capacity (WHC), oil holding capacity (OHC) and swelling capacity (SWC) were evaluated following the method of Robertson et al. [46] with some modification. Phosphate buffer

(1 M, pH 6.3, 25 mL) or commercial olive oil were added to 250 mg of dry sample, stirred thoroughly and left at room temperature for 1 h. The residue was weighed after centrifugation at 3000× *g* for 5 min. For SWC analysis, 0.1 g of sample was hydrated in 10 mL of distilled water in a calibrated cylinder (15 cm diameter) at room temperature. After equilibration for 18 h, the bed volume was documented. The WHC was expressed as amount of water (g) held per sample (g); the OHC was expressed as amount of oil (g) held per sample (g), while the SWC was expressed as mL/g of sample.

Colour analysis of MPP powder was measured according the method of Bolumar et al. [47] with some modification. The colour was assessed using a handheld colour spectrophotometer (NS800, 3nh, China). Before each set of measurements, the instrument was calibrated using a white ceramic tile. The measurement was with CIE Lab system; where L* denotes lightness on a 0 to 100 scale from black to white; a* denotes (+) red or (−) green and b* denotes (+) yellow or (−) blue.

2.6. Processing and Evaluation of Dried Chinese Sausage with Added Mango Peel Pectin

2.6.1. Dried Chinese Sausage Formulation

Chinese sausage ingredients (CTRL) obtained from Chiang Mai Livestock Product Research and Development Centre consisted of pork, fat, sugar, sodium nitrite, sodium erythorbate and water at (%*w/w*) 60.0, 20.0, 12.0, 1.2, 0.1 and 6.7, respectively. Pork and fat were ground and then mixed with all ingredients. Pectin powder was added at a level of 5%, 10% and 15% (*w/w*) fat replacement. It was firstly dissolved with 2 g of clean water and then mixed with the prepared ingredients for 10 min with cutter mixer (QS600, Baicheng, China). After that, the ingredients were added in dried pork sausage casing and dried in a hot air-oven at 60 ± 5 °C for 48 h. The sausages were left to cool at room temperature, packed in vacuum nylon bag and stored at 4 ± 1 °C.

2.6.2. Colour Evaluation

Chinese sausages added with 0%, 5%, 10% and 15% (*w/w*) of MPP were sliced into 10-mm thickness. The colour measurement was repeated 10 times at different parts of the sausage surface using a handheld colour spectrophotometer (NS800, 3nh, China). Before each set of measurements, the instrument was calibrated using a white ceramic tile. The measurement was with CIE Lab system; where L* denotes lightness on a 0 to 100 scale from black to white; a* denotes (+) red or (−) green; and b* denotes (+) yellow or (−) blue. To compare the overall colour changes between the MPP-supplemented Chinese sausage samples and the CTRL, the total colour differences (ΔE) between the samples (L*, a*, b*) and the CTRL (L_0*, a_0*, b_0*) were calculated as presented below [47,48];

$$\Delta E_{ab} = \sqrt{(L* - L_0*)^2 + (a* - a_0*)^2 + (b* - b_0*)^2} \quad (5)$$

2.6.3. Texture Profile Analysis

The sausages sliced for colour measurement were also used for Texture Profile Analysis (TPA) using a TA-TX2 texture analyser (Stable Micro Systems Ltd., Godalming, UK), attached with a 50-kg load cell. A 50-mm diameter compression cylindrical aluminium probe was used to compress a cylindrical shape of the sausage, which was compressed twice to 30% of the original height of the sausage at a compression rate of 1.0 mm/s at room temperature. The TPA settings were as follows: pretest speed: 2.0 mm/s; test speed: 1.0 mm/s; post-test speed: 2.0 mm/s; target mode distance: 3.0 mm; trigger force: 5 g; trigger type: auto and data acquisition rate: 200 points per sec. (pps). The delay between the first and second compression was 5 sec. The TPA analysis was carried out at ambient temperature (25 °C), and the analysis was completely operated within 17 sec. Six measurements were assessed for each sample in the same lot. A force-time graph was generated and textural parameters, including hardness, cohesiveness, springiness, gumminess and chewiness, were calculated with software provided along with the instrument [49].

2.6.4. Sensory Test

Sensory evaluation of the Chinese sausage products was operated following the modified procedures by Siddaiah et al. [50] using a panel of 12 individuals from Chiang Mai Livestock Product research and Development centre, who had been trained for the sensory assessment of Chinese sausage and with a minimum of 5-years-experience in meat processing. All the panels were assured that they understood the definitions of appearance, juiciness, springiness, firmness, colour and overall acceptability before the panel tested the Chinese sausage. Preparation of the meat products for testing occurred in a kitchen separated from the evaluation room, eliminating possible interference of fried odour. According to routine sensory evaluation of Chiang Mai Livestock Product and Development Centre, the sausage samples were sliced into 7-mm thickness and then fried with palm oil for 3 min in a low heat. Each panellist was given two pieces of each sample for evaluation on nine-point hedonic scale (1 = strongly dislike and 9 = strongly like).

2.7. Statistical Analysis

All experiments were operated with at least triplicate samples for each test. Data was analysed using one-way analysis of variance and Duncan's test. Differences in values were considered significant when the p value was <0.05. All statistical analysis was performed using SPSS program (version 22.0, Armonk, NY, USA).

3. Results and Discussion

3.1. Scanning Electron Microscope

SEM was performed to characterise the surface of commercial citrus pectin (Figure 1a) and our MPP (Figure 1b) samples by visualising their structures and morphology. The images demonstrate that the pectin particles are of distinct shapes, Nam Dok Mai MPP illustrates pellets to bulky and rough particles, which differs greatly from the shape of the commercial pectin, which has a comparatively smooth surface. Nevertheless, the MPP particles extracted using MAE 700 watts are crumblier in shape and with more porous surfaces. Begum et al. [51] reported that the dehydrated pectin obtained from jackfruit freeze-dried and spray-dried had high solubility due to their high porosities, smaller particle size and higher surface area. Thus, pectin particles with more porous structures usually have a better solubility than particles with the rigid structure and lower porosity, thereby increasing solution viscosity [52]. The porous quantity of pectin was correlated with water holding capacity and led to the low hardness property of low-fat frankfurter sausage [18]. The dietary ingredients influence the high binding ability and water holding capacity of meat product [6]. According to dielectric mode of action, microwave is in fact more efficient than other extraction methods due to the strong formation of vapour in polar substances created by the electromagnetic field [53]. Heat vapour modifies the cell wall matrix and leads to the severing of parenchymal cells, which rapidly and extensively opens the skin tissues, thus increasing the interaction between the extracting agent and the plant material during the extraction process [31]. Furthermore, the images of the MPP (Figure 1b) suggest a rough, ruptured and wrinkled surface, which could be due to the sudden increase of temperature in the MAE process. Similarly, Liew et al. [54] reported that the coarse surface of the extracted pectin using MAE could be due to the rapid rise in temperature. Sources of raw materials as well as modes of extraction could largely influence morphology of the resulted pectin [28]. Regarding commercial citrus pectin morphology, the surface showed multilaminate structures and was fluffy with a smooth surface [55], which was considerably different from the MPP surface. From their high porosity, MPP is appropriate for fat replacer in Chinese sausage.

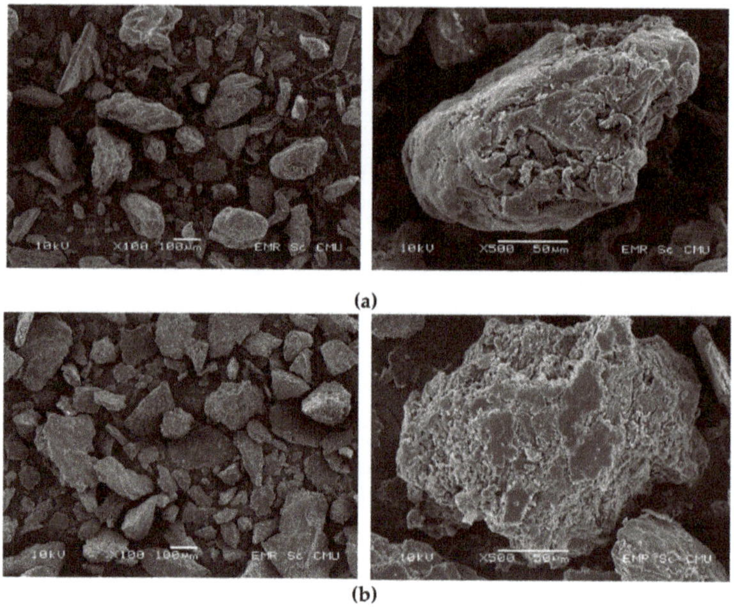

Figure 1. The SEM images of commercial pectin (citrus) (**a**) and pectin obtained using microwave-assisted extraction (MAE) from peel of Nam Dok Mai mango at 700 watts (**b**). The images were viewed at ×100 (left) and ×500 (right).

3.2. FT-IR Analysis of MPP

The FT-IR analysis is generally used to evaluate the conformation of pectin bands in the standard region usually between 1000 and 2000 cm^{-1} for the major chemical and functional groups [56]. Figure 2 illustrates the FT-IR region ranging from 900 to 4000 cm^{-1} of MPP. These demonstrate the similarities of the transmittance (%T) patterns in pectin extracted from different source materials. An individual peak at around 3400 cm^{-1} is likely due to the stretching of the hydroxyl groups, whereas a small peak at around 3000 cm^{-1} indicates C–H stretching of the CH_2 groups [42]. The strong absorption at 1730–1760 cm^{-1} represents the characteristic of esterified pectin, arising from the ester carbonyl-stretching band, and peaks at 1600–1630 cm^{-1} and 1400–1450 cm^{-1} are due to the antisymmetric and symmetric stretching frequencies of the ionic carboxyl groups [57]. The region at wavenumbers between 1500 and 1800 cm^{-1} is associated with the assessment of the degree of methylation [58]. Thus, the peak at around 1730 cm^{-1} in the pectin spectra corresponds to a higher DE value [59]. The region between 950 and 1200 cm^{-1} is accordingly referred to as the 'fingerprint' for carbohydrates, especially sugar composition [60]. The intense peaks relate to the characteristics of pectin polysaccharides (polygalacturonic acid) performed at 962, 1024, 1099, 1156 and 1223 cm^{-1}, which are assigned to C–O bending, C–C stretching, C–O stretching, C–H stretching and C–O stretching, respectively [59]. FT-IR analysis verified that the extracted constituent was pectin. Similar band patterns were detected in pectin extracted from Sam-pee mango [22], banana peel [61] and lime peel [59].

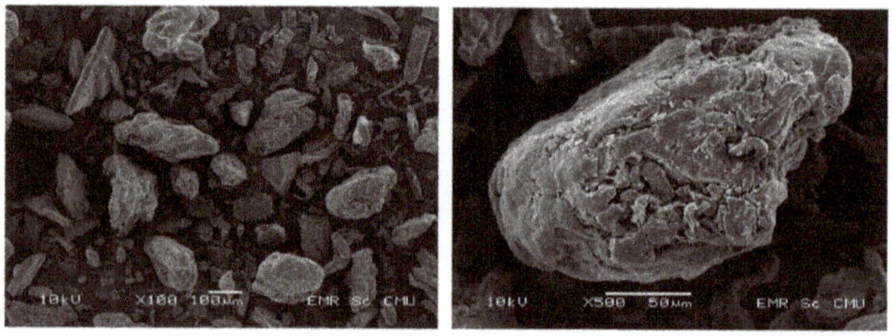

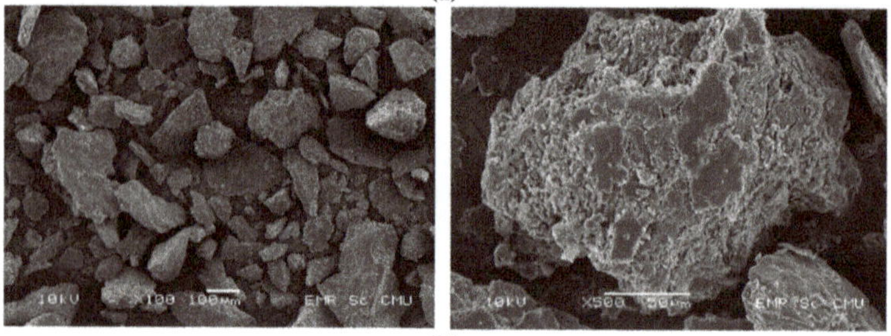

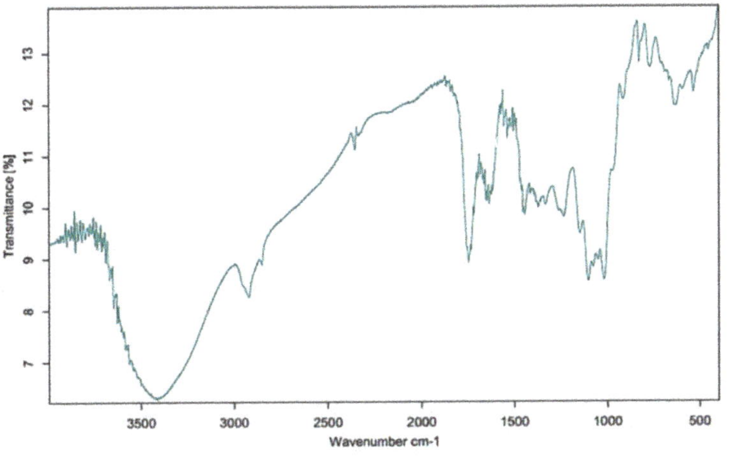

Figure 2. The FT-IR spectra of pectin extracted from Nam Dok Mai mango peel using MAE at 700 watts, from 900 to 4000 cm^{-1} (x axis). %T is the percentage of transmittance (y axis).

3.3. Characterisation of Mango Peel Pectin

Table 1 illustrates the characterisation of Nam Dok Mai MPP extracted by conventional and MAE techniques. The average yield of MPP extraction operated by conventional heating was approximately 0.80% [22], which was dramatically low when compared to the quantity of pectin extracted by MAE at 700 watts (13.85%). Microwave extraction gave better pectin recovery when compared to conventional extraction. Similarly, MAE was reported to be an applicable mode of extraction for high yield pectin recovery in grapefruit (27.81%) and navel orange peel (18.13%) [39,62]. Microwave heating is indeed more efficient than other extraction methods due to the intense formation of vapour in polar substances generated by the electromagnetic field [53]. Heat vapour modifies the cell wall matrix and leads to the severing of parenchymal cells, which rapidly and extensively break down cell membrane, thus increasing the interaction between the extracting agent and the plant material during the extraction process [31]. In addition, microwave energy also results in the inactivation of the pectinase [53].

Table 1. Qualities and functionalities of mango peel pectin extracted by conventional and MAE techniques.

Extraction Techniques	Qualities of Pectin							Functionalities of Pectin		
	Pectin Yield (%)	L*	a*	b*	Eq.W (mg/mol)	Mox (%)	DE (%)	SWC (mL/g sample)	WHC (g water/g sample)	OHC (g oil/g sample)
MAE 700	13.85 ± 0.51	36.33 ± 1.11	5.25 ± 1.05	11.26 ± 2.13	1485.78 ± 0.74	19.33 ± 0.04	77.19 ± 0.72	24.16 ± 0.22	9.60 ± 0.46	0.81 ± 0.04
Conventional	0.80 ± 0.06 [34]	36.88 ± 0.18	5.00 ± 0.20	11.39 ± 0.62	657.89 ± 47.33 [34]	13.90 ± 2.10 [34]	68.90 ± 3.70 [34]	25.50 ± 0.61	11.10 ± 0.23	1.04 ± 0.05
ANOVA Test		Ns	Ns	ns				ns	ns	ns

Data are expressed as mean ± standard deviation, n = 3; MAE 700 = microwave-assisted extraction at 700 watts; qualities and functionalities of mango peel pectin was according to our observation and those reported by Sommano et al. [34]. Eq.W = equivalent weight; Mox = methoxyl content; DE = degree of esterification; SWC = swelling capacity; WHC = water holding capacity and OHC = oil holding capacity.

Colour of pectin is an essential parameter as it influences the appearance of the formulated products. The colours of MPP obtained from conventional and MAE technique are shown in Table 1. L*, a* and b* values of MPP extracted by both techniques are not significantly different ($p > 0.05$).

Comparing the lightness (L*), our extracted MPPs were slightly darker than that of commercial citrus peel pectin extracted using the same extraction method [59]. To this end, pigmentation of the biomass could play an important role as the pigments cannot be removed by extraction steps. Moreover, nonenzymatic browning reactions, i.e., Maillard reaction and caramelisation, are also influenced by heating and might be of great contribution to the pectin colour [63]. In addition, high pigmented pectin may be a result of bound polyphenols [64] or other water-soluble pigments. Different extraction conditions (time and temperature) could also affect pectin colour [65].

Equivalent weight (Eq.W) of pectin is an indicator of gel-forming ability. The greater the Eq.W, the higher the gel-forming ability achieved [66]. The Eq.W of the MPP was about 1400 mg/mol, which was two-fold higher than that of the conventional extraction. The values are comparable with citrus pectin illustrated ranges of Eq.W between 635.63 to 2219.39 mg/mol depending on the extracting methods [59]. Pectin recovered by MAE seems to give a higher Eq.W than that of the conventional heating. The lower Eq.W could be due to higher partial degradation of pectin, thus the variation of Eq.W value depended upon the amount of free acid [67]. Consequently, it can be indicated that the heating of microwave does less damage to the pectin structure than that of the conventional method.

Methoxyl (Mox) content is an essential indicator of pectin setting time, their sensitivity to polyvalent cations and their beneficial properties in the preparation of low solid gels, films and fibres [68]. Moreover, Mox also represents the pectin distribution ability in water and gel ability [69,70]. Pectin extracted by MAE at 700 watts gave 19.33% Mox, which was significantly higher than that of the conventional extraction (13.90%) [22]. Commercially, a high methoxyl pectin (generally at 8%–11% Mox) can form gels at a high sugar content (>65% sugar), while a low methoxyl pectin with less than 7% Mox can form gels at a lower sugar content [71]. In this study, MPP was classified as high methyl pectin due to the higher Mox (>8%); therefore, it needs a very high sugar content (>65%) to suit high methoxyl pectins [59].

DE is a significant molecular index for pectin classification that defines the extent to which carboxyl groups in pectin molecules exist as the methyl ester [72]. The DE value of pectin extracted by MAE from Nam Dok Mai mango peel is 77.19%, which is higher than using a conventional heating method (68.90%). In a similar study, MAE of pectin from lime albedo, pulp and flavedo produced higher DE values than those of conventionally extracted pectin [28]. According to the DE values, MPP extracted by MAE can be classified as high methoxyl pectin as DE > 50% [36]. In addition, the pectin would have a rapid-set gel formation at DE > 72% [69].

Swelling capacity (SWC), water holding capacity (WHC) and oil holding capacity (OHC) of MPP extracted using MAE 700 watts are illustrated in Table 1. In comparison with the conventional MPP, SWC, WHC and OHC increased but were not significantly different from the MAE technique. It was advised this was due to the lower degree of esterification, the greater WHC and other physical properties the pectin demonstrated [73]. SWC elucidates how much the fibre matrix swells when water is absorbed. The high SWC is correlated with the amount of soluble dietary fibre, especially pectin [74]. The SWC value acquired for MPP (24.16 mL/g sample) is greater than those obtained for other fruit fibre, including those from passion fruit pulp, peel and seeds (7.2 mL/g sample) [75] or cocoa pod husks (5.81 mL/g sample) [76]. This value defines the structural characteristics and chemical composition of the fibre that play an important role in the kinetics of water uptake [77].

WHC is the ability of a moist material to hold water when subjected to an external centrifugal gravity force or compression. The value consists of the sum of linked water, hydrodynamic water and physically trapped water, the latter of which contributes most to this capacity [78,79]. WHC of MPP was 9.60 g water/g sample. High WHC value demonstrates that MPP has potential applications in products that require high viscosity and texture improvement, such as cooked meat or bakery [74,80]. To this, Boulos et al. [81] explained that water molecules either as free or bound forms react with carbohydrates in the association of heat. The linear molecules such as amylose and amylopectin are realigned into an immobile monolayer to form a precipitate or a gel, a phenomenon known as retrogradation. This therefore increases the viscosity ability of the carbohydrate.

OHC is a physical property associated with the chemical structure of plant polysaccharides and depends on surface properties, overall charge density, thickness and the hydrophobic nature of the fibre particle [82]. Our MPP showed a considerably lower OHC (0.81 g oil/g sample) than other fruit and vegetable-derived fibres, such as passion fruit albedo, 2.03 g oil/g sample [74], pomegranate bagasse, 5.9 g oil/g sample [83] or ripe kiwi 6.00 g oil/g sample [84]. As a result of its low OHC, the extractable MPP has potential ingredients for fried products since it would not provide a greasy sensation [74,85].

3.4. Physical Quality Assessments of Formulated Dried Chinese Sausage

3.4.1. Colour

Lightness (L*), redness (a*), and yellowness (b*) are considered the most informative parameters for quality assessment of product [86]. Surface colour of the dried sausage supplemented with MPP is illustrated in Table 2. From the result, it can be described that the higher concentration of the MPP added to the sausage, the lower the value of lightness. Our result also illustrates that the redness and yellowness of the sausage increases in all formulated products, and the colour intensity is higher with the increasing concentrations of the MPP. The result is correspondent with the report of Sarıçoban et al. [87] who found that the carotenoids as a food additive improved the redness in meat batters. Compared with the CTRL (0% pectin), ΔE values of the formulated products were significantly distinct from the CTRL ($p < 0.05$) ranging from 9.91–5.55 from the highest to the lowest concentrations, respectively. For this, it is possible that MPP could increase product colour intensity. The results corresponded well with that of Almeidat et al. [88], who advised that fat replacement with a high amount of amorphous cellulose (75% and 100%) in emulsified cooked sausage reduced the surface lightness of the product. Regardless of the product mouthfeel, it might be a promising option to adjust the colour of sausage by adding differently treated MPP [89]. The other textural enhancements such as protein isolate and starch however, affected colour of the meat product differently. Moreover, the protein isolate from pea can enhance cod sausage colour towards higher b*(yellowness) depending on ingredient mixtures and their concentrations [5]. Likewise, the addition of quinoa flour in frankfurter sausage significantly increased colour intensity of the product [90]. On the contrary, the resistance starch addition had no influence on the sausage colour [91]. In this study, addition of MPP had considerably altered the colour of Chinese sausage due to the bioactive compounds, especially carotenoid consisting in ripe mango peel [34,92,93]. This is quite beneficial for the use of dietary fibre of this kind as a functional ingredient.

Table 2. Colour of dried Chinese sausage with added mango peel pectin at different levels.

Percentage of Pectin	L*	a*	b*	ΔE	
0 (CTRL)	53.60 ± 7.44 ^{a}	6.69 ± 2.40 ^{c}	7.27 ± 1.32 ^{c}	-	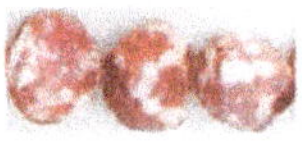
5	52.88 ± 2.87 ^{a}	9.36 ± 0.80 ^{b}	11.25 ± 0.62 ^{b}	5.55 ± 1.02 ^{a}	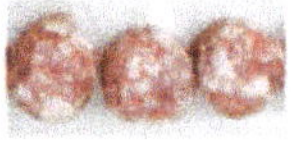
10	55.42 ± 1.82 ^{a}	10.46 ± 0.69 ^{ab}	13.31 ± 0.85 ^{a}	7.59 ± 0.74 ^{b}	
15	50.37 ± 3.81 ^{a}	12.16 ± 1.17 ^{a}	14.14 ± 0.74 ^{a}	9.91 ± 2.12 ^{c}	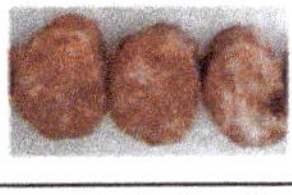

Data are expressed as mean ± standard deviation, $n = 10$; mean values with the same lowercase superscript letter of the same colour value are not significantly different ($p > 0.05$).

3.4.2. Texture

The force-deformation curves of the formulated samples are represented in Figure 3. The textural behaviour of sausages with MPP concentrations of 0%, 5%, 10% and 15% (*w/w*) are shown in Table 3. The hardness is the maximum peak force (F_1) during the first compression cycle required to compress a food between the molar teeth [94]. From the result, the hardness of all formulated samples was not significantly different ($p > 0.05$), whereas springiness, cohesiveness, gumminess and chewiness were lower in treatments with the pectin fibres ($p < 0.05$). The CTRL had the highest hardness value of 15.87 N followed by adding 5%, 10% and 15% (*w/w*) pectin powder with the values of 13.15, 12.89 and 12.70 N, respectively. Cierach et al. [95] described that the hardness in the sausages was related to their fat content. The higher MPP added to the sausage formula, the smaller the slope of the first peak obtained (hardness) (Figure 3). These differences in hardness profiles could be due to the binding ability and water holding capacity of fat and MPP mixture [6]. It could be obviously seen that the texture of sausages with addition of MPP were softer. This could be in association with gel strength of pectin quantity under compression [96]. According to Campagnol et al. [97], the hardness of fermented sausages added with amorphous cellulose as fat replacement at levels of 50%, 75% and 100% (*w/w*) was not significantly different from the control.

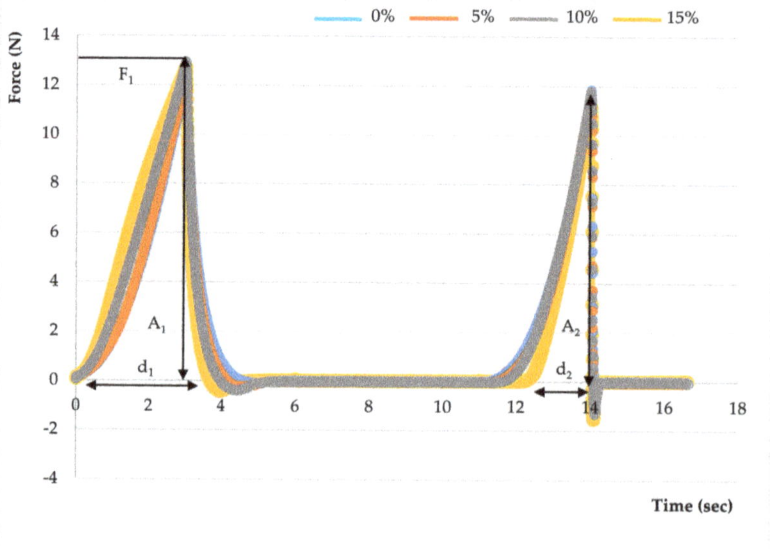

Figure 3. Texture profile of dried Chinese sausage with added mango peel pectin at different levels. Where; Hardness: F_1; Cohesiveness: A_2/A_1; Springiness: d_2/d_1; Gumminess: Hardness × Cohesiveness and Chewiness: Gumminess × Springiness.

Table 3. Texture profile analysis of dried Chinese sausage with added mango peel pectin at different levels.

Texture Characteristics	Percentage of Mango Peel Pectin			
	0 (CTRL)	5	10	15
Hardness (N)	15.87 ± 3.45 ^{a}	13.15 ± 0.66 ^{a}	12.89 ± 2.26 ^{a}	12.70 ± 1.48 ^{a}
Springiness (mm)	1.00 ± 0.01 ^{a}	0.84 ± 0.07 ^{b}	0.78 ± 0.06 ^{b}	0.61 ± 0.04 ^{c}
Cohesiveness	0.50 ± 0.01 ^{a}	0.39 ± 0.06 ^{b}	0.37 ± 0.03 ^{b}	0.28 ± 0.02 ^{c}
Gumminess (N)	7.92 ± 1.78 ^{a}	5.15 ± 0.99 ^{b}	4.78 ± 0.99 ^{b}	3.54 ± 0.52 ^{b}
Chewiness (N.mm)	7.94 ± 1.80 ^{a}	4.40 ± 1.15 ^{b}	3.78 ± 1.04 ^{bc}	2.18 ± 0.41 ^{c}

Data are expressed as mean ± standard deviation, $n = 6$; mean values with the same lowercase superscript letter of the same texture characteristic are not significantly different ($p > 0.05$).

The springiness is a textural parameter, which is correlated with elasticity of the sample. For texture profile analysis, springiness is associated with reversible ability of food after the end of first bite and the begin of the second bite. If springiness is high, it requires more mastication energy in the mouth [98]. The springiness values of four sausage samples are also represented in Table 3. There was a significant difference in the springiness values of all treatments of the sausages ($p < 0.05$). The sausage added with 15% (*w/w*) MPP showed the lowest springiness value compared with other samples. The higher concentration of MPP added, the lower springiness value obtained. Zapata and Pava [90] reported that quinoa flour supplementation had no significant influence on the springiness of frankfurter sausage. Whereas the higher MPP concentration negatively affected the springiness of Chinese sausage samples because of the gelling characteristic [12].

The cohesiveness (consistency) indicates the strength of internal bonds making up the body of food and the degree to which a food can be deformed before it ruptures (breaks) [99]. Cohesiveness is defined as the ratio of the positive force area during the second compression to that of the first compression. It also indicates the ability of the product to hold together [96]. The cohesiveness values of the sausage samples were in the ranges of 0.28 to 0.50. The highest and lowest values obtained were for 0% and 15% (*w/w*) of the pectin supplementation, respectively. Garcia-Santos et al. [91] revealed that the sausage with the addition of resistant starch had a low value of cohesiveness (0.50–0.70). Choe et al. [100] also reported the cohesiveness values of sausages supplemented with wheat fibre for the reduction of fat ranged from 0.27 to 0.34. Troutt et al. [101] found that the addition of three-ingredient combinations of Polydextrose®, potato starch and either sugar beet, oat or pea fibre reduced cohesiveness of beef patties. While quinoa flour had no noticeably effect on the cohesiveness value of frankfurter sausage [90]. The more supplementation of MPP in the Chinese sausage, the lower the value of cohesiveness ($p < 0.05$) because gelling was formed at higher concentration.

Gumminess is defined as the product of hardness and cohesiveness. It is a characteristic of semisolid foods with a low degree of hardness and high degree of cohesiveness. From Table 3, it can be seen that higher amount of MPP resulted in the lower values of gumminess; however, the values of Chinese sausages supplemented with MPP were not significantly distinguished ($p > 0.05$). The higher gumminess has also ascended from the higher hardness value [98]. Regarding research by Cardoso et al. [102], the gumminess value of cod frankfurter sausage remarkably increased ($p < 0.05$) with pea protein supplementation. Méndez-Zamora et al. [18] also found the gumminess of frankfurter sausages replacing fat with inulin and pectin was lower when a higher amount of pectin was added. In this research, Chinese sausage samples supplemented with MPP represented both visco-elastic and gumminess behaviour from the pectin attribute.

Chewiness is a measure of energy required to masticate the food and is normally reported for solid foods. It is defined as the product of gumminess and springiness [96]. The chewiness value of four Chinese sausage samples varied from 2.18 to 7.94 N. There was a significant difference in the value of all sausage treatments ($p < 0.05$). Similarly, higher amount of MPP powder supplemented in the sausage also led to a lower value of chewiness. Feng et al. [103] found statistical differences of gumminess between low-fat Chinese sausages supplemented with Mesona Blumes gum or rice starch mixed gels ($p < 0.05$). Cardoso et al. [102] reported that the chewiness value of cod sausage statistically increased ($p < 0.05$) with pea protein and carrageenan integration. The results could be due to the absence of a water content adjustment, causing moisture to decrease while protein and carbohydrate contents increased. On the other hand, the chewiness value of the Chinese sausages with MPP additive noticeably descended with the higher pectin levels. Because of the presence of high water content in the sausages, it could enhance swelling and gelling properties of the pectin.

Regarding all texture results, MPP influences on texture attributes of Chinese sausage due to their functional characteristics of pectin are used as gelling and texture modified agent in meat products [12]. Consequently, Chinese sausage supplemented with a low amount of the pectin has considerably similar texture properties of the conventional sausage.

3.4.3. Sensory Evaluation

Sensory evaluation can assist food scientists in instructively gaining a distinct understanding of the consequences of reformulation low-fat meat processes. Table 4 represents the acceptance of the sensory attributes of Chinese sausages added with MPP. Each sample was evaluated by 12 trained panels (sex: 6 females, 6 males; age = 25–40 years). The addition of pectin in the levels of 5% and 10% (*w/w*) slightly influenced ($p > 0.05$) the sensory attributes compared with the CTRL. However, the maximum pectin amount (15%(*w/w*)) shows the least acceptance scores in all parameters. Regarding overall acceptability, five percentage of the pectin was the favourite treatment because of its juiciness and appearance. With similar texture attributes (Table 3), the low pectin level added in sausage was more accepted than higher levels. Méndez-Zamora et al. [18] reported that higher levels of pectin added in low-fat frankfurter sausage affected the flavour and odour. Rahman et al. [104] reported that fish sausages with higher starch content had given higher sensorial hardness. Lin and Huang [105] revealed that the konjac or gellan gum additive could improve the firmness of low-fat frankfurter sausage owing to the reduction of fat. Feng et al. [103] found the Mesona Blumes gum or rice starch mixed gels still exhibited the properties of juiciness, facilitating a better overall acceptability of the low-fat Chinese sausage. From sensory evaluation of low-fat Chinese sausage added MPP results, it can be primarily summarised that MPP at high concentrations had dramatically influenced Chinese sausage sensory attributes after sample preparation by pan-frying.

Table 4. Sensory analysis of dried Chinese sausage with added mango peel pectin at different levels with 9-points hedonic scale scoring.

Parameters	Percentage of Mango Peel Pectin (*w/w*)			
	0%	5%	10%	15%
Appearance	7.42 ± 2.15 ^{a}	7.08 ± 1.08 ^{a}	5.83 ± 1.53 ^{ab}	4.92 ± 1.83 ^{b}
Juiciness	8.33 ± 0.89 ^{a}	6.92 ± 1.16 ^{a}	6.83 ± 0.94 ^{a}	5.42 ± 1.56 ^{a}
Springiness	6.75 ± 1.66 ^{a}	6.75 ± 1.82 ^{a}	6.17 ± 1.53 ^{ab}	4.00 ± 2.45 ^{b}
Firmness	6.08 ± 2.02 ^{a}	6.17 ± 1.70 ^{a}	5.92 ± 1.56 ^{a}	4.00 ± 2.00 ^{b}
Colour	5.58 ± 2.23 ^{a}	6.00 ± 2.00 ^{b}	5.83 ± 1.70 ^{b}	4.08 ± 2.68 ^{c}
Overall acceptability	6.58 ± 1.68 ^{a}	6.58 ± 1.56 ^{a}	6.00 ± 1.41 ^{a}	3.08 ± 1.88 ^{b}
Prepared sausage for sensory				

Data are expressed as mean ± standard deviation, $n = 12$; mean values with the same lowercase superscript letter of the same parameter are not significantly different ($p > 0.05$).

4. Conclusions

The microwave-assisted extraction technique evaluated in this study has successfully proven to be a complementary method for the extraction of mango pectin. Consequently, we achieved a significantly greater pectin yield from peel of Nam Dok Mai mango with the MAE 700 watts. The equivalent weight, methoxyl content and degree of esterification of MPP processed higher than that of the conventional method. The substitution of 5% pectin to fat content in the Chinese sausage could enhance colour and conserve the physical qualities as well as sensory attribute. In conclusion, MPP can be utilised in the low-fat Chinese sausage formula as a novel fat replacer.

Author Contributions: Conceptualisation, S.R.S.; Methodology, M.W. and S.R.S.; Formal Analysis, M.W.; Investigation, M.W. and S.R.S.; Writing-Original Draft Preparation, M.W.; Writing-Review & Editing, S.R.S. and M.W.; Supervision, S.R.S. and T.T.; Project Administration, M.W.; Funding Acquisition, K.J. and P.R. All authors have read and agreed to the published version of the manuscript.

Funding: This research work was partially supported by Chiang Mai University and the research grant by Industrial Research and Technology Capacity Development Program (IRTC), Chiang Mai University and Heng heng seng lee hor, co. Ltd., Thailand.

Acknowledgments: We would like to thank staff at Chiang Mai Livestock Product Research and Development Centre for Chinese sausage recipe and sensory evaluation.

Conflicts of Interest: The authors declare no conflict of interest.

References

1. Jiménez-Colmenero, F.; Cofrades, S.; López-López, I.; Ruiz-Capillas, C.; Pintado, T.; Solas, M.T. Technological and sensory characteristics of reduced/low-fat, low-salt frankfurters as affected by the addition of konjac and seaweed. *Meat Sci.* **2010**, *84*, 356–363. [CrossRef]
2. Ritzoulis, C.; Petridis, D.; Derlikis, E.M.; Fytianos, K.; Asteriou, P. Utilization of inverse water-in-oil emulsions as fat replacers in frankfurter model sausages: Influence of fat emulsion content on the organoleptic and mechanical properties. *J. Texture Stud.* **2010**, *41*, 62–74. [CrossRef]
3. Carrapiso, A. Effect of fat content on flavor release from sausages. *Food Chem.* **2007**, *103*, 396–403. [CrossRef]
4. Wang, Q.; Zhao, X.; Ren, Y.; Fan, E.; Chang, H.; Wu, H. Effects of high pressure treatment and temperature on lipid oxidation and fatty acid composition of yak (*Poephagus grunniens*) body fat. *Meat Sci.* **2013**, *94*, 489–494. [CrossRef] [PubMed]
5. Cardoso, C.; Mendes, R.; Nunes, M. Development of a healthy low-fat fish sausage containing dietary fibre. *Int. J. Food Sci. Technol.* **2008**, *43*, 276–283. [CrossRef]
6. Choi, Y.-S.; Kim, H.-W.; Hwang, K.E.; Song, D.H.; Choi, J.-H.; Lee, M.-A.; Chung, H.-J.; Kim, C.-J. Physicochemical properties and sensory characteristics of reduced-fat frankfurters with pork back fat replaced by dietary fiber extracted from *makgeolli* lees. *Meat Sci.* **2014**, *96*, 892–900. [CrossRef] [PubMed]
7. Rezende, N.V.; Benassi, M.T.; Vissotto, F.Z.; Augusto, P.P.C.; Grossmann, M.V.E. Mixture design applied for the partial replacement of fat with fibre in sucrose-free chocolates. *LWT-Food Sci. Technol.* **2015**, *62*, 598–604. [CrossRef]
8. Koutsopoulos, D.A.; Koutsimanis, G.E.; Bloukas, J.G. Effect of carrageenan level and packaging during ripening on processing and quality characteristics of low-fat fermented sausages produced with olive oil. *Meat Sci.* **2008**, *79*, 188–197. [CrossRef]
9. Heinz, G.; Hautzinger, P. *Meat Processing Technology for Small to Medium Scale Producers*; Food and Agricultural Organization of the United Nations: Bangkok, Thailand, 2007.
10. Tan, F.-J.; Liao, F.-Y.; Jhan, Y.-J.; Liu, D.-C. Effect of replacing pork backfat with yams (*Dioscorea alata*) on quality characteristics of Chinese sausage. *J. Food Eng.* **2007**, *79*, 858–863. [CrossRef]
11. Lin, K.-W.; Huang, C.-Y. Physicochemical and textural properties of ultrasound-degraded konjac flour and their influences on the quality of low-fat Chinese-style sausage. *Meat Sci.* **2008**, *79*, 615–622. [CrossRef]
12. Biswas, A.; Kumar, V.; Bhosle, S.; Sahoo, J.; Chatli, M.K. Dietary fibers as functional ingredients in meat products and their role in human health. *Int. J. Livest. Prod.* **2011**, *2*, 45–54.
13. Villegas, D.; Handford, M.; Alcalde, J.A.; Perez-Donoso, A. Exogenous application of pectin-derived oligosaccharides to grape berries modifies anthocyanin accumulation, composition and gene expression. *Plant Physiol. Biochem.* **2016**, *104*, 125–133. [CrossRef] [PubMed]
14. Ajila, C.M.; Jaganmohan Rao, L.; Prasada Rao, U.J.S. Characterization of bioactive compounds from raw and ripe *Mangifera indica* L. peel extracts. *Food Chem. Toxicol.* **2010**, *48*, 3406–3411. [CrossRef] [PubMed]
15. Macagnan, F.; Santos, L.; Roberto, B.; Moura, F.; Bizzani, M.; Silva, L. Biological properties of apple pomace, orange bagasse and passion fruit peel as alternative source of dietary fibre. *Bioact. Carbohydr. Diet. Fibre* **2015**, *6*, 1–6. [CrossRef]
16. Mohnen, D. Pectin structure and biosynthesis. *Curr. Opin. Plant Biol.* **2008**, *11*, 266–277. [CrossRef]
17. Güzel, M.; Akpınar, Ö. Valorisation of fruit by-products: Production characterization of pectins from fruit peels. *Food Bioprod. Process.* **2019**, *115*, 126–133. [CrossRef]
18. Méndez-Zamora, G.; García-Macías, J.A.; Santellano-Estrada, E.; Chávez-Martínez, A.; Durán-Meléndez, L.A.; Silva-Vázquez, R.; Quintero-Ramos, A. Fat reduction in the formulation of frankfurter sausages using inulin and pectin. *Food Sci. Technol.* **2015**, *35*, 25–31. [CrossRef]
19. Verma, A.; Banerjee, R. Dietary fibre as functional ingredient in meat products: A novel approach for healthy living—A review. *J. Food Sci. Technol.* **2010**, *47*, 247–257. [CrossRef]

20. Silva-Vazquez, R.; Flores-Giron, E.; Quintero-Ramos, A.; Hume, M.E.; Méndez-Zamora, G. Effect of inulin and pectin on physicochemical characteristics and emulsion stability of meat batters. *CyTA-J. Food* **2018**, *16*, 306–310. [CrossRef]
21. Ajila, C.M.; Prasada Rao, U.J.S. Mango peel dietary fibre: Composition and associated bound phenolics. *J. Funct. Foods* **2013**, *5*, 444–450. [CrossRef]
22. Sommano, S.; Ounamornmas, P.; Nisoa, M.; Sriwattana, S.; Page, P.; Colelli, G. Characterisation and physiochemical properties of mango peel pectin extracted by conventional and phase control microwave-assisted extractions. *Int. Food Res. J.* **2018**, *25*, 2657–2665.
23. Garcia-Magana Mde, L.; Garcia, H.S.; Bello-Perez, L.A.; Sayago-Ayerdi, S.G.; de Oca, M.M. Functional properties and dietary fiber characterization of mango processing by-products (*Mangifera indica* L., cv Ataulfo and Tommy Atkins). *Plant Foods Hum. Nutr.* **2013**, *68*, 254–258. [CrossRef] [PubMed]
24. Manthey, J.A.; Perkins-Veazie, P. Influences of harvest date and location on the levels of beta-carotene, ascorbic acid, total phenols, the in vitro antioxidant capacity, and phenolic profiles of five commercial varieties of mango (*Mangifera indica* L.). *J. Agric. Food Chem.* **2009**, *57*, 10825–108230. [CrossRef] [PubMed]
25. Schieber, A.; Berardini, N.; Carle, R. Identification of flavonol and xanthone glycosides from mango (*Mangifera indica* L. Cv. "Tommy Atkins") peels by high-performance liquid chromatography-electrospray ionization mass spectrometry. *J. Agric. Food Chem.* **2003**, *51*, 5006–5011. [CrossRef]
26. Nagel, A.; Sirisakulwat, S.; Carle, R.; Neidhart, S. An acetate-hydroxide gradient for the quantitation of the neutral sugar and uronic acid profile of pectins by HPAEC-PAD without postcolumn pH adjustment. *J. Agric. Food Chem.* **2014**, *62*, 2037–2048. [CrossRef]
27. Panouillé, M.; Ralet, M.C.; Bonnin, E.; Thibault, J.F. Recovery and reuse of trimmings and pulps from fruit and vegetable processing. In *Handbook of Waste Management and Co-Product Recovery in Food Processing*; Woodhead Publishing Limited: Cambridge, UK, 2007; pp. 417–447.
28. Fishman, M.L.; Chau, H.K.; Hoagland, P.D.; Hotchkiss, A.T. Microwave-assisted extraction of lime pectin. *Food Hydrocoll.* **2006**, *20*, 1170–1177. [CrossRef]
29. Adetunji, L.R.; Adekunle, A.; Orsat, V.; Raghavan, V. Advances in the pectin production process using novel extraction techniques: A review. *Food Hydrocoll.* **2017**, *62*, 239–250. [CrossRef]
30. Guolin, H.; Jeffrey, S.; Kai, Z.; Xiaolan, H. Application of ionic liquids in the microwave-assisted extraction of pectin from lemon peels. *J. Anal. Methods Chem.* 2012. [CrossRef]
31. Wang, S.; Chen, F.; Wu, J.; Wang, Z.; Liao, X.; Hu, X. Optimization of pectin extraction assisted by microwave from apple pomace using response surface methodology. *J. Food Eng.* **2007**, *78*, 693–700. [CrossRef]
32. Swamy, G.J.; Muthukumarappan, K. Optimization of continuous and intermittent microwave extraction of pectin from banana peels. *Food Chem.* **2017**, *220*, 108–114. [CrossRef]
33. Matharu, A.; Houghton, J.; Lucas-Torres, C.; Moreno, A. Acid-free microwave-assisted hydrothermal extraction of pectin and porous cellulose from mango peel waste—Towards a zero waste mango biorefinery. *Green Chem.* **2016**, *18*, 5280–5287. [CrossRef]
34. Sommano, S.; Ounamornmas, P.; Nisoa, M.; Sriwattana, S. Bioactive functionality of pectin from peels of seven Thai mango cultivars. *Acta Hortic.* **2018**. [CrossRef]
35. Košťálová, Z.; Aguedo, M.; Hromádková, Z. Microwave-assisted extraction of pectin from unutilized pumpkin biomass. *Chem. Eng. Proces.* **2016**, *102*, 9–15. [CrossRef]
36. Hosseini, S.S.; Khodaiyan, F.; Yarmand, M.S. Optimization of microwave assisted extraction of pectin from sour orange peel and its physicochemical properties. *Carbohydr. Polym.* **2016**, *140*, 59–65. [CrossRef]
37. Chaiwarit, T.; Masavang, S.; Mahe, J.; Sommano, S.; Ruksiriwanich, W.; Brachais, C.-H.; Chambin, O.; Jantrawut, P. Mango (cv. Nam Dokmai) peel as a source of pectin and its potential use as a film-forming polymer. *Food Hydrocoll.* **2020**, *102*, 105611. [CrossRef]
38. Pandit, S.G.; Vijayanand, P.; Kulkarni, S.G. Pectic principles of mango peel from mango processing waste as influenced by microwave energy. *LWT-Food Sci. Technol.* **2015**, *64*, 1010–1014. [CrossRef]
39. Bagherian, H.; Zokaee Ashtiani, F.; Fouladitajar, A.; Mohtashamy, M. Comparisons between conventional, microwave- and ultrasound-assisted methods for extraction of pectin from grapefruit. *Chem. Eng. Process.* **2011**, *50*, 1237–1243. [CrossRef]
40. Maran, J.P.; Swathi, K.; Jeevitha, P.; Jayalakshmi, J.; Ashvini, G. Microwave-assisted extraction of pectic polysaccharide from waste mango peel. *Carbohydr. Polym.* **2015**, *123*, 67–71. [CrossRef]

41. Guandalini, B.B.V.; Rodrigues, N.P.; Marczak, L.D.F. Sequential extraction of phenolics and pectin from mango peel assisted by ultrasound. *Food Res. Int.* **2019**, *119*, 455–461. [CrossRef]
42. Jiang, Y.; Du, Y.; Zhu, X.; Xiong, H.; Woo, M.W.; Hu, J. Physicochemical and comparative properties of pectins extracted from *Akebia trifoliata* var. australis peel. *Carbohydr. Polym.* **2012**, *87*, 1663–1669. [CrossRef]
43. Ranganna, S. *Hand Book of Analysis and Quality Control for Fruits and Vegetable Products*, 2nd ed.; McGraw Hill Publishing Co., Ltd.: New Delhi, India, 1995.
44. Ranganna, S. *Handbook of Analysis and Quality Control for Fruit and Vegetable Products*; Tata Mac Graw Hill: New Delhi, India, 1986.
45. Pinheiro, E.S.R.; Silva, I.M.D.A.; Gonzaga, L.V.; Amante, E.R.; Teófilo, R.F.; Ferreira, M.M.C.; Amboni, R.D. Optimization of extraction of high-ester pectin from passion fruit peel (*Passiflora edulis* flavicarpa) with citric acid by using response surface methodology. *Bioresour. Technol.* **2008**, *99*, 5561–5566. [CrossRef]
46. Robertson, J.A.; de Monredon, F.D.; Dysseler, P.; Guillon, F.; Amado, R.; Thibault, J.-F. Hydration properties of dietary fibre and resistant starch: A european collaborative study. *LWT-Food Sci. Technol.* **2000**, *33*, 72–79. [CrossRef]
47. Bolumar, T.; Toepfl, S.; Heinz, V. Fat reduction and replacement in dry-cured fermented sausage by using high pressure processing meat as fat replacer and olive oil. *Pol. J. Food Nutr. Sci.* **2015**, *65*, 175–182. [CrossRef]
48. Femenia, A.; Lefebvre, A.C.; Thebaudin, J.; Robertson, J.A.; Bourgeois, C.M. Physical and sensory properties of model foods supplemented with cauliflower fiber. *J. Food Sci.* **2006**, *62*, 635–639. [CrossRef]
49. Rahman, M.S.; Al-Farsi, S.A. Instrumental texture profile analysis (TPA) of date flesh as a function of moisture content. *J. Food Eng.* **2005**, *66*, 505–511. [CrossRef]
50. Siddaiah, D.; Sagar Reddy, G.V.; Raju, C.V.; Chandrasekhar, T.C. Changes in lipids, proteins and kamaboko forming ability of silver carp (*Hypophthalmichthys molitrix*) mince during frozen storage. *Food Res. Int.* **2001**, *34*, 47–53. [CrossRef]
51. Begum, R.; Yusof, Y.A.; Aziz, M.G.; Uddin, M.B. Structural and functional properties of pectin extracted from jackfruit (*Artocarpus heterophyllus*) waste: Effects of drying. *Int. J. Food Prop.* **2017**, *20*, S190–S201. [CrossRef]
52. Tamnak, S.; Mirhosseini, H.; Tan, C.P.; Ghazali, H.M.; Muhammad, K. Physicochemical properties, rheological behavior and morphology of pectin-pea protein isolate mixtures and conjugates in aqueous system and oil in water emulsion. *Food Hydrocoll.* **2016**, *56*, 405–416. [CrossRef]
53. Kratchanova, M.; Pavlova, E.; Panchev, I. The Effect of microwave heating of fresh orange peels on the fruit tissue and quality of extracted pectin. *Carbohydr. Polym.* **2004**, *56*, 181–185. [CrossRef]
54. Liew, S.Q.; Ngoh, G.C.; Yusoff, R.; Teoh, W.H. Sequential ultrasound-microwave assisted acid extraction (UMAE) of pectin from pomelo peels. *Int. J. Biol. Macromol.* **2016**, *93*, 426–435. [CrossRef]
55. Jiang, Y.; Xu, Y.; Li, F.; Li, D.; Huang, Q. Pectin extracted from persimmon peel: A physicochemical characterization and emulsifying properties evaluation. *Food Hydrocoll.* **2020**, *101*. [CrossRef]
56. Kačuráková, M.; Capek, P.; Sasinková, V.; Wellner, N.; Ebringerová, A. FT-IR study of plant cell wall model compounds: Pectic polysaccharides and hemicelluloses. *Carbohydr. Polym.* **2000**, *43*, 195–203. [CrossRef]
57. Posé, S.; Kirby, A.R.; Mercado, J.A.; Morris, V.J.; Quesada, M.A. Structural characterization of cell wall pectin fractions in ripe strawberry fruits using AFM. *Carbohydr. Polym.* **2012**, *88*, 882–890. [CrossRef]
58. Abid, M.; Cheikhrouhou, S.; Renard, C.; Sylvie, B.; Cuvelier, G.; Attia, H.; Ayadi, M.A. Characterization of pectins extracted from pomegranate peel and their gelling properties. *Food Chem.* **2016**, *215*, 318–325. [CrossRef] [PubMed]
59. Rodsamran, P.; Sothornvit, R. Microwave heating extraction of pectin from lime peel: Characterization and properties compared with the conventional heating method. *Food Chem.* **2019**, *278*, 364–372. [CrossRef]
60. Černá, M.; Barros, A.S.; Nunes, A.; Rocha, S.M.; Delgadillo, I.; Čopíková, J.; Coimbra, M.A. Use of FT-IR spectroscopy as a tool for the analysis of polysaccharide food additives. *Carbohydr. Polym.* **2003**, *51*, 383–389. [CrossRef]
61. Oliveira, T.Í.S.; Rosa, M.F.; Cavalcante, F.L.; Pereira, P.H.F.; Moates, G.K.; Wellner, N.; Mazzetto, S.E.; Waldron, K.W.; Azeredo, H.M.C. Optimization of pectin extraction from banana peels with citric acid by using response surface methodology. *Food Chem.* **2016**, *198*, 113–118. [CrossRef]
62. Guo, X.; Han, D.; Xi, H.; Rao, L.; Liao, X.; Hu, X.; Wu, J. Extraction of pectin from navel orange peel assisted by ultra-high pressure, microwave or traditional heating: A comparison. *Carbohydr. Polym.* **2012**, *88*, 441–448. [CrossRef]

63. Manzocco, L.; Calligaris, S.; Mastrocola, D.; Nicoli, M.C.; Lerici, C.R. Review of non-enzymatic browning and antioxidant capacity in processed foods. *Trends Food Sci. Technol.* **2000**, *11*, 340–346. [CrossRef]
64. Baississe, S.; Ghannem, H.; Fahloul, D.; Lekbir, A. Comparison of structure and emulsifying activity of pectin extracted from apple pomace and apricot pulp. *World J. Dairy Food Sci.* **2010**, *5*, 79–84.
65. Nguyen, B.M.N.; Pirak, T.; Yildiz, F. Physicochemical properties and antioxidant activities of white dragon fruit peel pectin extracted with conventional and ultrasound-assisted extraction. *Cogent Food Agric.* **2019**, *5*, 1633076. [CrossRef]
66. Vaclavik, V.A.; Christian, E.W. *Essentials in Food Science*, 3rd ed.; Springer Science and Business: New York, NY, USA, 2008.
67. Azad, M.A.K.; Ali, M.; Akter, M.; Rahman, M.J. Isolation and characterization of pectin extracted from lemon pomace during ripening. *J. Food Nutr. Sci.* **2014**, *2*, 30–35. [CrossRef]
68. Ranajit, K.S.; Yoga, N.A.P.P.; Asrul, A. Opimized extraction condition and characterization of pectin from kaffir lime (*Citrus hystrix*). *Res. J. Agric. For. Sci.* **2013**, *1*, 1–11.
69. Israel, K.A.; Baguio, S.F.; Diasanta, M.D.B.; Lizardo, R.C.; Dizon, E.; Mejico, M.I.F. Extraction and characterization of pectin from Saba banana (*Musa 'saba'*(*Musa acuminata* × *Musa balbisiana*)) peel wastes: A preliminary study. *Int. Food Res. J.* **2015**, *22*, 202–207.
70. Constenla, D.; Lozano, J. Kinetic model of pectin demethylation. *Lat. Am. Appl. Res.* **2003**, *33*, 91–95.
71. Rouse, A.H.; Atkins, C.D.; Moore, E.L. The occurrence and evaluation of pectin in component parts of valencia oranges during maturation. *Fla. State Hortic. Soc.* **1962**, *75*, 307–311.
72. Sudhir, D.Y.; Namrata, S.B.; Namrata, N.W.; Deepali, C.S. Extraction and Characterization of Pectin from sweet lime. In Proceedings of the 4th International Conference on Multidisciplinary Research & Practice, Ahmedabad, India, 22 December 2017; pp. 58–63.
73. Barrera, A.M.; Ramírez, J.A.; González-Cabriales, J.J.; Vázquez, M. Effect of pectins on the gelling properties of surimi from silver carp. *Food Hydrocoll.* **2002**, *16*, 441–447. [CrossRef]
74. López-Vargas, J.H.; Fernández-López, J.; Pérez-Álvarez, J.A.; Viuda-Martos, M. Chemical, physico-chemical, technological, antibacterial and antioxidant properties of dietary fiber powder obtained from yellow passion fruit (*Passiflora edulis* var. flavicarpa) co-products. *Food Res. Int.* **2013**, *51*, 756–763. [CrossRef]
75. Martínez, R.; Torres, P.; Meneses, M.A.; Figueroa, J.G.; Pérez-Álvarez, J.A.; Viuda-Martos, M. Chemical, technological and in vitro antioxidant properties of mango, guava, pineapple and passion fruit dietary fibre concentrate. *Food Chem.* **2012**, *135*, 1520–1526. [CrossRef]
76. Martínez, R.; Torres, P.; Meneses, M.A.; Figueroa, J.G.; Pérez-Álvarez, J.A.; Viuda-Martos, M. Chemical, technological and in vitro antioxidant properties of cocoa (*Theobroma cacao* L.) co-products. *Food Res. Int.* **2012**, *49*, 39–45. [CrossRef]
77. Figuerola, F.; Hurtado, M.L.; Estévez, A.M.; Chiffelle, I.; Asenjo, F. Fibre concentrates from apple pomace and citrus peel as potential fibre sources for food enrichment. *Food Chem.* **2005**, *91*, 395–401. [CrossRef]
78. Lan, G.; Chen, H.; Chen, S.; Tian, J. Chemical composition and physicochemical properties of dietary fiber from *Polygonatum odoratum* as affected by different processing methods. *Food Res. Int.* **2012**, *49*, 406–410. [CrossRef]
79. Chirinang, P.; Oonsivilai, R. Physicochemical properties, in vitro binding capacities for lard, cholesterol, bile acids and assessment of prebiotic potential of dietary fiber from cassava pulp. *Int. Food Res. J.* **2018**, *25*, S63–S74.
80. Jacometti, G.A.; Mello, L.R.P.F.; Nascimento, P.H.A.; Sueiro, A.C.; Yamashita, F.; Mali, S. The physicochemical properties of fibrous residues from the agro industry. *LWT-Food Sci. Technol.* **2015**, *62*, 138–143. [CrossRef]
81. Boulos, N.N.; Greenfield, H.; Wills, R.B.H. Water holding capacity of selected soluble and insoluble dietary fibre. *Int. J. Food Prop.* **2000**, *3*, 217–231. [CrossRef]
82. Fernández-López, J.; Sendra, E.; Navarro, C.; Sayas, E.; Viuda-Martos, M.; Pérez-Álvarez, J. Storage stability of a high dietary fibre powder from orange by-products. *Int. J. Food Sci. Technol.* **2009**, *44*, 748–756. [CrossRef]
83. Viuda-Martos, M.; Ruiz-Navajas, Y.; Martin-Sánchez, A.; Sánchez-Zapata, E.; Fernández-López, J.; Sendra, E.; Sayas-Barberá, E.; Navarro, C.; Pérez-Álvarez, J.A. Chemical, physico-chemical and functional properties of pomegranate (*Punica granatum* L.) bagasses powder co-product. *J. Food Eng.* **2012**, *110*, 220–224. [CrossRef]
84. Femenia, A.; Sastre-Serrano, G.; Simal, S.; Garau, M.C.; Eim, V.S.; Rosselló, C. Effects of air-drying temperature on the cell walls of kiwifruit processed at different stages of ripening. *LWT-Food Sci. Technol.* **2009**, *42*, 106–112. [CrossRef]

85. Moura, F.A.; Macagnan, F.T.; Santos, L.R.; Bizzani, M.; Petkowicz, C.L.O.; Silva, L.P. Characterization and physicochemical properties of pectins extracted from agroindustrial by-products. *J. Food Sci. Technol.* **2017**, *54*, 3111–3117. [CrossRef]
86. Mielnik, J.A.N.; Slinde, E. Sausage color measured by integrating sphere reflectance spectrophotometry when whole blood or blood cured by nitrite is added to sausages. *J. Food Sci.* **2006**, *48*, 1723–1725. [CrossRef]
87. Sarıçoban, C.; Özalp, B.; Yılmaz, M.T.; Özen, G.; Karakaya, M.; Akbulut, M. Characteristics of meat emulsion systems as influenced by different levels of lemon albedo. *Meat Sci.* **2008**, *80*, 599–606. [CrossRef] [PubMed]
88. Almeida, C.; Wagner, R.; Mascarin, L.; Zepka, L.; Campagnol, P. Production of low-fat emulsified cooked sausages using amorphous cellulose gel. *J. Food Qual.* **2014**, *37*, 437–443. [CrossRef]
89. Wang, Q.; Xiong, Z.; Li, G.; Zhao, X.; Wu, H.; Ren, Y. Tomato peel powder as fat replacement in low-fat sausages: Formulations with mechanically crushed powder exhibit higher stability than those with airflow ultra-micro crushed powder: Tomato particles as fat replacement on quality of sausages. *Eur. J. Lipid Sci. Technol.* **2015**, *118*, 175–184. [CrossRef]
90. Zapata, J.I.H.; Pava, G.C. Physicochemical analysis of frankfurter type sausages made with red tilapia fillet waste (*Oreochromis* sp.) and quinoa flour (*Chenopodium quinoa* W.). *Braz. J. Food Technol.* **2017**, *21*. [CrossRef]
91. Garcia-Santos, M.D.S.L.; ConceiÇÃO, F.S.; Villas Boas, F.; Salotti De Souza, B.M.; Barretto, A.C.D.S. Effect of the addition of resistant starch in sausage with fat reduction on the physicochemical and sensory properties. *Food Sci. Technol.* **2019**, *39*, 491–497. [CrossRef]
92. Karanjalker, G.; Kodthalu, S.K.R.; Dinesh, M.R.; Geetha, G.; Pavithra, K.; Ravishankar, K. Profiling of anthocyanins and carotenoids in fruit peel of different colored mango cultivars. *J. Food Sci. Technol.* **2018**, *55*, 4566–4577.
93. Ajila, C.; Naidu, A.; Bhat, S.G.; Prasada Rao, U. Bioactive compounds and antioxidant potential of mango peel extract. *Food Chem.* **2007**, *105*, 982–988. [CrossRef]
94. Szczesniak, A. Texture is a sensory property. *Food Qual. Prefer.* **2002**, *13*, 215–225. [CrossRef]
95. Cierach, M.; Modzelewska-Kapitula, M.; Szacilo, K. The influence of carrageenan on the properties of low-fat frankfurters. *Meat Sci.* **2009**, *82*, 295–299. [CrossRef]
96. Chandra, M.; Aswathnaryan, S. Texture profile analysis and functional properties of gelatin from the skin of three species of fresh water fish. *Int. J. Food Prop.* **2014**, *18*, 572–584. [CrossRef]
97. Campagnol, P.; Dos Santos, B.; Wagner, R.; Terra, N.; Pollonio, M. Amorphous cellulose gel as a fat substitute in fermented sausages. *Meat Sci.* **2011**, *90*, 36–42. [CrossRef] [PubMed]
98. Rahman, M.; Al-Mahrouqi, A. Instrumental texture profile analysis of gelatin gel extracted from grouper skin and commercial (bovine and porcine) gelatin gels. *Int. J. Food Sci. Nutr.* **2009**, *7*, 229–242. [CrossRef] [PubMed]
99. Radocaj, O.; Dimic, E.; Vujasinovic, V. Optimization of the texture of fat-based spread containing hull-less pumpkin (*Cucurbita pepo* L.) seed press-cake. *Acta Period. Technol.* **2011**, *42*, 131–143. [CrossRef]
100. Choe, J.H.; Kim, H.Y.; Lee, J.M.; Kim, Y.J.; Kim, C.J. Quality of frankfurter-type sausages with added pig skin and wheat fiber mixture as fat replacers. *Meat Sci.* **2013**, *93*, 849–854. [CrossRef]
101. Troutt, E.S.; Hunt, M.C.; Johnson, D.E.; Claus, J.R.; Kastner, C.L.; Kropf, D.H. Characteristics of low-fat ground beef containing texture-modifying ingredients. *J. Food Sci.* **1992**, *57*, 19–24. [CrossRef]
102. Cardoso, C.M.L.; Mendes, R.; Nunes, M.L. Instrumental texture and sensory characteristics of cod frankfurter sausages. *Int. J. Food Prop.* **2009**, *12*, 625–643. [CrossRef]
103. Feng, T.; Ye, R.; Zhuang, H.; Rong, Z.; Fang, Z.; Wang, Y.; Gu, Z.; Jin, Z. Physicochemical properties and sensory evaluation of *Mesona Blumes* gum/rice starch mixed gels as fat-substitutes in Chinese Cantonese-style sausage. *Food Res. Int.* **2013**, *50*, 85–93. [CrossRef]
104. Rahman, M.; Al-Waili, H.; Guizani, N.; Kasapis, S. Instrumental-sensory evaluation of texture for fish sausage and its storage stability. *Fish. Sci.* **2007**, *73*, 1166–1176. [CrossRef]
105. Lin, K.-W.; Huang, H.-Y. Konjac/gellan gum mixed gels improve the quality of reduced-fat frankfurters. *Meat Sci.* **2003**, *65*, 749–755. [CrossRef]

foods

MDPI

Article

Inclusion of Healthy Oils for Improving the Nutritional Characteristics of Dry-Fermented Deer Sausage

Márcio Vargas-Ramella [1,2], Paulo E. S. Munekata [2], Mohammed Gagaoua [3], Daniel Franco [2], Paulo C. B. Campagnol [4], Mirian Pateiro [2], Andrea Carla da Silva Barretto [5], Rubén Domínguez [2] and José M. Lorenzo [2,6,*]

1 Centro de Educação Superior da Região Sul—CERES da Universidade do Estado de Santa Catarina, Chapecó, Santa Catarina 89.800-000, Brazil; marcio.ramella@hotmail.com

2 Centro Tecnológico de la Carne de Galicia, Rúa Galicia N° 4, Parque Tecnológico de Galicia, San Cibrao das Viñas, 32900 Ourense, Spain; paulosichetti@ceteca.net (P.E.S.M.); danielfranco@ceteca.net (D.F.); mirianpateiro@ceteca.net (M.P.); rubendominguez@ceteca.net (R.D.)

3 Food Quality and Sensory Science Department, Teagasc Ashtown Food Research Centre, Ashtown, Dublin 15, Ireland; gmber2001@yahoo.fr

4 Departmento de Tecnologia e Ciência de Alimentos, Universidade Federal de Santa Maria, Santa Maria, Rio Grande do Sul CEP 97105-900, Brazil; paulocampagnol@gmail.com

5 Department of Food Technology and Engineering, UNESP—São Paulo State University, Street Cristóvão Colombo, 2265, Sao Jose do Rio Preto 15054-000, Brazil; andrea.carla@unesp.br

6 Área de Tecnología de los Alimentos, Facultad de Ciencias de Ourense, Universidad de Vigo, 32004 Ourense, Spain

* Correspondence: jmlorenzo@ceteca.net

Received: 7 September 2020; Accepted: 15 October 2020; Published: 18 October 2020

check for updates

Abstract: The influence of partial replacement of animal fat by healthy oils on composition, physicochemical, volatile, and sensory properties of dry-fermented deer sausage was evaluated. Four different batches were manufactured: the control was formulated with animal fat (18.2%), while in the reformulated batches the 50% of animal fat was substituted by olive, canola, and soy oil emulsions immobilized in Prosella gel. The reformulation resulted in a decrease of moisture and fat contents and an increase of protein and ash amount. Moreover, reformulated sausages were harder, darker, and had higher pH values. This fact is related to the lower moisture content in these samples. As expected, the fatty acid composition was changed by the reformulation. The use of soy and canola oils increased polyunsaturated fatty acids and omega-3 content and decreased n-6/n-3 ratio and saturated fatty acids. Thus, the use of these two oils presented the best nutritional benefits. The changes observed in the fatty acids reflected the fatty acid composition of the oils employed in the emulsions. Regarding volatile compounds (VOC), the replacement of animal fat by healthy emulsion gels increased the content of both total VOC and most of individual VOC. However, the lipid-derived VOC did not show this trend. Generally speaking, the control samples presented similar or higher VOC derived from lipid oxidation processes, which could be related to the natural antioxidant compounds present in the vegetable oils. Finally, all reformulated sausages presented higher consumer acceptability than control samples. In fact, the sausage reformulated with soy oil emulsion gel was the most preferred. Thus, as a general conclusion, the reformulation of deer sausages with soy emulsion gel improves both composition and sensory quality of the final product, which could be an excellent strategy to the elaboration of healthy fermented sausages.

Keywords: food reformulation; healthy meat product; game meat; fatty acids; volatile compounds; sensory properties

1. Introduction

Over the last number of years, the consumption of game meat has increased. The characteristic texture and taste of the meat from game, such as wild deer is appreciated by the consumer [1]. Spain is the second world producer of deer venison which mainly exports meat from hunted animals [2]. According to official data, the capture of cynegetic deer in Spain surpassed 144,000 animals (red deer) in 2018. The average live weights of these animals were 80 kg and resulted in 11,530 tons with an economic value of 25 million euros. In addition to red deer, 66,737 roe deer and 24,337 fallow deer were hunted in the same year [3]. Between 2012 and 2018, the red deer captures ranged between 144,061 and 182,458 animals per year [3]. However, it is important to highlight that deer meat is simply considered as a by-product of hunting [4]. Additionally, the consumption of both fresh meat and traditional meat products elaborated with game meat has increased [1]. It is due to their particular taste and excellent nutritional characteristics (low fat and cholesterol contents and high amounts of other essential nutrients) [2,5–8]. Additionally, taking into account health aspects, it also has high amount of PUFA (39–50 g/100 g fatty acids) [8] and long-chain n-3 fatty acids [2,7–9].

A wide range of meat products can be produced with deer meat, including burgers [10] dry-cured cecina, pâté [9], dry-cured loin [11], and dry-fermented sausages such as chorizo [1,12,13] and salchichón [4,12,14,15]. However, the market for game products is very restricted [12] and these products are labelled as "gourmet products" in the international market. Nevertheless, due to the very lean character of meat of deer, the use of pork backfat or pork meat with high fat contents are necessary in the dry-fermented sausages elaboration in order to ensure the correct dry-ripening process [4,14]. This strategy is necessary because fat limits the mobility of the water from the interior of the meat product to the environment, which cause a progressive, slow, and controlled drying process. In fact, the use of an inadequate amount of fat (a very lean product) causes a rapid initial drying, which creates a superficial crust that prevents the correct subsequent drying process, affecting all the changes that occur during ripening and yielding a poor-quality product. Thus, with the addition of animal fat, the health benefits of the special composition of deer meat are masked by the high values of saturated fat and cholesterol of the pork backfat. In this regard, cynegetic venison sausages have around 40–50% of fat [4]. In fact, in an article that studied the proteolysis, free fatty acids, and composition of commercial dry-fermented deer sausages, the authors reported values of fat between 32% and 53% of dry matter (DM) in chorizo and between 33% and 42% of dry matter salchichón [12]. Similarly, other authors reported fat values of 31–36% (DM) in deer chorizo [1,13] and between 19% and 40% (DM) in salchichón [14,15]. Additionally, the saturated fatty acids accounted for more than 40% of total fatty acids in deer salchichón [14].

It is known that the diets rich in fat (particularly saturated fat) increase the risk of overweighting and developing chronic disorders, particularly ischemic heart disease, stroke, and cancer [16–20]. Furthermore, the consumers, aware that diet plays a crucial role in their health, have increased the demand for more healthy meat products [21]. Therefore, following the international recommendations and consumers' demands, the meat industry has focused on two main strategies during the last decade.

A direct reduction of fat added in the sausages elaboration by using a higher proportion of lean meat is the first option. Several researchers studied this option [4,22–25]. However, the fat has a great influence on dry-fermented sausage quality. Sensory characteristics as appearance, tenderness, hardness, and overall palatability depend directly on the fat content [23,26]. The characteristic flavor of meat products, which is one of the most appreciated attributes for the consumers, are also related to fat. Additionally, fat is the precursor of the lipolysis and lipid oxidation reactions that modulate the release and formation of several volatile compounds, which are crucial in typical dry-cured aroma [27–30]. Thus, the numerous and complex reactions during ripening conferred the characteristic aroma of these products, and depend on the concentration and the olfactory threshold of each compound [31]. Moreover, as commented above, fat also exert a technological function during the dry-ripening process, because the granulated fat can facilitate the regular moisture release occurring

during ripening phase [32]. Thus, the fat reduction strategy tended to result in lower sensory quality and technological problems.

With the aim to mitigate fat reduction drawbacks, multiple authors used fat replacers such as konjac gel [32], fructooligosaccharides [31,33,34], oat, and chia emulsion gel [35], cellulose gel [30], or boiled quinoa [20]. However, in most of these studies, the use of fat replacers only affected the total fat content, but did not affect or had low effect in the fatty acid profile. Therefore, in order to improve the nutritional value of fat, the most recent studies replaced animal fat by oil-in-water emulsions. In this regard, several researches indicated that the replacement of animal fat by different non-animal lipids (vegetable or marine oils) resulted in a reduction of both total fat and SFA contents [9,36–39], and also reduces the cholesterol content in the final product [37].

However, the correct choice of oils as substitutes for animal fat is the most important point of the reformulation strategy. The nutritional recommendations indicate that the intake of saturated fats should be replaced by monounsaturated or polyunsaturated fatty acids (mainly omega-3). Nevertheless, not all oils rich in these fatty acids are good candidates for the reformulation of meat products. Various authors have observed that the use of chia [9,10,38,40], linseed [9,10,40], or fish oils [36,41,42] (oils with high omega-3 content) in the reformulation of meat products have resulted in a significant increase in oxidation rates and/or rancid and fishy flavors, which decreased the consumer acceptability and sensory properties. For this reason, these oils have been ruled out for the design of this experiment. Following the recommendations, and taking into account the composition of multiple oils, the use of olive oil has been considered, since it contains a high amount of oleic acid (about 70% of total fatty acids) [43]. In fact, the use of olive oil as animal fat replacer in other meat products were carried out in previous studies with promising results [37,38]. Moreover, soybean oil presented also a favorable fatty acids profile, with high polyunsaturated fatty acids amounts (about 60%) and intermediate values of omega-3 fatty acids (5–11%) [44]. In addition to their composition, there are few studies that used soybean oil as fat replacers in meat products [44–46], which demonstrated its viability as animal fat replacer. Finally, canola oil presents high monounsaturated (mainly oleic acid; 60%) and polyunsaturated fatty acids amounts (36%), with intermediate α-linolenic acid contents (about 10%) [47]. This oil was previously used in the reformulation of chicken meatballs [47] and sausages [48], and the results demonstrated a significant nutritional improvement without affecting the sensory characteristics of the products (presented the same consumer acceptability as control samples). Based on these aspects, the use of olive, canola, and soybean oils was proposed in the present study.

On the one hand, in cooked meat products such as Frankfurt type sausages [38,41], pâté [9,42] or in fresh products such as burgers [39,40,49], the use of oil-in-water emulsion is easy. On the other hand, the application of emulsion gels in dry-fermented or dry-ripened products is more difficult. Drying and ripening stages are characterized by multiple and complex physicochemical changes and the emulsion gels must be stable throughout the process [50]. In order to overcome these issues, the use of emulsion gels (healthy oils immobilized in a gel structure) has been proposed in recent years [51]. In this sense, the application of these emulsions as animal fat replacer can confer the distinctive characteristics of saturated fat but with favorable fatty acid profile. Recent techniques of converting liquid healthy oils into a solid-like gel were explored by multiple researchers. Different dry-fermented sausages were reformulated with healthy oil-in-water emulsion gels using konjac [36,50,52,53], whey protein powder [54], carrageenan [55], or oleogels [56], which is a promising alternative to produce healthier fermented sausages [30].

However, as far as the authors are aware, the use of a healthy oil stabilized into an alginate-wheat glucose-phosphate matrix as a functional ingredient and animal fat replacement in the development of healthy dry-fermented sausage has not been explored. Thus, the effect of dry-ripening process in its viability and stability was not studied.

On the other hand, very few studies examined the effect of pork backfat replacement by healthy oils in deer dry-fermented sausages. To this regard, only two manuscripts were found about the influence of partial pork meat replacement by olive oil organogel in the sensory [15] or composition and nutritional values [14] of deer dry-fermented sausages. In these studies, the authors used pork meat with high fat content (around 50%) instead of pork backfat as we used in the present manuscript. These different strategies gave totally different results in both, the composition and the sensory quality.

With the aforementioned in mind, the objective of this research was to design a technological strategy for the incorporation of healthy oil-in-water gelled emulsions as a partial animal fat replacer to improve nutritional characteristics of dry-fermented deer sausages. The chemical and physicochemical characteristics, nutritional properties, volatile release modifications, sensory properties, and consumer acceptability were assessed.

2. Materials and Methods

2.1. Preparation of Prosella Gel and Fatty Acid Characterization of Fat Sources

Elaboration of emulsion alginate-based hydrogels (Prosella gel) were carried out following the procedure described by Barros et al. [39]. Its preparation was a day before of sausages manufacture. Three different batches of emulsion hydrogels were obtained using olive oil, canola oil, and soy oil. The fatty acids composition of the fat sources used in the present study is shown in Table 1.

Table 1. Fatty acid composition (expressed as g/100 g of total fatty acids) of fat sources.

	Fat Source			
	Pork Backfat	**Olive Oil**	**Canola Oil**	**Soy Oil**
C14:0	1.19 ± 0.002	0.03 ± 0.003	0.06 ± 0.005	0.08 ± 0.002
C16:0	23.03 ± 0.16	11.73 ± 0.04	4.96 ± 0.07	10.50 ± 0.02
C16:1n-7	1.89 ± 0.07	0.84 ± 0.11	0.18 ± 0.01	0.09 ± 0.003
C17:0	0.23 ± 0.001	0.08 ± 0.000	0.05 ± 0.006	0.09 ± 0.001
C17:1n-7	0.19 ± 0.007	0.12 ± 0.003	0.06 ± 0.002	0.05 ± 0.001
C18:0	12.74 ± 0.22	3.40 ± 0.02	2.01 ± 0.05	3.31 ± 0.008
C18:1n-9	41.18 ± 0.04	72.16 ± 0.06	54.65 ± 0.002	24.99 ± 0.006
C18:1n-7	2.67 ± 0.005	2.21 ± 0.04	2.68 ± 0.11	1.55 ± 0.02
C18:2n-6	13.68 ± 0.23	7.70 ± 0.022	27.44 ± 0.05	52.10 ± 0.05
C20:0	0.24 ± 0.000	0.41 ± 0.004	0.48 ± 0.001	0.33 ± 0.001
C18:3n-6	0.04 ± 0.000	0.03 ± 0.001	0.33 ± 0.001	0.26 ± 0.001
C20:1n-9	0.87 ± 0.03	0.24 ± 0.003	0.97 ± 0.02	0.22 ± 0.001
C18:3n-3	0.66 ± 0.01	0.61 ± 0.000	5.30 ± 0.01	5.70 ± 0.02
C20:2n-6	0.64 ± 0.02	n.d.	0.05 ± 0.001	0.04 ± 0.002
C22:0	0.01 ± 0.001	0.14 ± 0.001	0.39 ± 0.000	0.47 ± 0.002
C20:3n-3	0.11 ± 0.004	0.24 ± 0.000	n.d.	n.d.
C20:4n-6	0.23 ± 0.04	n.d.	n.d.	n.d.
C24:0	n.d.	0.06 ± 0.002	0.15 ± 0.001	0.17 ± 0.003
SFA	37.61 ± 0.38	15.85 ± 0.07	8.14 ± 0.13	14.99 ± 0.03
MUFA	45.93 ± 0.05	75.33 ± 0.09	57.74 ± 0.10	26.69 ± 0.03
PUFA	15.59 ± 0.29	8.58 ± 0.02	33.15 ± 0.05	58.10 ± 0.06
n-6	14.75 ± 0.29	7.73 ± 0.02	27.84 ± 0.06	52.40 ± 0.05
n-3	0.77 ± 0.005	0.85 ± 0.001	5.30 ± 0.001	5.70 ± 0.02

SFA: saturated fatty acids; MUFA: monounsaturated fatty acids; PUFA: polyunsaturated fatty acids; n.d.: not detected.

2.2. Manufacture of Dry-Fermented Sausages

For the present research, four different batches of fermented sausages were manufactured in the pilot plant of the Meat Technology Center of Galicia (San Cibrao das Viñas, Ourense, Spain). In control (CON) batch was used 100% of pork backfat as fat source. In the other ones, 50% of pork backfat was replaced by Prosella gel containing olive (OLI), canola (CAN), or soy (SOY) oils. An identical formula was used for all batches, except for fat source. The deer sausage formulation included lean deer meat (740 $g \cdot kg^{-1}$), fat source (182 $g \cdot kg^{-1}$), and a supplement "542 Salchichon" (46 $g \cdot kg^{-1}$) from Laboratorios

Ceylamix (Valencia, Spain). No starter culture was added. All manufacture conditions were previously published by Bis-Souza et al. [31].

The manufacturing process was replicated with the same ingredients, formulation, and methods in two different months. Ten samples from each batch and elaboration were taken after 45 days of ripening.

2.3. Physicochemical, Lipid Oxidation, and Composition Analysis

The physicochemical (pH, color, and texture) and proximate composition was evaluated following the procedure described by Lorenzo et al. [36], while for lipid oxidation determination, the TBARS index was evaluated [57].

2.4. Fatty Acids Analysis

Total fat was extracted from 10 g of sample [39]. Then, the fatty acids were transesterified, and analyzed by gas chromatography-FID technique (Agilent Technologies, Santa Clara, CA, USA), following the conditions reported by Barros et al. [39]. The results were expressed as g/100 g of total identified fatty acids.

2.5. Volatile Compounds Analysis

The volatile compounds of 1 g of sample were analyzed using SPME-gas chromatography-mass spectrometry technique (Agilent Technologies, Santa Clara, CA, USA), following the procedure described by Domínguez et al. [28]. The volatile results were expressed as area units per gram of sample (AU × 10^4/g of sample).

2.6. Sensory Analysis of Deer Sausages

The quantitative-descriptive analysis (QDA) was conducted with 15 trained panellists selected from the Meat Technology Centre of Galicia. The sausages were cut into slices (≈5 mm thick), coded with three numbers, and presented at room temperature. Nine sensory traits of deer sausages, grouped according to appearance (meat color and fat color), flavor (black pepper flavor, rancid flavor, and global flavor), odor, taste, and texture (hardness and chewiness) were assessed according to the methodology proposed by ISO regulations [58–60]. The intensity of every attribute was rated on a structured scale from 0 (sensation not perceived) to 9 (maximum of the sensation).

Sensory acceptability analysis was conducted by 68 consumers (29–40 years and from both genders) from Ourense (Spain) (Ethical Committee for Sensorial Analysis of Centro Tecnolóxico da Carne; approved number: SEN.CTC.20.001). The samples were evaluated to determine whether the panellist liked or disliked of the reformulated sausages in comparison to Control. The samples were three-digit coded and presented individually in unwrapped oblique slices approximately 5 mm thick on a small plate. The samples were randomly presented to consumers [61]. Consumers evaluated the deer sausages by the acceptance test using a 7-point hedonic scale ("1—disliked much"; "7—liked much"). Additionally, consumers also ordered the sausages according to their preference (structured 4-point scale: 1 = most favorite and 4 = less favorite).

2.7. Statistical Analysis

A total of 80 samples were used in the present study (10 sausages × 4 batches × 2 manufacture process replicates). Shapiro–Wilk analysis was used to test the normal distribution and variance homogeneity. Data from chemical composition, physicochemical parameters, fatty acids, and volatile compounds were examined using a one-way ANOVA analysis. Duncan's test was used for the determination of differences between least squares means ($p < 0.05$), while correlations between variables were evaluated with the Pearson's linear coefficient. All statistical analyses was achieved using IBM SPSS statistics for Windows (version 19.0. IBM Corp., New York, NY, USA).

The XLSTAT-Sensory version 2018 (Addinsoft SARL, Paris, France) software was utilized to determinate all the sensory data. For QDA analysis, a 2-way Mixed Model ANOVA was conducted (panellist and treatment). Dependent variables were intensity ratings corresponding to each sensory attribute. Duncans' mean separation tests were used for post hoc analyses ($\alpha = 0.05$). In addition, Friedman two-way ANOVA, was used to analyze preference data. Principal component analysis (PCA) was applied for the significantly different attributes and it was conducted to evaluate the deer sausages and, in this way, create the attribute maps.

3. Results and Discussion

3.1. Proximate Composition, Color, and Texture Parameters of Deer Sausages

The use of emulsion hydrogels (olive, canola, and soy oils) resulted in a significant decrease of moisture ($p < 0.001$) (Table 2). The CON samples presented a final moisture content of 39.79%, while in the experimental batches the moisture content varied between 27.69% and 29.90%. The moisture values found in the control samples were similar to those reported by other authors, who found moisture values about 34–45% in dry-fermented sausages [30,54]. In a similar way, the moisture values of experimental batches agree with those described by other authors (around 25–30%) also in dry-cured sausages [20,27,36,55,56].

Table 2. Effect of fat source on physicochemical and composition parameters of dry-fermented deer sausage.

	Fat Source				Sig.
	CON	OLI	CAN	SOY	
Composition (g/100 g)					
Moisture	39.79 ± 0.72 ^{c}	29.68 ± 0.41 ^{b}	29.90 ± 0.43 ^{b}	27.69 ± 0.38 ^{a}	***
Fat (dry matter)	37.62 ± 0.52 ^{b}	32.53 ± 0.56 ^{a}	32.30 ± 0.35 ^{a}	32.54 ± 0.42 ^{a}	***
Protein (dry matter)	47.39 ± 0.44 ^{a}	49.60 ± 0.32 ^{b}	49.53 ± 0.32 ^{b}	49.43 ± 0.29 ^{b}	***
Ashes (dry matter)	7.24 ± 0.07 ^{a}	8.77 ± 0.21 ^{b}	8.63 ± 0.06 ^{b}	8.59 ± 0.06 ^{b}	***
pH	4.84 ± 0.02 ^{a}	5.04 ± 0.01 ^{b}	5.04 ± 0.03 ^{b}	5.12 ± 0.02 ^{c}	***
Color Parameters					
L^*	40.81 ± 0.63 ^{c}	30.77 ± 0.40 ^{a}	30.97 ± 0.46 ^{a}	33.01 ± 0.59 ^{b}	***
a^*	5.99 ± 0.19 ^{b}	6.02 ± 0.27 ^{b}	6.07 ± 0.32 ^{b}	5.08 ± 0.20 ^{a}	*
b^*	8.49 ± 0.25 ^{b}	5.17 ± 0.31 ^{a}	4.81 ± 0.23 ^{a}	5.03 ± 0.26 ^{a}	***
Texture Parameters					
Hardness (N)	61.98 ± 3.10 ^{a}	279.47 ± 6.91 ^{b}	271.56 ± 6.37 ^{b}	316.18 ± 6.57 ^{c}	***
Springiness (mm)	0.45 ± 0.01 ^{a}	0.54 ± 0.005 ^{b,c}	0.54 ± 0.01 ^{c}	0.52 ± 0.005 ^{b}	***
Cohesiveness	0.29 ± 0.01 ^{a}	0.34 ± 0.005 ^{b}	0.34 ± 0.003 ^{b}	0.35 ± 0.003 ^{b}	***
Gumminess (N)	18.36 ± 1.05 ^{a}	96.07 ± 3.23 ^{b}	92.83 ± 2.38 ^{b}	109.54 ± 2.25 ^{c}	***
Chewiness (N·mm)	8.40 ± 0.62 ^{a}	51.44 ± 1.81 ^{b}	50.24 ± 1.37 ^{b}	56.49 ± 1.35 ^{c}	***

$^{a–d}$ Mean values in the same row (corresponding to the same parameter) not followed by a common letter differ significantly ($p < 0.05$; Duncan test); Sig.: significance: * ($p < 0.05$), *** ($p < 0.001$); SEM: standard error of the mean. Treatments: CON: sausages prepared 100% pork fat; OLI: sausages reformulated with 50% of pork fat replaced by olive oil; CAN: sausages reformulated with 50% of pork fat replaced by canola oil; SOY: sausages reformulated with 50% of pork fat replaced by soy oil.

Some authors also reported the lowest moisture amounts in reformulated dry-fermented sausages, in which animal fat was replaced by encapsulated oil-in-konjac matrix [36]. Additionally, the samples of dry-cured sausage reformulated with linseed oil oleogel (made with sterols) showed lower moisture content than the control samples [56]. In contrast, other researchers reported that the addition of oil-in-cellulose gel [30], beeswax oleogel [56], oil-in-konjac matrix [50], chia and oat emulsion gels [35], chia and olive oils structured in emulsion gels [62], and hazelnut oil pre-emulsified with whey protein [54] as animal fat replacers resulted in a significant increase of moisture content. It is well known that the drying-ripening process is vital for the progressive and adequate dehydration

process of the product. Therefore, the differences observed between studies could be related to the different ripening conditions (time, temperature, relative humidity, air speed, etc.) as well as to the characteristics of the product (amount of fat, type of meat, casing size, etc.) or the behavior of the different fat substitutes (oleogels or emulsions, encapsulated oils, etc.).

The moisture differences found between CON and experimental batches suggest a faster drying process in sausages with animal fat replacement. It is well known that animal fat generates a barrier and reduces water loss during the drying step. In contrast, the oil-in-water emulsion gels used in the elaboration of OLI, CAN, and SOY samples contain 56% of water, which favors the drying process. Thus, our results showed that the barrier effect of Prosella gels is very low in comparison with animal fat. This aspect is an advantage for the manufacturer. In this sense, because the drying process is more intense and faster, the reformulated sausages can be produced in less time, with the economic and technological benefit that this result implies. Moreover, these variations in the final moisture between control and reformulated sausages will influence other parameters such as pH, texture, and sensorial characteristics.

Fat content (on a dry matter basis) showed a significant reduction with animal fat replacement. CON samples presented 37.62% of fat, while in the reformulated sausages were about 32.5%. Thus, the fat reduction between CON and reformulated samples were about 13–14%. Our values were similar to those reported for similar dry-fermented sausages [12,20,56]. As expected, the replacement of pork backfat, which contained about 80% of total fat [63], by oil-in-water emulsions that had only 37.2% of oil resulted in a significantly decrease of fat in the reformulated samples. These findings were reported by several authors, who proved that the use of gelled oils as animal fat replacers caused a total fat reduction in other similar dry-ripening sausages, such as sucuk [54], salchichón [36,56], fuet [62], and other types of fermented sausages [20,30,55].

In contrast, the animal fat replacement significantly enhanced the content of protein and ash. These results agree with those reported by other authors, who found an increase in protein and ash content when animal fat was replaced by boiled quinoa [20], amorphous cellulose gel [30], encapsulated oil in konjac matrix [36], or emulsion gel using protein isolate and gelatin [62]. These variations reflect the decrease in fat content, which results in a higher proportion of protein and ash in the dry matter of the reformulated sausages. Additionally, the variation in the ash content in the present study may be related to the addition of Prosella powder in the oil-in-water emulsion. This fact was also reported in a previous study in which beef burgers reformulated with oil emulsions using Prosella gel presented higher ash content than control samples [39]. Similarly, a study in which animal fat was replaced by chia and oat emulsion gels concluded that the concentration of minerals increased when emulsion gel was used in the fresh sausage formulation [35].

The reformulation also affected the pH of dry-fermented deer sausage. The samples with oil-in-water emulsions had higher pH values ($p < 0.001$) than CON sausages. The sausages from SOY batch presented the highest values (5.12; $p < 0.05$), followed by OLI and CAN (5.04) and finally the lowest values were observed in CON (4.48). Despite the pH changes, these pH values are consistent with those obtained in similar products [14,22,36,55,56]. Our pH level reached values less than 5.15 in all samples because of fermentation process and the acidification caused by lactic acid bacteria [30] and thus increased the microbial stability of the final product. The changes observed between the samples of different formulations could be related to the differences in moisture content among batches. In fact, as reported in a previous study [56], a negative and significant correlation was observed among pH and moisture values ($r = -0.667$; $p < 0.01$). Similarly, other authors also found that the animal fat replacement by emulsions gels resulted in higher pH values in the final product [35,36,62].

The instrumental color data obtained in this research was similar to those described by other authors in deer sausage [14]. As could be seen in Table 2, animal fat replacement by oil-in-water emulsions affected all color parameters. In this regard, the reformulated sausages were darker ($p < 0.001$) than CON samples. The highest L^* values were observed in CON samples (40.81) followed by SOY (33.01) and finally OLI and CAN, with similar values (about 31). The same trend was observed by several authors in dry-fermented sausages reformulated with oil emulsions [20,24,36,62]. It is well-known that the fat amount has a strong influence on the L^* values, since the higher the fat content in sausages, the higher the L^* values [22]. Thus, the lightness reduction in reformulated sausages can be explained by the reduction of animal fat because this fat is white, and it provides the brilliant aspect of dry-fermented sausages [36]. Another possible explanation is that the higher dehydration process during ripening in the reformulated sausages resulted in lower L^* values [24].

Regarding a^* values, only SOY samples showed lower values (5.08; $p < 0.05$) than the other three batches (between 5.99 and 6.07). Despite these differences, it is worth mentioning that the values of all batches were similar. This means that the strategy of animal replacement by oil-in-water emulsions stabilized in Prosella gels gives rise to products that maintain the characteristic red color of the sausage.

As occurs with L^* values, the yellowness (b^*) also decreased as the animal fat was replaced by emulsion gels. In this sense, b^* values of reformulated sausages were similar (between 4.81 and 5.17), while CON samples presented the highest values (8.49; $p < 0.001$). The same trend was observed by Lorenzo et al. [36], who found that the replacement of animal fat by encapsulated fish oil in konjac matrix produced a decreased in b^* values. In contrast, other authors described that the use of emulsions or oleogels as fat replacers increased or did not affect the yellowness of the final product [54,56,62]. As reported by other authors, the differences between studies could be due to the characteristic color of the oleogelators or emulsifiers used in the production of the gels as well as the amount of oil used in the emulsion or the oleogel and their natural color. For example, the use of beeswax linseed oleogel in the reformulation of sausage resulted in higher b^* values, and the authors attributed this fact to the yellow color of both oil and beeswax [56].

Texture parameters were also influenced by fat replacement. All parameters increased as the animal fat was substituted by oil-in-water emulsion gels. In accordance with our results, the formation of harder structures has been found as fat content decreased in the product [20,22,24]. Focusing on the hardness, which is the most important parameter from the consumer's point of view, this parameter increased from 61.98 N in CON samples to about 275 N in OLI and CAN and 316 N in SOY samples. Similarly to our findings, the replacement of pork backfat by encapsulated fish oil in konjac matrix of pork sausage produced a progressive increase in hardness. Conversely, the use of beeswax oleogel or protein isolate/gelatin emulsion decreased the hardness of fuet [62]. Other authors also reported a decrease in the hardness values using oil-in-water emulsions in konjac matrix in different dry-fermented sausages [32,52]. It is well known that the dehydration process during ripening and the final moisture content is the most important parameter in the textural characteristics of this type of product. Thus, the differences observed between batches in the present research, as well as between the data of different studies could be attributed to the moisture content. In agreement with other studies about salchichón [22,36,56] and fuet [62], a significant and negative correlation between hardness and moisture content was observed ($r = -0.904$; $p < 0.01$). Similarly, springiness ($r = -0.595$; $p < 0.01$), cohesiveness ($r = -0.614$; $p < 0.01$), gumminess ($r = -0.881$; $p < 0.01$), and chewiness ($r = -0.863$; $p < 0.01$) also showed a significant and negative correlation with moisture. Thus, the present results corroborate that the moisture content is the most important factor that influences the texture parameters of the dry-fermented sausages.

3.2. Fatty Acids Composition of Deer Sausages

The reformulation of deer sausages significantly affected the content of fatty acids (Table 3). It is important to note that in all samples, monounsaturated fatty acids (MUFA) were the most abundant fatty acids, followed by saturated fatty acids (SFA) and polyunsaturated fatty acids (PUFA). Individually, in CON, OLI, and CAN samples the majority was the C18:1n-9, followed by C16:0, C18:2n-6, and C18:0. This profile was previously reported in similar dry-fermented sausages [14,36,55,56]. In contrast, in SOY samples the order was C18:1n-9 > C18:2n-6 > C16:0 > C18:0.

Table 3. Effect of fat source on fatty acid composition (expressed as g/100 g of total fatty acids) of dry-fermented deer sausage.

	Fat Source				Sig.
	CON	OLI	CAN	SOY	
C14:0	1.32 ± 0.01 [d]	1.02 ± 0.01 [b]	1.00 ± 0.01 [a]	1.08 ± 0.01 [c]	***
C15:0	0.09 ± 0.002 [b]	0.08 ± 0.001 [a]	0.08 ± 0.001 [a]	0.10 ± 0.002 [b]	***
C16:0	22.68 ± 0.05 [d]	19.10 ± 0.08 [c]	17.28 ± 0.06 [a]	18.67 ± 0.07 [b]	***
C16:1n-7	2.12 ± 0.01 [d]	1.75 ± 0.01 [c]	1.65 ± 0.01 [b]	1.62 ± 0.01 [a]	***
C17:0	0.41 ± 0.002 [c]	0.33 ± 0.002 [a]	0.34 ± 0.002 [a]	0.36 ± 0.003 [b]	***
C17:1n-7	0.30 ± 0.003 [c]	0.25 ± 0.002 [a]	0.26 ± 0.001 [b]	0.26 ± 0.002 [b]	***
C18:0	12.12 ± 0.05 [d]	9.79 ± 0.05 [b]	8.92 ± 0.04 [a]	10.02 ± 0.05 [c]	***
9t-C18:1	0.29 ± 0.01 [c]	0.26 ± 0.01 [b]	0.22 ± 0.01 [a]	0.26 ± 0.01 [b]	***
C18:1n-9	39.69 ± 0.07 [b]	48.14 ± 0.12 [d]	46.89 ± 0.13 [c]	35.40 ± 0.12 [a]	***
C18:1n-7	2.58 ± 0.02 [b]	2.41 ± 0.02 [a]	2.74 ± 0.01 [c]	2.36 ± 0.03 [a]	***
C18:2n-6	14.29 ± 0.09 [b]	12.90 ± 0.09 [a]	14.90 ± 0.11 [c]	24.11 ± 0.19 [d]	***
C20:0	0.22 ± 0.001 [a]	0.27 ± 0.001 [b]	0.31 ± 0.002 [d]	0.29 ± 0.001 [c]	***
C18:3n-6	0.05 ± 0.001 [a]	0.06 ± 0.001 [a]	0.14 ± 0.002 [c]	0.13 ± 0.001 [b]	***
C20:1n-9	0.93 ± 0.003 [c]	0.73 ± 0.003 [a]	1.02 ± 0.003 [d]	0.77 ± 0.003 [b]	***
C18:3n-3	0.67 ± 0.01 [a]	0.87 ± 0.01 [b]	2.20 ± 0.04 [c]	2.43 ± 0.03 [d]	***
C20:2n-6	0.62 ± 0.004 [d]	0.45 ± 0.003 [c]	0.40 ± 0.004 [b]	0.36 ± 0.002 [a]	***
C22:0	0.06 ± 0.001 [a]	0.09 ± 0.001 [b]	0.15 ± 0.001 [c]	0.21 ± 0.002 [d]	***
C20:3n-6	0.13 ± 0.002 [c]	0.12 ± 0.002 [b]	0.11 ± 0.002 [a]	0.11 ± 0.001 [a]	***
C20:3n-3	0.09 ± 0.001 [c]	0.07 ± 0.001 [b]	0.06 ± 0.001 [a]	0.06 ± 0.001 [a]	***
C20:4n-6	0.58 ± 0.01 [a]	0.63 ± 0.02 [b]	0.61 ± 0.01 [a, b]	0.63 ± 0.01 [b]	*
C20:5n-3	0.09 ± 0.004	0.09 ± 0.004	0.09 ± 0.01	0.10 ± 0.004	ns
C22:5n-6	0.12 ± 0.002 [b]	0.10 ± 0.002 [a]	0.09 ± 0.002 [a]	0.09 ± 0.001 [a]	***
C22:5n-3	0.18 ± 0.004 [a]	0.20 ± 0.01 [b]	0.20 ± 0.005 [b]	0.20 ± 0.003 [b]	**
C22:6n-3	0.04 ± 0.001 [a, b]	0.04 ± 0.001 [a, b]	0.04 ± 0.001 [a]	0.04 ± 0.000 [b]	*
SFA	37.08 ± 0.09 [c]	30.81 ± 0.14 [b]	28.23 ± 0.10 [a]	30.91 ± 0.12 [b]	***
MUFA	45.68 ± 0.08 [b]	53.34 ± 0.11 [d]	52.65 ± 0.13 [c]	40.49 ± 0.14 [a]	***
PUFA	16.94 ± 0.12 [b]	15.58 ± 0.13 [a]	18.91 ± 0.17 [c]	28.34 ± 0.22 [d]	***
n-6	15.79 ± 0.11 [b]	14.24 ± 0.11 [a]	16.26 ± 0.13 [c]	25.43 ± 0.19 [d]	***
n-3	1.08 ± 0.01 [a]	1.27 ± 0.02 [b]	2.58 ± 0.04 [c]	2.83 ± 0.03 [d]	***
n-6/n-3	14.66 ± 0.07 [d]	11.20 ± 0.08 [c]	6.31 ± 0.05 [a]	9.00 ± 0.04 [b]	***

[a–d] Mean values in the same row (corresponding to the same parameter) not followed by a common letter differ significantly ($p < 0.05$; Duncan test); Sig.: significance: * ($p < 0.05$), ** ($p < 0.01$), *** ($p < 0.001$), ns (not significant); SEM: standard error of the mean; SFA: saturated fatty acids; MUFA: monounsaturated fatty acids; PUFA: polyunsaturated fatty acids. Treatments: CON: sausages prepared 100% pork fat; OLI: sausages reformulated with 50% of pork fat replaced by olive oil; CAN: sausages reformulated with 50% of pork fat replaced by canola oil; SOY: sausages reformulated with 50% of pork fat replaced by soy oil.

Regarding SFA content, the replacement of pork backfat by oil-in-water emulsions reduced SFA between 16.6% and 23.8%. Except for the contents of C20:0 and C22:0 (minor fatty acids; < 0.5% of total fatty acids), all individual SFA showed a significant reduction with the animal fat replacement. These findings were also observed by other authors, in which reformulated sausages with emulsified and gelled oils presented significantly lower SFA values than those formulated with pork fat [50,55,56,62].

The MUFA content varied among the batches. The samples from OLI and CAN had significantly higher values of total MUFA than CON, due mainly to the higher amounts of C18:1n-9 in these batches. However, the content of MUFA in SOY sausages was lower ($p < 0.05$) than CON samples.

As occurs in MUFA, PUFA content also varied depending on the batch. In comparison with CON samples, total PUFA decreased in OLI samples and increased in CAN and SOY sausages. These differences are mainly due to the different contents of C18:2n-6 and to a lesser extent of C18:3n-3 between pork backfat and oils. In this regard, it is important to note that the improvement in the fatty acids profile of the reformulated samples are also related to the high content of C18:3n-3 when fat was replacement by oil-in-water emulsions. This increase was more pronounced in the CAN and SOY sausages. As a result of these variations, significant differences were also observed in the content of total n-3 and n-6 fatty acids. Thus, following the same trend described for C18:3n-3, the content of n-3 fatty acids increased in all reformulated batches. The highest concentration was observed in the SOY (2.83 g/100 g) batch followed by CAN (2.58 g/100 g), OLI (1.27 g/100 g), and finally CON (1.08 g/100 g). Similarly, the content of total n-6 was affected by the variation in the amount of C18:2n-6. In SOY (25.43 g/100 g) and CAN (16.26 g/100 g) samples the total n-6 content increased, while in the OLI (14.24 g/100 g) sausages n-6 value decreased in comparison with CON samples (15.79 g/100 g).

As commented above, several studies investigated the effect of the addition of oil-in-water emulsion as animal fat replacer in dry-fermented sausages [36,50,54,55,62]. Generally speaking, in all studies the authors found a significant reduction of SFA content and a progressive increased of MUFA and/or PUFA. However, the intensity of this effect depends on multiple factors, such as the level of animal fat replacement (partial or total replacement), the amount of oil used in the emulsions, and the type of oil used in the sausage manufacture.

Thus, as a general conclusion for fatty acid composition, due to the low fat content of deer meat [7], dry-fermented sausages reflected the fatty acid composition of the fat or oil used in their manufacture. Therefore, the differences in fatty acids commented above are due to the fat/oil composition. The same trend was reported by other authors who observed that the fatty acid composition of the oil used in the sausage manufacture conditioned the fatty acids content of the final product [50,55,62].

Finally, due to the variations in total n-6 and n-3 fatty acids, the inclusion of emulsion gel in sausages significantly decreased of the n-6/n-3 ratio. This ratio is of great interest, because unbalanced diets with high n-6/n-3 ratio are associated with an increased risk of developing cardiovascular diseases, cancer and depressive disorders [64], while the consumption of n-3 has a protective effect against these diseases. Following nutritional recommendations, this ratio should be less than 4 [65], being the optimal value 1 [64]. In our study, all samples presented higher values than those recommended. The use of canola oil in the formulation of oil-in-water emulsion showed the best n-6/n-3 values (6.31), followed by SOY (9.00), OLI (11.20), and CON (14.66). Although these values were higher than recommended, the inclusion of emulsion oils resulted in a significant improve in the nutritional quality of the deer sausages.

3.3. Volatile Compounds of Deer Sausages

The reformulation significantly affected the levels of several volatiles organic compounds (VOC) of deer sausages (Table 4). After 45 days, a total of 84 compounds were identified in dry-fermented sausage samples. These compounds were grouped into seven chemical classes: terpenoids and benzene-derived compounds (34), aldehydes (16), alcohols (11), esters (8), ketones (6), acids (5), and others (4).

Terpenoids and benzene-derived compounds were the predominant volatiles in all formulations, representing 71.3% of total VOC in CON samples and about 73% of total VOC in the reformulated sausages. In agreement with our findings, previous studies in salchichón showed that terpenes were the predominant class [22,27,28], representing between 58% and 71% of total VOC. The major terpene was *o*-cymene followed by 1R-α-pinene and β-phellandrene. Moreover, significant contents of thujene (both, α- and β-thujene), β-myrcene, α-phellandrene, 3-carene, D-limonene, terpinene (both, α- and δ-terpinene), safrole, and caryophyllene were also found in salchichón samples.

Table 4. Effect of fat source on volatile compounds (expressed as AU·10^4/g) of dry-fermented deer sausage.

Compound Information			Fat Source				Sig.
Name	**LRI**	**m/z**	**Control**	**Olive Oil**	**Canola Oil**	**Soy Oil**	
Glycidol	468	44	28.76 ± 1.33 [a]	70.80 ± 3.51 [c]	60.79 ± 2.83 [b]	65.44 ± 1.67 [b,c]	***
1-Propanol	546	59	9.19 ± 1.10 [d]	5.29 ± 0.50 [b]	7.25 ± 0.34 [c]	3.21 ± 0.16 [a]	***
2-Butanol, (R)-	583	45	5.13 ± 0.47	4.48 ± 0.43	5.54 ± 0.41	5.25 ± 0.35	ns
1-Butanol	696	56	6.59 ± 0.49 [b]	4.66 ± 0.37 [a]	5.26 ± 0.54 [a]	4.05 ± 0.25 [a]	***
(R)-(-)-2-Pentanol	744	45	3.04 ± 0.18 [b]	1.56 ± 0.11 [a]	1.38 ± 0.08 [a]	7.04 ± 0.37 [c]	***
1-Pentanol	851	70	18.45 ± 1.26 [c]	9.67 ± 0.36 [a]	13.75 ± 1.05 [b]	14.67 ± 0.73 [b]	***
2,3-Butanediol	922	45	119.2 ± 9.13 [a]	412.7 ± 16.80 [c]	327.6 ± 16.69 [b]	489.4 ± 24.39[d]	***
1-Hexanol	971	56	32.61 ± 2.01 [c]	14.80 ± 0.72 [a]	23.00 ± 2.30 [b]	21.07 ± 0.93 [b]	***
1-Octen-3-ol	1080	57	35.24 ± 1.83 [b]	26.94 ± 0.91 [a]	36.76 ± 1.14 [b,c]	39.84 ± 1.18 [c]	***
1-Octanol	1165	84	7.11 ± 0.34 [a]	12.44 ± 1.34 [b]	11.61 ± 0.36 [b]	10.92 ± 0.86 [b]	***
Linalool	1185	71	26.64 ± 0.92 [a]	44.08 ± 2.32 [b]	49.90 ± 1.73 [c]	43.37 ± 2.35 [b]	***
Total Alcohols			292.0 ± 11.09 [a]	607.4 ± 15.72 [c]	542.8 ± 17.04 [b]	704.2 ± 22.78[d]	***
Propanal	496	58	45.25 ± 3.23 [b]	87.61 ± 3.25 [c]	10.88 ± 1.24 [a]	6.94 ± 0.40 [a]	***
Propanal, 2-methyl-	529	72	10.01 ± 0.72 [a]	32.23 ± 1.10 [c]	29.77 ± 1.37 [b,c]	27.70 ± 0.74 [b]	***
Butanal, 3-methyl-	640	58	46.75 ± 1.96 [a]	163.04 ± 3.68 [c]	149.72 ± 7.49 [c]	135.07 ± 4.91 [b]	***
Butanal, 2-methyl-	654	57	23.55 ± 1.20 [a]	83.42 ± 2.64 [c]	74.17 ± 2.90 [b]	69.61 ± 3.39 [b]	***
Pentanal	717	58	35.68 ± 3.17 [b]	26.32 ± 1.06 [a]	34.51 ± 1.33 [b]	44.89 ± 2.26 [c]	***
2-Butenal, 2-methyl-	801	84	3.08 ± 0.19 [a]	4.83 ± 0.26 [b]	7.11 ± 0.45 [c]	4.39 ± 0.28 [b]	***
Hexanal	872	56	249.3 ± 27.99 [b]	83.76 ± 4.08 [a]	209.8 ± 18.21 [b]	209.9 ± 13.97 [b]	***
Methylal	903	45	192.2 ± 13.73	157.1 ± 9.73	183.9 ± 7.64	174.9 ± 6.28	ns
Heptanal	993	70	13.85 ± 1.53 [c]	7.82 ± 0.39 [a]	11.69 ± 0.86 [b,c]	9.96 ± 0.66 [a,b]	***
Hexanal, 5-methyl-	993	55	9.11 ± 1.44 [c]	3.83 ± 0.19 [a]	6.28 ± 0.40 [b]	4.66 ± 0.26 [a,b]	***
Methional	1022	48	13.31 ± 0.87 [a]	32.75 ± 0.97 [b]	35.97 ± 1.26 [c]	31.29 ± 0.59 [b]	***
Benzaldehyde	1072	106	99.77 ± 7.53 [a]	125.2 ± 2.69 [b]	152.8 ± 12.04 [c]	111.6 ± 4.72 [a b]	***
Benzeneacetaldehyde	1155	91	57.85 ± 3.38 [a]	58.51 ± 3.43 [a]	89.15 ± 3.89 [b]	85.82 ± 5.59 [b]	***
2-Octenal, (E)-	1160	55	12.98 ± 3.53	11.17 ± 0.41	13.49 ± 1.06	11.11 ± 0.62	ns
Nonanal	1187	98	7.94 ± 0.61 [c]	5.84 ± 0.24 [a]	7.45 ± 0.49 [b,c]	6.32 ± 0.28 [a b]	**
3-Isopropylbenzaldehyde	1321	148	2.95 ± 0.20 [a]	3.55 ± 0.15 [b]	3.66 ± 0.18 [b]	3.37 ± 0.20 [a b]	*
Total Aldehydes			823.5 ± 27.86 [a]	887.0 ± 19.57 [a b]	1020 ± 26.97 [c]	937.6 ± 23.93 [b]	***

Table 4. *Cont.*

Compound Information			Fat Source				Sig.
Name	LRI	m/z	Control	Olive Oil	Canola Oil	Soy Oil	
Carbon disulfide	504	76	18.95 ± 2.78 b	19.23 ± 2.08 b	12.76 ± 0.88 a	13.19 ± 1.31 a	*
Furan, 2-pentyl-	1065	81	14.09 ± 0.92 a,b	11.85 ± 0.64 a	15.77 ± 1.07 b	15.79 ± 0.55 b	**
1,3-Benzenediol, monobenzoate	1072	77	39.70 ± 1.48 a	56.20 ± 1.71 b	67.41 ± 2.62 c	54.23 ± 1.83 b	***
Hexane, 2,4,4-trimethyl-	1110	57	51.81 ± 3.49 b	24.91 ± 2.33 a	25.13 ± 1.62 a	31.42 ± 2.90 a	***
Total Others			124.5 ± 4.06	112.2 ± 2.93	121.0 ± 3.65	114.6 ± 4.35	ns
2,3-Butanedione	563	86	14.39 ± 1.09 a,b	17.62 ± 1.09 c	11.79 ± 0.45 a	16.51 ± 0.96 b c	***
2-Butanone	569	72	32.67 ± 2.16 a	58.90 ± 2.54 c	60.79 ± 1.61 c	45.88 ± 1.52 b	***
3-Pentanone	722	57	18.34 ± 1.30 b	10.49 ± 1.46 a	16.93 ± 1.03 b	16.74 ± 0.95 b	***
Acetoin	785	45	76.61 ± 3.67 a	151.9 ± 6.34 c	132.3 ± 4.82 b	138.8 ± 3.58 b c	***
2-Heptanone	985	58	3.58 ± 0.94 b	1.27 ± 0.05 a	1.90 ± 0.17 a	6.31 ± 0.25 c	***
Butyrolactone	1072	86	17.11 ± 0.57 a	36.26 ± 1.01 b	41.24 ± 1.51 c	37.04 ± 1.06 b	***
Total Ketones			162.7 ± 4.96 a	276.4 ± 9.39 b	265.0 ± 6.57 b	261.2 ± 5.47 b	***
Acetic acid	676	60	1513 ± 40.31 a	2088 ± 69.46 b	2013 ± 44.52 b	1964 ± 58.87 b	***
Butanoic acid	932	60	341.6 ± 13.55 a	578.7 ± 25.65 c	683.7 ± 23.27d	513.7 ± 14.79 b	***
Butanoic acid, 3-methyl-	986	60	18.87 ± 1.02 a	34.33 ± 1.72 b	37.57 ± 1.33 b	36.75 ± 1.46 b	***
Pentanoic acid	1029	60	13.40 ± 1.15 a	17.94 ± 0.52 b	22.04 ± 0.70 c	20.46 ± 0.97 c	***
Hexanoic acid	1117	60	55.30 ± 2.41 a	65.52 ± 2.51 b	68.80 ± 2.49 b c	74.99 ± 3.03 c	***
Total Acids			1943 ± 46.64 a	2785 ± 85.03 b,c	2825 ± 55.68 c	2610 ± 71.52 b	***
Propanoic acid, ethyl ester	728	57	28.23 ± 1.93 a	36.39 ± 1.83 b	52.58 ± 2.42 c	26.61 ± 1.05 a	***
n-Propyl acetate	736	61	2.96 ± 0.17 a	3.96 ± 0.31 b	5.72 ± 0.26 c	2.37 ± 0.07 a	***
Propanoic acid, 2-methyl-, ethyl ester	800	71	4.74 ± 0.31 a	6.60 ± 0.35 b	7.01 ± 0.39 b	4.81 ± 0.18 a	***
Butanoic acid, ethyl ester	860	88	126.8 ± 7.95 a	178.3 ± 8.66 c	198.8 ± 9.69 c	155.2 ± 4.68 b	***
Butanoic acid, 3-methyl-, ethyl ester	925	88	12.87 ± 0.54 a	20.66 ± 1.18 b	21.67 ± 1.21 b	20.41 ± 0.62 b	***
Pentanoic acid, ethyl ester	978	88	10.15 ± 0.93 a	11.67 ± 0.96 a	15.01 ± 1.73 b	11.40 ± 0.62 a	*
Octanoic acid, ethyl ester	1250	88	17.66 ± 0.81 a	19.99 ± 0.72 b	22.12 ± 0.75 b	20.09 ± 0.72 b	**
Decanoic acid, ethyl ester	1397	88	7.17 ± 0.34d	4.92 ± 0.17 b	5.63 ± 0.18 c	3.82 ± 0.16 a	***
Total Esters			210.6 ± 9.73 a	282.5 ± 11.66 c	328.5 ± 11.68d	244.8 ± 6.75 b	***

Table 4. *Cont.*

Compound Information			Fat Source				Sig.
Name	LRI	m/z	Control	Olive Oil	Canola Oil	Soy Oil	
Benzene, 1,3-dimethyl-	939	91	5.13 ± 0.19 [a]	5.51 ± 0.09 [a,b]	5.34 ± 0.16 [a,b]	5.67 ± 0.09 [b]	*
α-Thujene	990	92	380.1 ± 14.23 [a]	622.9 ± 30.47 [b]	596.3 ± 31.77 [b]	579.3 ± 26.52 [b]	***
1R-α-Pinene	998	93	1208 ± 53.60 [a]	1745 ± 58.31 [b]	1728 ± 72.11 [b]	1688 ± 74.63 [b]	***
Camphene	1018	93	42.92 ± 3.36 [a]	57.17 ± 3.08 [b]	53.59 ± 2.65 [b]	49.58 ± 2.28 [a,b]	**
(+)-Camphene	1018	121	41.92 ± 3.27 [a]	54.49 ± 2.88 [b]	51.71 ± 2.42 [b]	47.52 ± 2.51 [a,b]	*
β-Thujene	1046	136	398.5 ± 17.13 [a]	658.7 ± 28.11 [b]	634.3 ± 29.62 [b]	605.2 ± 29.37 [b]	***
Pseudolimonene	1048	121	192.0 ± 10.53 [a]	233.9 ± 11.56 [b]	241.9 ± 13.38 [b]	229.7 ± 14.21 [b]	*
β-Myrcene	1058	93	380.4 ± 20.94 [a]	692.0 ± 23.94 [b]	725.1 ± 28.23 [b]	692.2 ± 31.09 [b]	***
α-Phellandrene	1075	93	471.7 ± 27.49 [a]	980.3 ± 34.30 [b]	1025 ± 57.47 [b]	934.0 ± 41.24 [b]	***
3-Carene	1078	121	373.6 ± 23.83 [a]	685.0 ± 30.62 [b]	661.0 ± 31.60 [b]	712.9 ± 36.69 [b]	***
α-Terpinene	1087	121	236.4 ± 20.23 [a]	416.7 ± 17.56 [b]	465.2 ± 15.54 [c]	397.7 ± 14.10 [b]	***
D-Limonene	1098	79	460.9 ± 8.99 [a]	707.0 ± 18.02 [b]	770.7 ± 27.49 [c]	682.5 ± 22.11 [b]	***
o-Cymene	1102	119	2074 ± 78.00 [a]	2702 ± 103.42 [c]	2657 ± 85.9 [b c]	2426 ± 95.25 [b]	***
β-Phellandrene	1103	93	940.3 ± 57.34 [a]	1418 ± 56.98 [b]	1592 ± 58.51 [c]	1484 ± 53.9 [b c]	***
Eucalyptol	1109	154	12.35 ± 0.33 [a]	14.70 ± 0.20 [b]	16.13 ± 0.35 [c]	14.56 ± 0.25 [b]	***
β-Ocimene	1114	91	5.76 ± 0.50 [a]	9.41 ± 0.86 [b]	12.11 ± 0.53 [c]	10.42 ± 0.47 [b c]	***
δ-Terpinene	1125	77	147.0 ± 9.96 [a]	228.4 ± 11.26 [b]	233.5 ± 14.97 [b]	229.8 ± 7.73 [b]	***
(+)-4-Carene	1152	121	137.1 ± 9.85 [a]	237.7 ± 12.49 [b]	261.5 ± 16.38 [b]	246.0 ± 9.91 [b]	***
Benzyl alcohol	1162	79	66.16 ± 2.94 [a]	78.58 ± 2.00 [b]	102.1 ± 3.00[d]	90.14 ± 3.31 [c]	***
m-Cymenene	1167	117	98.41 ± 5.89 [a]	124.1 ± 7.87 [b]	117.5 ± 7.13 [b]	97.66 ± 4.68 [a]	**
trans-4-Thujanol	1196	71	16.67 ± 0.60 [a]	27.56 ± 0.84 [b]	32.04 ± 1.14 [c]	27.23 ± 1.02 [b]	***
p-Cresol	1223	107	7.64 ± 0.39 [a]	11.17 ± 0.39 [b]	11.71 ± 0.58 [b]	11.78 ± 0.49 [b]	***
Phenylethyl Alcohol	1226	91	31.21 ± 1.19 [a]	37.32 ± 2.55 [b,c]	39.09 ± 1.35 [c]	32.83 ± 1.39 [a,b]	**
α-Phellandren-8-ol	1251	91	31.67 ± 1.40 [a]	38.76 ± 0.98 [b c]	41.97 ± 0.95 [c]	37.12 ± 1.35 [b]	***
Terpinen-4-ol	1254	71	187.4 ± 6.69 [a]	253.4 ± 7.80 [b]	291.8 ± 6.59 [c]	250.2 ± 9.24 [b]	***
α-Terpineol	1271	121	9.73 ± 0.35 [a]	15.19 ± 0.43 [b]	16.90 ± 0.42 [c]	14.54 ± 0.61 [b]	***
Safrole	1340	162	430.0 ± 14.41 [a]	679.0 ± 16.97 [c]	712.3 ± 13.91 [c]	624.4 ± 23.68 [b]	***
δ-Elemene	1357	121	25.12 ± 0.90 [a]	57.76 ± 2.18 [b c]	62.67 ± 2.92 [c]	54.74 ± 3.48 [b]	***
α-Cubebene	1365	161	9.03 ± 0.55 [a]	14.77 ± 0.50 [b]	17.54 ± 0.65 [c]	15.52 ± 0.83 [b]	***

Table 4. *Cont.*

Compound Information			Fat Source				Sig.
Name	LRI	m/z	Control	Olive Oil	Canola Oil	Soy Oil	
Copaene	1388	161	93.26 ± 5.16 [a]	147.7 ± 5.24 [b]	174.6 ± 5.65 [c]	155.7 ± 7.62 [b]	***
Methyleugenol	1422	178	12.88 ± 0.42 [a]	23.53 ± 0.60 [c]	25.74 ± 1.01 [d]	21.35 ± 0.82 [b]	***
Caryophyllene	1430	133	258.8 ± 9.19 [a]	451.8 ± 16.22 [b]	481.3 ± 20.21 [b]	455.6 ± 21.50 [b]	***
Myristicin	1493	192	31.03 ± 0.99 [a]	51.40 ± 1.23 [b]	56.88 ± 1.07 [c]	49.53 ± 1.69 [b]	***
Elemicin	1502	208	3.94 ± 0.13 [a]	7.17 ± 0.14 [c]	6.87 ± 0.13 [c]	5.92 ± 0.17 [b]	***
Total Terpenoids and Benzene-Derive Compounds			8821 ± 303.2 [a]	13,490 ± 334.8 [b]	13,923 ± 407.1 [b]	12,981 ± 388.7 [b]	***
TOTAL COMPOUNDS			12,378 ± 324.4 [a]	18,441 ± 373.7 [b,c]	19,027 ± 446.6 [c]	17,854 ± 369.8 [b]	***

[a–d] Mean values in the same row (corresponding to the same parameter) not followed by a common letter differ significantly ($p < 0.05$; Duncan test); Sig.: significance: * ($p < 0.05$), ** ($p < 0.01$), *** ($p < 0.001$), ns (not significant); SEM: standard error of the mean; LRI: linear retention index calculated for DB-624 capillary column (J&W scientific: 30 m × 0.25 mm id, 1.4 μm film thickness) installed on a gas chromatograph equipped with a mass selective detector; *m/z*: quantifier ion. Treatments: CON: sausages prepared 100% pork fat; OLI: sausages reformulated with 50% of pork fat replaced by olive oil; CAN: sausages reformulated with 50% of pork fat replaced by canola oil; SOY: sausages reformulated with 50% of pork fat replaced by soy oil.

The same individual terpenes were previously reported by other authors in salchichón samples [22,27,28,31,66] and in other dry-cured sausages formulated with pepper [20]. A previous study found these terpenes in pepper [67,68]. However, some dry-cured products without added spices presented terpenes such as D-limonene, α- and β-pinene, or 3-carene, which could be related to the diet of the animals [68,69]. With these results it is easy to conclude that the VOC derived from spices clearly dominated the volatile profile of sausage, while the VOC derived from lipid oxidation, microbial metabolism, or other physicochemical changes contribute to a lesser extent to the total volatile compounds of this type of sausage. These terpenes have been described to contribute with fresh, menthol, herbal, and lemon notes [27,28,70].

Generally speaking, the reformulated sausages presented higher individual and total terpenoids and benzene-derive compounds ($p < 0.001$) (about 13,000 AU·10^4/g) than CON samples (8821 AU·10^4/g). Similarly, other authors reported that the reduced-fat sausages resulted in a higher terpenoid content [20,30]. This fact could be related to the fat that acts as solvent for these compounds, delaying their release [25]. Additionally, in the present study the content of terpenes among reformulated (OLI, CAN, and SOY) batches were similar (except some individual terpene). The higher VOC could be related to the low moisture content in the reformulated sausages. This fact is corroborated by the negative and significant correlation observed between pH and moisture values ($r = -0.674$; $p < 0.001$).

The second most important VOC in sausage samples were acids, representing about 15% of total VOC in all samples. Among them, acetic acid (1513 AU·10^4/g in CON and about 2000 AU·10^4/g in the other batches) was the major organic acid, followed by butanoic (341–683 AU·10^4/g) and hexanoic acids (55.3–74.9 AU·10^4/g). All these compounds were found in previous studies of salchichón [27,28,31,66]. Moreover, in agreement with our results, some authors reported that acetic acid was the most important organic acid in dry-fermented sausages [20,28,67]. The most probable origin of butanoic, pentanoic, and hexanoic acids in dry-fermented sausages is the carbohydrate fermentation induced by microorganisms, such as lactic bacteria and staphylococci [66,70]. However, the origin of butanoic acid, 3 methyl is the oxidation of its Streker aldehyde (butanal, 3 methyl). The acetic acid gives notes of ripened aroma, while butanoic acid gives cheese notes [27]. It is well known that the acids found in this research (butanoic, acetic, and butanoic, 3-methyl), together with 1-octen-3-ol have a vital role in the development of typical aroma of sausages [28,67]. As occurs in the terpenoids section, the reformulated sausages presented higher amounts of total and individual acids than CON samples. In this case, also a significant and negative correlation was observed between moisture and acids ($r = -0.657$; $p < 0.001$), which could explain the differences found in the VOC amounts.

The aldehydes content represents 4.8–5.4% of total VOC in reformulated samples and 6.6% in CON samples. In this group, the most abundant compound was the hexanal (except for OLI samples, in which the butanal, 3-methyl had the highest amounts). Other authors also found that hexanal was the most important aldehyde in dry-fermented sausages [22,27,30,36], while benzaldehyde or benzenacetaldehyde were found the most abundant aldehydes in other research [28,31]. In this study, other important aldehydes found in deer sausage samples were butanal, 2-methyl, butanal, 3-methyl, methylal, benzaldehyde, and benzeneacetaldehyde. The origin of the aldehydes could be grouped in two main routes. The linear aldehydes derived mainly from lipid oxidation [28,29], while the branched aldehydes are related to amino acid degradation and proteolysis [28,71]. The contents of pentanal, hexanal, or 2-octenal, depends on linoleic, linolenic, and arachidonic fatty acids, while the content of heptanal and nonanal derived from oleic fatty acid [29]. As could be seen in Table 4, the hexanal content was not influenced by the replacement of animal fat by emulsions containing canola or soy oils, while the samples reformulated with the olive oil presented a significantly lower value than the other batches. A similar trend was observed in the contents of pentanal, heptanal, and nonanal. Thus, OLI samples presented a significantly lower content of lipid-derived VOC than the other three batches (with similar values among them), which could be related to the high content of tocopherols in olive oil. It is well-known that the presence of tocopherols inhibit the lipid oxidation process [29], thus this fact could explain the lower values of these aldehydes in OLI samples. Moreover, it is also important to note that, besides the high unsaturated level of oils used in the present study, the VOC derived from lipid oxidation process did not have a significant increase in comparison with CON samples, which could also be related to the presence of natural antioxidants (phenolic compounds) in the vegetable oils. In contrast with our results, the reformulation of salchichón samples with encapsulated fish oil immobilized in konjac matrix resulted in a significant increase of lipid-derived aldehydes [36]. In another study, the use of emulsion gel with olive and chia oil mixture as animal fat replacer also increased the amounts of aldehydes in the fuet samples [62].

On the other hand, contrary to observed results for linear aldehydes, branched aldehydes and cycloaldehydes increased significantly as the animal fat was replaced by the emulsion gels. The origin of propanal, 2-methyl, butanal, 2-methyl and butanal, 3-methyl is the deamination-decarboxylation of the amino acids valine, isoleucine, and leucine, respectively [28,71]. The main route for the formation of benzaldehyde and benzeneacetaldehyde is the Streker degradation of some amino acids such as leucine or phenylalanine [72].

Regarding the aroma notes, linear aldehydes contributed with sweet, floral, grassy, and fruity notes. Hexanal presents a rancid aroma at high amounts, while in low content it gives a pleasant grassy aroma [28], and linear aldehydes derived from oleic acid oxidation had pleasant meaty notes [70]. The two cycloaldehydes detected in salchichón samples contributed with floral, acorn, and bitter almonds notes (benzaldehyde) and with acorn, rancid, and pungent aroma (benzeneacetaldehdye) [27,28]. Finally, propanal, 2-methyl had pungent and nutty odor and butanal, 3-methyl presented acorn-like, salty, fruity, and cheesy aroma [28]. Additionally, it is important to note that due to their low odor thresholds, aldehydes are one of the main VOC that contributed to the typical sausage aroma. Particularly, the content of butanal, 3-methyl was reported as important VOC that impart a characteristic "ripened flavor" [28,70].

Alcohols represented between 2.35% and 3.94% of total VOC. The major alcohol was the 2,3-butanediol, followed by glycidol, linalool, 1-octen-3-ol, and 1-hexanol. These compounds and 1-pentanol or 1-octanol are commonly detected in fermented sausage samples [20,27,28,31]. The alcohols of fermented sausages are mainly generated from the reduction of aldehydes [66]. As commented above, the 1-octen-3-ol is described as an important VOC that contributed to characteristic aroma of dry-cured products [28,71]. This compound is derived from the oxidation process of linoleic acid [28,29]. Similarly, 1-pentanol is derived from the degradation of hydroperoxides and 1-hexanol from reduction of hexanal, while the 1-octanol arises from oleic acid oxidation [28,29,67]. As occurs in other VOC, the content of total alcohols was higher in reformulated than in CON sausages, mainly due to the higher amounts of 2,3-butanediol, glycidol, and linalool in these samples. A similar trend was described in another study in which the authors found 2,3-butanediol only in fat reduced sausages [20]. In contrast, the content of lipid-derived alcohols did not show a clear trend, and CON samples presented the highest values of some of them such as 1-hexanol of 1-pentanol. This fact is in accordance with the content of linear aldehydes, in which OLI samples presented the lowest values of lipid oxidation VOC. Despite this result, a significant and negative correlation was observed between moisture and total alcohols ($r = -0.855$; $p < 0.001$), which could explain the differences found in the VOC amounts.

The content of ketones and esters were similar (about 1.5% of total VOC). The most abundant ketone was the acetoin, followed by 2-butanone, butyrolactone, 3-pentanone, and 2,3-butanedione. Other authors also reported in salchichón that acetoin was the major ketone [28,31]. The origin of linear ketones is lipid oxidation of free fatty acids, while acetoin are formed through Maillard reactions [28]. Some compounds of this class are 2-butanone and 2-heptanone (give a characteristic blue cheese aroma and has an intense odor) and the acetoin (buttery and sweet odor and a very low odor threshold) [28]. However, high amounts of butyrolactone in sausage were previously reported [28]. This VOC contributes with pleasant butter, fatty, creamy, fruity, and coconut-like nuances. Regarding esters, butanoic acid, ethyl ester was the most abundant, followed by the propanoic acid, ethyl ester. The main origin of esters in meat product is the esterification of carboxylic acids and alcohols, while the low molecular weight esters can be also a product of carbohydrate metabolism [28]. In the present study, seven out of eight detected esters were ethyl esters. It is well-known that ethyl esters have lower odor thresholds than methyl esters, thus they have more impact on the overall aroma of the product. Additionally, ethyl esters contributed to a proper fermented sausage odor and mask rancid notes [70]. In both cases (ketones and esters), the total amount was higher in reformulated samples than in control samples.

Finally, due to all-mentioned differences between samples, the total VOC compounds were significantly higher (17,854–19,027 AU·10^4/g) in reformulated samples than in CON samples (12,378 AU·10^4/g). Additionally, a negative and significant correlation between moisture and total VOC was also observed ($r = -0.723$; $p < 0.001$). With this in mind, it could be concluded that the higher VOC in reformulated samples are related to the lower moisture content in these samples, which determined that the volatile compounds are proportionally in greater amount in the reformulated batches than in the control batch. Conversely, the lipid-derived VOC did not show this trend, and, in some cases, the CON samples presented significantly higher VOC content of that derived from lipid oxidation processes, which could be related to the natural antioxidant compounds present in the vegetable oils.

3.4. Sensory Analysis of Deer Sausages

The results of the descriptive sensorial analysis for the evaluated attributes in deer sausage are shown in Figure 1. From all attributes, only meat color, odor, chewiness, and rancid flavor were influenced by the reformulation. To this regard, the samples from CON batch presented the lowest scores for meat color attribute, which could be explained by lower L^* values (darker) of the reformulated samples, as previously discussed in the color parameters section.

The "odor" was significantly higher in the reformulated sausages (mainly CAN and SOY samples) than in the CON group. This fact could be explained by the lower values of total VOC found in CON samples in comparison with the values found in reformulated samples. Similarly, the sensory analysis revealed significantly higher "rancid flavor" in CON than in reformulated sausages, which could be related to the highest amounts of lipid-derived volatiles in these samples. As discussed in the volatile section, in most cases, the content of volatiles from lipid oxidation was higher in the control samples, which is clearly reflected in the scores of the sensory panel. As occurs in our research, Alejandre et al. [55] also found differences in odor between control and sausages formulated with linseed emulsion gel as animal fat replacer. Moreover, another study also reported that the replacement of animal fat by amorphous cellulose gel in dry-fermented sausages resulted in positive scores for aromatic attributes than control samples [30].

On the other hand, although panelists found higher "black pepper flavor" in reformulated samples than in CON, these differences were not significant. These results are also supported by the volatile analysis, due to presented higher amounts of terpenoids and benzene-derived compounds in reformulated sausages, which are mainly derived from spices.

It is important to note that panelists described higher scores of "hardness" in reformulated samples than in CON sausages ($p > 0.05$). This fact is related to the clear differences found in the texture analysis, and due to the lowest value of moisture in reformulated sausages. Contrary to our results, a previous study in dry-fermented sausage found that the reformulation process with oil-in-water emulsion immobilized in konjac gel decreased all sensorial parameters [50]. These authors found that the use of oil-in-konjac matrix resulted in a softer structure in meat product, decreasing the firmness scores, which justified the lower sensorial scores in the reformulated sausages. Similarly, other authors also observed that fuet (a fermented sausage similar to salchichón) reformulated with oleogel or emulsion gel presented lower scores for all attributes than samples formulated with animal fat [62].

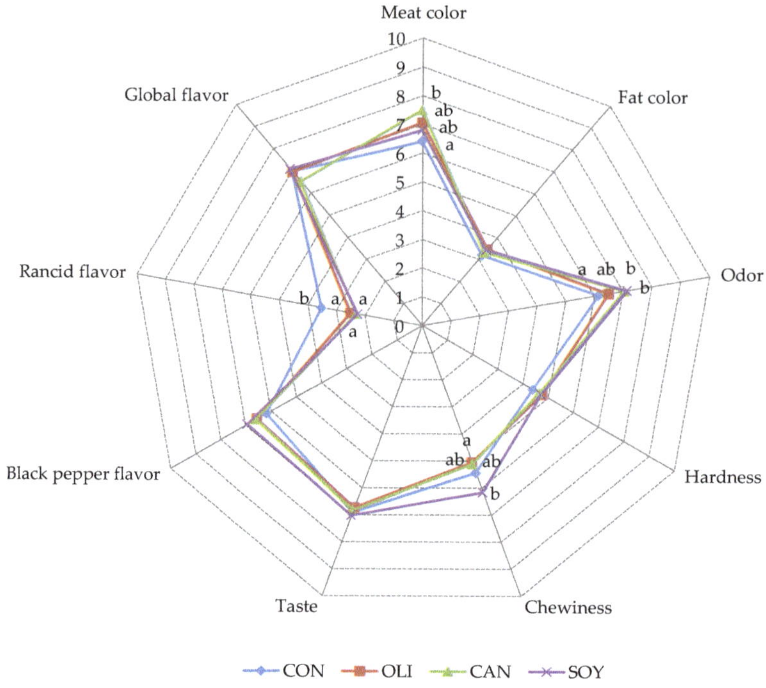

Figure 1. Effect of pork fat replacement by vegetable oil emulsion on sensory characteristics of dry-fermented deer sausages. [a,b] Mean values in the same row (corresponding to the same parameter) not followed by a common letter differ significantly ($p < 0.05$; Duncan test); Treatments: CON: sausages prepared 100% pork fat; OLI: sausages reformulated with 50% of pork fat replaced by olive oil; CAN: sausages reformulated with 50% of pork fat replaced by canola oil; SOY: sausages reformulated with 50% of pork fat replaced by soy oil.

Regarding acceptance test, all batches obtained a score higher than four (acceptability limit; Figure 2A). Sausages from CON batch presented the lowest sores (4.53), while reformulated samples with olive and canola oils had intermediate values (4.8 both) and SOY samples the highest values (5.40) (significantly higher than control). Although the results show higher acceptability scores in all reformulated samples, it should be noted that the acceptability of the sausages with olive oil and canola did not show significant differences with the control samples, while the reformulated sausages with soy oil obtained significantly higher values than the control samples. With these results, it could be concluded that the reformulation of deer sausage using oil-in-water emulsions immobilized with Prosella gel does not affect or increase acceptance of the final product. Additionally, there are no significant differences among reformulated samples. Our findings agree with the results reported in another study with deer fermented sausage, who found that the replacement of up to 25% of pork meat (with 50% of fat) by olive oil immobilized in a protein concentrate organogel resulted in a similar appearance and odor to the control samples [14]. Other authors also found that the inclusion of amorphous cellulose gel up to 50% did not influence the sensory properties of fermented sausages. In contrast, the use of oil-in-konjac matrix in chorizo [50], emulsion or oleogel in fuet [62], and oleogels as replacer of 40% of fat in salchichón [56] resulted in a significant decrease in consumers acceptance.

An acceptable differentiation between batches was possible with PCA (Figure 2B). The attribute map of the sausages showed 75.48% of total variability (F1 and F2 explained 38.44% and 37.04% of total variability, respectively). The attributes more influenced on F1 were fat color, taste and global flavor, while black pepper flavor, rancid flavor, hardness, odor, and meat color had higher weight in F2.

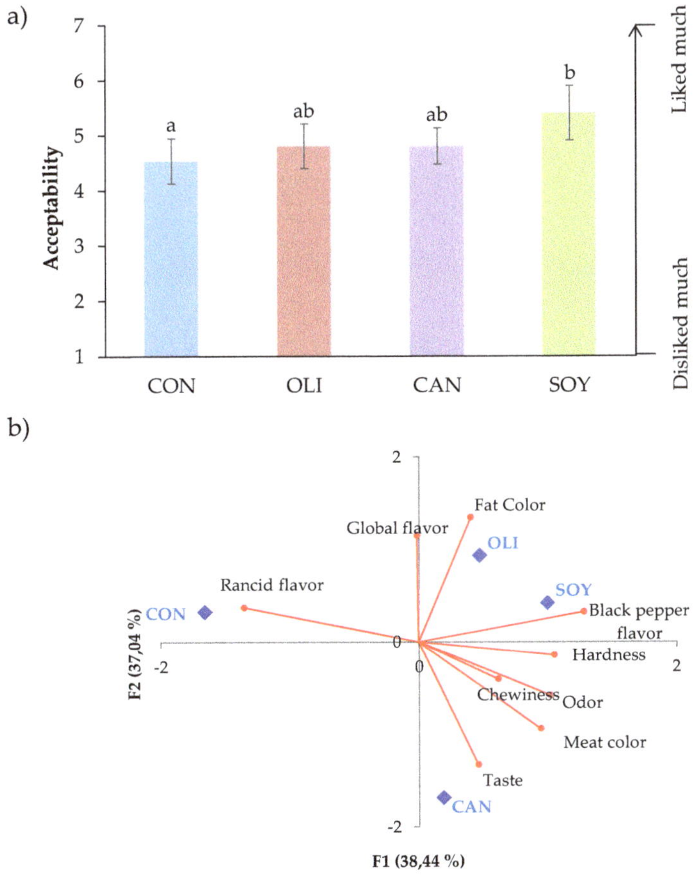

Figure 2. Global acceptance of dry-fermented deer sausages reformulated with healthy oils (**A**) and projection of the sensory attributes and sample batch in the plane defined by the first two components (**B**). [a–b] The bars of the Figure 2A with different letter differ significantly ($p < 0.05$; Duncan test).

The spatial separation showed that batches were separated in two different groups. One of them represented by CON samples and the second one by sausages from OLI, SOY, and CAN batches.

Finally, regarding preference ordination of the different treatments, there were significant differences (Figure 3). Freedman test showed that less than 8% of consumers would choose CON as the most preferred. In contrast, SOY sausages obtained the highest scores, being chosen by 60% of consumers. Intermediate values were obtained by OLI and CAN samples, chosen by 13% and 20% of consumers, respectively.

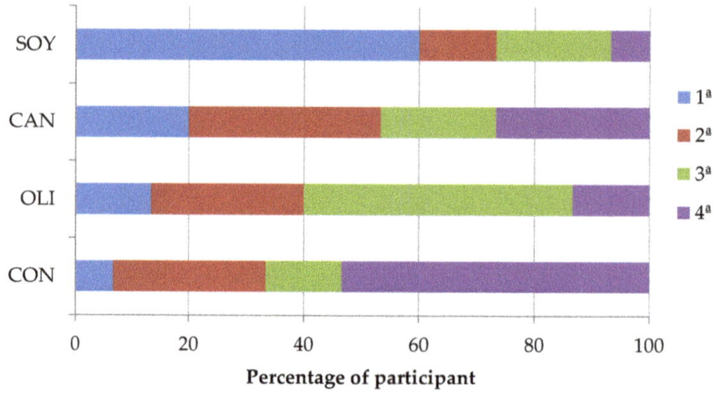

Figure 3. Preference attributed by the panelist on dry-fermented deer sausages.

The sensory characteristics indicated that the use of soy oil emulsion immobilized in Prosella gel resulted in higher consumer acceptability of the final product than control samples. Thus, vegetable oil emulsion immobilized in Prosella gel could be used as a suitable alternative of animal fat in the reformulation of fermented sausages.

4. Conclusions

The use of vegetable oils structured into emulsion gel shows a strong potential for application in the meat industry, especially for the development of healthier dry-fermented sausages. The strategy of replacing animal fat by oils gives rise products with good color and texture characteristics, improves the chemical composition and fatty acids profile, increases the odor characteristics of the final product, and inhibits the lipid oxidation process. Additionally, the sensory results indicated that reformulated sausages presented higher acceptability than traditional samples and the use of soy oil is the most preferred reformulation strategy for the consumers. It is important to note that, from technological point of view, the use of vegetable oil emulsions immobilized in Prosella gel resulted in faster drying process, which could be economically interesting for the manufacturers. Thus, as a general conclusion, the use of vegetable oils emulsions in Prosella gel as animal fat replacer in dry-fermented sausages is technologically viable and improves the composition and nutritional quality, oxidation stability, and sensory properties of the final product.

Author Contributions: Conceptualization, M.P., R.D. and J.M.L.; formal analysis, M.V.-R., M.P. and R.D.; writing—original draft preparation, R.D.; writing—review and editing, M.P., M.G., D.F., P.C.B.C., P.E.S.M., A.C.d.S.B., R.D. and J.M.L. All authors have read and agreed to the published version of the manuscript.

Funding: This work was supported by Ministerio de Economía y Competitividad (Retos-Colaboración program; Project INCYDEN, RTC-2016-5327-2).

Acknowledgments: Authors (J.M.L., M.P., D.F., P.E.S.M., P.C.B.C. and R.D.) are members of the HealthyMeat network, funded by CYTED (ref. 119RT0568). Thanks to GAIN (Axencia Galega de Innovación) for supporting this research (grant number IN607A2019/01). Paulo E. S. Munekata acknowledges postdoctoral fellowship support from Ministry of Economy and Competitiveness (MINECO, Spain) "Juan de la Cierva" program (FJCI-2016-29486). Special thanks to Cárnicas Dibe S.L. for the samples supplied for this research.

Conflicts of Interest: The authors declare no conflict of interest.

References

1. García-Ruiz, A.; Mariscal, C.; González-Vinas, M.A.; Soriano, A. Influence of hunting-season stage and ripening conditions on microbiological, physicochemical and sensory characteristics of venison (*Cervus Elaphus*) chorizo sausages. *Ital. J. Food Sci.* **2010**, *22*, 386–394.

2. Serrano, M.P.; Maggiolino, A.; Pateiro, M.; Landete-Castillejos, T.; Domínguez, R.; García, A.; Franco, D.; Gallego, L.; De Palo, P.; Lorenzo, J.M. Carcass Characteristics and Meat Quality of Deer. In *More than Beef, Pork and Chicken—The Production, Processing, and Quality Traits of Other Sources of Meat for Human Diet*; Lorenzo, J.M., Munekata, P.E.S., Barba, F., Toldrá, F., Eds.; Springer International Publishing: Cham, Switzerland, 2019; pp. 227–268. ISBN 978-3-030-05483-0.
3. Ministerio de Agricultura. Anuarios de Estadística Forestal. Available online: https://www.mapa.gob.es/es/desarrollo-rural/estadisticas/forestal_anuarios_todos.aspx (accessed on 18 April 2020).
4. Utrilla, M.C.; Soriano, A.; García Ruiz, A. Determination of the optimal fat amount in dry-ripened venison sausage. *Ital. J. Food Sci.* **2015**, *27*, 409–415.
5. Serrano, M.P.; Maggiolino, A.; Lorenzo, J.M.; De Palo, P.; García, A.; Landete-Castillejos, T.; Gambín, P.; Cappelli, J.; Domínguez, R.; Pérez-Barbería, F.J.; et al. Meat quality of farmed red deer fed a balanced diet: Effects of supplementation with copper bolus on different muscles. *Animal* **2019**, *13*, 888–896. [CrossRef] [PubMed]
6. Maggiolino, A.; Pateiro, M.; Serrano, M.P.; Landete-Castillejos, T.; Domínguez, R.; García, A.; Gallego, L.; De Palo, P.; Lorenzo, J.M. Carcass and meat quality characteristics from Iberian wild red deer (*Cervus elaphus*) hunted at different ages. *J. Sci. Food Agric.* **2019**, *99*, 1938–1945. [CrossRef] [PubMed]
7. Lorenzo, J.M.; Maggiolino, A.; Gallego, L.; Pateiro, M.; Serrano, M.P.; Domínguez, R.; García, A.; Landete-Castillejos, T.; De Palo, P. Effect of age on nutritional properties of Iberian wild red deer meat. *J. Sci. Food Agric.* **2019**, *99*, 1561–1567. [CrossRef] [PubMed]
8. Serrano, M.P.; De Palo, P.; Maggiolino, A.; Pateiro, M.; Gallego, L.; Dominguez, R.; García, A.; Landete-Castillejos, T.; Lorenzo, J.M. Seasonal variations of carcass characteristics, meat quality and nutrition value in Iberian wild red deer. *Span. J. Agric. Res.* **2020**, in press. [CrossRef]
9. Vargas-Ramella, M.; Pateiro, M.; Barba, F.J.; Franco, D.; Campagnol, P.C.B.; Munekata, P.E.S.; Tomasevic, I.; Domínguez, R.; Lorenzo, J.M. Microencapsulation of healthier oils to enhance the physicochemical and nutritional properties of deer pâté. *LWT* **2020**, *125*, 109223. [CrossRef]
10. Vargas-Ramella, M.; Munekata, P.E.S.; Pateiro, M.; Franco, D.; Campagnol, P.C.B.; Tomasevic, I.; Domínguez, R.; Lorenzo, J.M. Physicochemical composition and nutritional properties of deer burger enhanced with healthier oils. *Foods* **2020**, *9*, 571. [CrossRef]
11. Vargas-Ramella, M.; Domínguez, R.; Pateiro, M.; Franco, D.; Barba, F.J.; Lorenzo, J.M. Chemical and physico-chemical changes during the dry-cured processing of deer loin. *Int. J. Food Sci. Technol.* **2020**, *55*, 1025–1031. [CrossRef]
12. Soriano, A.; Cruz, B.; Gómez, L.; Mariscal, C.; García Ruiz, A. Proteolysis, physicochemical characteristics and free fatty acid composition of dry sausages made with deer (*Cervus elaphus*) or wild boar (*Sus scrofa*) meat: A preliminary study. *Food Chem.* **2006**, *96*, 173–184. [CrossRef]
13. Soriano, A.; Mariscal, C.; Utrilla, M.C.; García-Ruiz, A. Free fatty acids and lipid oxidation in venison chorizo sausages made at different stages of the hunting season and under different ripening conditions. *Ital. J. Food Sci.* **2010**, *22*, 274–283.
14. Utrilla, M.C.; García Ruiz, A.; Soriano, A. Effect of partial replacement of pork meat with an olive oil organogel on the physicochemical and sensory quality of dry-ripened venison sausages. *Meat Sci.* **2014**, *97*, 575–582. [CrossRef] [PubMed]
15. Utrilla, M.C.; García Ruiz, A.; Soriano, A. Effect of partial replacement of pork meat with olive oil on the sensory quality of dry-ripened venison sausage. *Ital. J. Food Sci.* **2015**, *27*, 443–449.
16. WHO. The Top 10 Causes of Death. Available online: https://www.who.int/news-room/fact-sheets/detail/the-top-10-causes-of-death (accessed on 28 September 2020).
17. Belc, N.; Smeu, I.; Macri, A.; Vallauri, D.; Flynn, K. Reformulating foods to meet current scientific knowledge about salt, sugar and fats. *Trends Food Sci. Technol.* **2019**, *84*, 25–28. [CrossRef]
18. WHO. *Diet, Nutrition, and the Prevention of Chronic Diseases: Report of a Joint WHO/FAO Expert Consultation*, 1st ed.; World Health Organization: Geneva, Switzerland, 2003.
19. WHO. Obesity and Overweight. Available online: https://www.who.int/news-room/fact-sheets/detail/obesity-and-overweight (accessed on 28 September 2020).
20. Fernández-Diez, A.; Caro, I.; Castro, A.; Salvá, B.K.; Ramos, D.D.; Mateo, J. Partial fat replacement by boiled quinoa on the quality characteristics of a dry-cured sausage. *J. Food Sci.* **2016**, *81*, C1891–C1898. [CrossRef]

21. Domínguez, R.; Gullón, P.; Pateiro, M.; Munekata, P.E.S.; Zhang, W.; Lorenzo, J.M. Tomato as potential source of natural additives for meat industry. A review. *Antioxidants* **2020**, *9*, 73. [CrossRef] [PubMed]
22. Fonseca, S.; Gómez, M.; Domínguez, R.; Lorenzo, J.M. Physicochemical and sensory properties of Celta dry-ripened "salchichón" as affected by fat content. *Grasas y Aceites* **2015**, *66*, e059.
23. Gómez, M.; Lorenzo, J.M. Effect of fat level on physicochemical, volatile compounds and sensory characteristics of dry-ripened "chorizo" from Celta pig breed. *Meat Sci.* **2013**, *95*, 658–666. [CrossRef]
24. Liaros, N.G.; Katsanidis, E.; Bloukas, J.G. Effect of the ripening time under vacuum and packaging film permeability on processing and quality characteristics of low-fat fermented sausages. *Meat Sci.* **2009**, *83*, 589–598. [CrossRef]
25. Lorenzo, J.M.; Montes, R.; Purriños, L.; Franco, D. Effect of pork fat addition on the volatile compounds of foal dry-cured sausage. *Meat Sci.* **2012**, *91*, 506–512. [CrossRef]
26. Lorenzo, J.M.; Temperán, S.; Bermúdez, R.; Cobas, N.; Purriños, L. Changes in physico-chemical, microbiological, textural and sensory attributes during ripening of dry-cured foal salchichón. *Meat Sci.* **2012**, *90*, 194–198. [CrossRef] [PubMed]
27. Domínguez, R.; Agregán, R.; Lorenzo, J.M. Role of commercial starter cultures on microbiological, physicochemical characteristics, volatile compounds and sensory properties of dry-cured foal sausage. *Asian Pac. J. Trop. Dis.* **2016**, *6*, 396–403. [CrossRef]
28. Domínguez, R.; Purriños, L.; Pérez-Santaescolástica, C.; Pateiro, M.; Barba, F.J.; Tomasevic, I.; Campagnol, P.C.B.; Lorenzo, J.M. Characterization of volatile compounds of dry-cured meat products using HS-SPME-GC/MS technique. *Food Anal. Methods* **2019**, *12*, 1263–1284. [CrossRef]
29. Domínguez, R.; Pateiro, M.; Gagaoua, M.; Barba, F.J.; Zhang, W.; Lorenzo, J.M. A comprehensive review on lipid oxidation in meat and meat products. *Antioxidants* **2019**, *8*, 429. [CrossRef]
30. Campagnol, P.C.B.; dos Santos, B.A.; Wagner, R.; Terra, N.N.; Rodrigues Pollonio, M.A. Amorphous cellulose gel as a fat substitute in fermented sausages. *Meat Sci.* **2012**, *90*, 36–42. [CrossRef] [PubMed]
31. Bis-Souza, C.V.; Pateiro, M.; Domínguez, R.; Lorenzo, J.M.; Penna, A.L.B.; da Silva Barretto, A.C. Volatile profile of fermented sausages with commercial probiotic strains and fructooligosaccharides. *J. Food Sci. Technol.* **2019**, *56*, 5465–5473. [CrossRef] [PubMed]
32. Ruiz-Capillas, C.; Triki, M.; Herrero, A.M.; Rodriguez-Salas, L.; Jiménez-Colmenero, F. Konjac gel as pork backfat replacer in dry fermented sausages: Processing and quality characteristics. *Meat Sci.* **2012**, *92*, 144–150. [CrossRef]
33. Santos, B.A.; Campagnol, P.C.B.; Pacheco, M.T.B.; Pollonio, M.A.R. Fructooligosaccharides as a fat replacer in fermented cooked sausages. *Int. J. Food Sci. Technol.* **2012**, *47*, 1183–1192. [CrossRef]
34. Bis-Souza, C.V.; Pateiro, M.; Domínguez, R.; Penna, A.L.B.; Lorenzo, J.M.; Silva Barretto, A.C. Impact of fructooligosaccharides and probiotic strains on the quality parameters of low-fat Spanish Salchichón. *Meat Sci.* **2020**, *159*, 107936. [CrossRef]
35. Pintado, T.; Herrero, A.M.; Jiménez-Colmenero, F.; Pasqualin Cavalheiro, C.; Ruiz-Capillas, C. Chia and oat emulsion gels as new animal fat replacers and healthy bioactive sources in fresh sausage formulation. *Meat Sci.* **2018**, *135*, 6–13. [CrossRef]
36. Lorenzo, J.M.; Munekata, P.E.S.; Pateiro, M.; Campagnol, P.C.B.; Domínguez, R. Healthy Spanish salchichón enriched with encapsulated n − 3 long chain fatty acids in konjac glucomannan matrix. *Food Res. Int.* **2016**, *89*, 289–295. [CrossRef] [PubMed]
37. Domínguez, R.; Agregán, R.; Gonçalves, A.; Lorenzo, J.M. Effect of fat replacement by olive oil on the physico-chemical properties, fatty acids, cholesterol and tocopherol content of pâté. *Grasas y Aceites* **2016**, *67*, e133.
38. de Carvalho, F.A.L.; Munekata, P.E.S.; Pateiro, M.; Campagnol, P.C.B.; Domínguez, R.; Trindade, M.A.; Lorenzo, J.M. Effect of replacing backfat with vegetable oils during the shelf-life of cooked lamb sausages. *LWT* **2020**, *122*, 109052. [CrossRef]
39. Barros, J.C.; Munekata, P.E.S.; de Carvalho, F.A.L.; Pateiro, M.; Barba, F.J.; Domínguez, R.; Trindade, M.A.; Lorenzo, J.M. Use of tiger nut (*Cyperus esculentus* L.) oil emulsion as animal fat replacement in beef burgers. *Foods* **2020**, *9*, 44. [CrossRef] [PubMed]
40. Heck, R.T.; Vendruscolo, R.G.; de Araújo Etchepare, M.; Cichoski, A.J.; de Menezes, C.R.; Barin, J.S.; Lorenzo, J.M.; Wagner, R.; Campagnol, P.C.B. Is it possible to produce a low-fat burger with a healthy n-6/n-3 PUFA ratio without affecting the technological and sensory properties? *Meat Sci.* **2017**, *130*, 16–25. [CrossRef]

41. Domínguez, R.; Pateiro, M.; Agregán, R.; Lorenzo, J.M. Effect of the partial replacement of pork backfat by microencapsulated fish oil or mixed fish and olive oil on the quality of frankfurter type sausage. *J. Food Sci. Technol.* **2017**, *54*, 26–37. [CrossRef]
42. Domínguez, R.; Pateiro, M.; Munekata, P.E.S.; Campagnol, P.C.B.; Lorenzo, J.M. Influence of partial pork backfat replacement by fish oil on nutritional and technological properties of liver pâté. *Eur. J. Lipid Sci. Technol.* **2017**, *119*, 1600178. [CrossRef]
43. Zamuz, S.; Purriños, L.; Tomasevic, I.; Domínguez, R.; Brnčić, M.; Barba, F.J.; Lorenzo, J.M. Consumer acceptance and quality parameters of the commercial olive oils manufactured with cultivars grown in Galicia (NW Spain). *Foods* **2020**, *9*, 427. [CrossRef]
44. de Souza Paglarini, C.; de Figueiredo Furtado, G.; Honório, A.R.; Mokarzel, L.; da Silva Vidal, V.A.; Ribeiro, A.P.B.; Cunha, R.L.; Pollonio, M.A.R. Functional emulsion gels as pork back fat replacers in Bologna sausage. *Food Struct.* **2019**, *20*, 100105. [CrossRef]
45. Paglarini, C.d.S.; Furtado, G.d.F.; Biachi, J.P.; Vidal, V.A.S.; Martini, S.; Forte, M.B.S.; Cunha, R.L.; Pollonio, M.A.R. Functional emulsion gels with potential application in meat products. *J. Food Eng.* **2018**, *222*, 29–37. [CrossRef]
46. de Souza Paglarini, C.; Vidal, V.A.S.; Ribeiro, W.; Badan Ribeiro, A.P.; Bernardinelli, O.D.; Herrero, A.M.; Ruiz-Capillas, C.; Sabadini, E.; Rodrigues Pollonio, M.A. Using inulin-based emulsion gels as fat substitute in salt reduced Bologna sausage. *J. Sci. Food Agric.* **2020**. [CrossRef]
47. Vieira, A.S.P.; De Souza, X.R.; Rodrigues, E.C.; Sousa, D.C. Replacement of animal fat by canola oil in chicken meatball. *Rev. Bras. Cienc. Avic.* **2019**, *21*, 21. [CrossRef]
48. Baek, K.H.; Utama, D.T.; Lee, S.K.S.G.; An, B.K.; Lee, S.K.S.G. Effects of replacing pork back fat with canola and flaxseed oils on physicochemical properties of emulsion sausages from spent layer meat. *AsianAustralas. J. Anim. Sci.* **2016**, *29*, 865–871. [CrossRef] [PubMed]
49. Heck, R.T.; Saldaña, E.; Lorenzo, J.M.; Correa, L.P.; Fagundes, M.B.; Cichoski, A.J.; de Menezes, C.R.; Wagner, R.; Campagnol, P.C.B. Hydrogelled emulsion from chia and linseed oils: A promising strategy to produce low-fat burgers with a healthier lipid profile. *Meat Sci.* **2019**, *156*, 174–182. [CrossRef]
50. Jiménez-Colmenero, F.; Triki, M.; Herrero, A.M.; Rodríguez-Salas, L.; Ruiz-Capillas, C. Healthy oil combination stabilized in a konjac matrix as pork fat replacement in low-fat, PUFA-enriched, dry fermented sausages. *LWT - Food Sci. Technol.* **2013**, *51*, 158–163. [CrossRef]
51. Domínguez, R.; Munekata, P.E.S.; Pateiro, M.; López-Fernández, O.; Lorenzo, J.M. Immobilization of oils using hydrogels as strategy to replace animal fats and improve the healthiness of meat products. *Curr. Opin. Food Sci.* **2020**, in press.
52. Triki, M.; Herrero, A.M.; Rodríguez-Salas, L.; Jiménez-Colmenero, F.; Ruiz-Capillas, C. Chilled storage characteristics of low-fat, n-3 PUFA-enriched dry fermented sausage reformulated with a healthy oil combination stabilized in a konjac matrix. *Food Control* **2013**, *31*, 158–165. [CrossRef]
53. Munekata, P.E.S.; Domínguez, R.; Franco, D.; Bermúdez, R.; Trindade, M.A.; Lorenzo, J.M. Effect of natural antioxidants in Spanish salchichón elaborated with encapsulated n-3 long chain fatty acids in konjac glucomannan matrix. *Meat Sci.* **2017**, *124*, 54–60. [CrossRef]
54. Yıldız-Turp, G.; Serdaroğlu, M.; Yildiz-Turp, G.; Serdaroğlu, M. Effect of replacing beef fat with hazelnut oil on quality characteristics of sucuk – A Turkish fermented sausage. *Meat Sci.* **2008**, *78*, 447–454. [CrossRef]
55. Alejandre, M.; Poyato, C.; Ansorena, D.; Astiasarán, I. Linseed oil gelled emulsion: A successful fat replacer in dry fermented sausages. *Meat Sci.* **2016**, *121*, 107–113. [CrossRef]
56. Franco, D.; Martins, A.J.; López-Pedrouso, M.; Cerqueira, M.A.; Purriños, L.; Pastrana, L.M.; Vicente, A.A.; Zapata, C.; Lorenzo, J.M. Evaluation of linseed oil oleogels to partially replace pork backfat in fermented sausages. *J. Sci. Food Agric.* **2020**, *100*, 218–224. [CrossRef]
57. Tarladgis, B.G.; Watts, B.M.; Younathan, M.T.; Dugan, L. A distillation method for the quantitative determination of malonaldehyde in rancid foods. *J. Am. Oil Chem. Soc.* **1960**, *37*, 44–48. [CrossRef]
58. International Organization for Standarization. *Sensory Analysis—Methodology—Method of Investigating Sensitivity of Taste: ISO 3972*; International Organization for Standarization: Geneva, Switzerland, 1991.
59. International Organization for Standarization. *Sensory Analysis—Methodology—Texture Profile: ISO 11036*; International Organization for Standarization: Geneva, Switzerland, 1994.

60. International Organization for Standarization. *Sensory Analysis—Methodology—Initiation and Training of Assessors in the Detection and Recognition of Odours: ISO 5496*; International Organization for Standarization: Geneva, Switzerland, 2006.
61. Macfie, H.J.; Bratchell, N.; Greenhoff, K.; Vallis, L.V. Designs to balance the effect of order of presentation and first-order carry-over effects in hall tests. *J. Sens. Stud.* **1989**, *4*, 129–148. [CrossRef]
62. Pintado, T.; Cofrades, S. Quality characteristics of healthy dry fermented sausages formulated with a mixture of olive and chia oil structured in oleogel or emulsion gel as animal fat replacer. *Foods* **2020**, *9*, 830. [CrossRef] [PubMed]
63. Alves, L.A.A.S.; Lorenzo, J.M.; Gonçalves, C.A.A.; Santos, B.A.; Heck, R.T.; Cichoski, A.J.; Campagnol, P.C.B. Production of healthier bologna type sausages using pork skin and green banana flour as a fat replacers. *Meat Sci.* **2016**, *121*, 73–78. [CrossRef] [PubMed]
64. Marventano, S.; Kolacz, P.; Castellano, S.; Galvano, F.; Buscemi, S.; Mistretta, A.; Grosso, G. A review of recent evidence in human studies of n-3 and n-6 PUFA intake on cardiovascular disease, cancer, and depressive disorders: Does the ratio really matter? *Int. J. Food Sci. Nutr.* **2015**, *66*, 611–622. [CrossRef]
65. Simopoulos, A.P. Evolutionary aspects of the dietary Omega-6:Omega-3 fatty acid ratio: Medical implications. *World Rev. Nutr. Diet.* **2009**, *100*, 1–21. [PubMed]
66. Lorenzo, J.M.; Gómez, M.; Purriños, L.; Fonseca, S. Effect of commercial starter cultures on volatile compound profile and sensory characteristics of dry-cured foal sausage. *J. Sci. Food Agric.* **2016**, *96*, 1194–1201. [CrossRef] [PubMed]
67. Montanari, C.; Gatto, V.; Torriani, S.; Barbieri, F.; Bargossi, E.; Lanciotti, R.; Grazia, L.; Magnani, R.; Tabanelli, G.; Gardini, F. Effects of the diameter on physico-chemical, microbiological and volatile profile in dry fermented sausages produced with two different starter cultures. *Food Biosci.* **2018**, *22*, 9–18. [CrossRef]
68. Marušić, N.; Vidaček, S.; Janči, T.; Petrak, T.; Medić, H. Determination of volatile compounds and quality parameters of traditional Istrian dry-cured ham. *Meat Sci.* **2014**, *96*, 1409–1416. [CrossRef]
69. Petričević, S.; Marušić Radovčić, N.; Lukić, K.; Listeš, E.; Medić, H. Differentiation of dry-cured hams from different processing methods by means of volatile compounds, physico-chemical and sensory analysis. *Meat Sci.* **2018**, *137*, 217–227. [CrossRef] [PubMed]
70. Andrade, M.J.; Córdoba, J.J.; Casado, E.M.; Córdoba, M.G.; Rodríguez, M. Effect of selected strains of Debaryomyces hansenii on the volatile compound production of dry fermented sausage "salchichón". *Meat Sci.* **2010**, *85*, 256–264. [CrossRef]
71. Narváez-Rivas, M.; Gallardo, E.; León-Camacho, M. Analysis of volatile compounds from Iberian hams: A review. *Grasas y Aceites* **2012**, *63*, 432–454.
72. Lorenzo, J.M.; Carballo, J. Changes in physico-chemical properties and volatile compounds throughout the manufacturing process of dry-cured foal loin. *Meat Sci.* **2015**, *99*, 44–51. [CrossRef] [PubMed]

foods

MDPI

Article

Punica granatum and *Citrus* spp. Extract Mix Affects Spoilage Microorganisms Growth Rate in Vacuum-Packaged Cooked Sausages Made from Pork Meat, Emmer Wheat (*Triticum dicoccum* Schübler), Almond (*Prunus dulcis* Mill.) and Hazelnut (*Corylus avellana* L.)

David Ranucci [1,2], Rossana Roila [2], Egon Andoni [3], Paolo Braconi [1] and Raffaella Branciari [1,2,*]

1 Centro Interuniversitario per l'Ambiente (CIPLA), University of Perugia, Via Enrico dal Pozzo, 06123 Perugia, Italy; david.ranucci@unipg.it (D.R.); paolo.braconi@unipg.it (P.B.)
2 Department of Veterinary Medicine, University of Perugia, Via San Costanzo 4, 06126 Perugia, Italy; rossana.roila@studenti.unipg.it
3 Faculty of Veterinary Medicine, Universiteti Bujqësor i Tiranës, Kodër Kamëz, SH1, 1000 Tiranë, Albania; eandoni@ubt.edu.al
* Correspondence: raffaella.branciari@unipg.it

Received: 18 November 2019; Accepted: 9 December 2019; Published: 10 December 2019

check for updates

Abstract: Sausage made from pork meat, emmer wheat (*Triticum dicoccum* Schübler), almond (*Prunus dulcis* Mill.), and hazelnut (*Corylus avellana* L.) was integrated with a mix of *Punica granatum* and *Citrus* spp. extracts to evaluate the possible effects on the growth and oxidation of spoilage microorganisms. Two concentrations of the mix were added, respectively, during sausage-making, and the final products were compared with a control group, without the extract mix, during storage. The use of the mix, especially at 10 g/1000 g of the whole ingredients, delayed the pH drop and thiobarbituric acid-reactive substances (TBARs) value during storage. Total viable count, lactic acid bacteria and psychrotrophic microbial counts were also affected, as the extract mix lowered the maximum growth rate of the microbial population considered. The sensory analyses revealed an improvement in the shelf-life of 6 and 16 days, respectively, when 5‰ and 10‰ of the mix were used.

Keywords: meat products; antioxidant; antimicrobials; shelf-life; plant extracts; pomegranate

1. Introduction

The increased interest in food with healthy properties has led to a many studies on meat products in which meat is integrated or substituted at different levels with other ingredients, such as fibres, cereals and nuts [1–4]. The presence of cereals or products derived from them is reported in traditional meat products [5–7] and historically confirmed even form the Ancient Roman period [8]. Despite the presence of these meat products in the market, scarce information is available on their shelf-life and safety, especially if a high percentage of non-meat ingredients is added [8–10]. Moreover, the presence of non-meat ingredients may affect the microbial growth and oxidation and, consequently, the product shelf-life [8]. Additives are usually adopted to improve shelf-life and food safety, but consumers interested in healthier meat request products without synthetic additives (i.e., nitrites) or with natural substitutes that could increase aspects of both commercial stability and safety [4,11,12]. This study evaluated the effects of different concentrations of a commercial mix of pomegranate (*Punica granatum* L.) and *Citrus* spp. extracts (Naturmix WM®, MEC Import, Rome, Italy) on the growth of spoilage microorganisms and thereby the shelf-life of vacuum-packaged, cooked sausages made from pork meat, emmer wheat (*Triticum dicoccum* Schübler), almond (*Prunus dulcis* Mill.), and hazelnut (*Corylus avellana* L.), obtained from an Imperial Roman recipe. The extract mix was chosen to maintain a philological approach to this "historical" sausage, based on the evidence that both botanical species were present in the Mediterranean area during the Roman Age [13,14].

2. Materials and Methods

2.1. Sausage Production and Experimental Design

Sausages were produced in a local factory from shoulder pork meat (51% by weight of the ingredients), boiled emmer wheat (30%), equal parts of chopped peeled roasted almonds and hazelnuts (total of 15%), Squid Brand fish sauce (Thai Fishsauce Factory, Co., Ltd., Bangkok, Thailand) at 1%, salt and black pepper (3%). A control batch (C group) was produced without the mix addition when combining the ingredients, and two experimental groups were produced with 5 g/1000 g (MIX5 group) and 10 g/1000 g (MIX10 group), respectively, of the pomegranate and *Citrus* spp. mix. The meat was ground, as described in Ranucci et al. [8], and mixed with the remaining ingredients. Lamb small intestine casings were filled using a Piston stuffer (RL 15 IDRV, AMB Food Tech Ltd., Bologna, Italy) to obtain sausages of 25 g each (12 cm long by 1.5 cm in diameter) that were cooked in an oven (Self Cooking Centre, Rational AG Italia, Mestre, Italy) until reaching a core temperature of 72 °C for 1 min (registered with thermocouples inserted at the core of the product). The products were then cooled to 2 °C in a chiller (Tecnodom, Ristotecno Ltd., Gubbio, Italy), sealed under vacuum (A/PP 95, Seven Distribuzione, Città di Castello, Italy) in packs of five sausages and pasteurised on the surface inside the oven at 75 °C for 5 min (vacuum-sealed pre-cooked ready-to-eat meat product). The products were cooled again and then stored at 4 ± 1 °C.

Two batches were produced under the same conditions using the same recipe (three groups per batch). Five packs of products from each batch were immediately sampled for analytical determination after the final cool down (day 1 = T1), followed by samplings of the other packs after 7 days and every 15 days (five packs per batch at the following times: T2 = 7 days; T3 = 15 days; T4 = 30 days; T5 = 45 days; T6 = 60 days). From each pack, two sausages were randomly chosen for analytical determination, to obtain 10 sausages per group. Furthermore, 15 packs were collected for sensory analyses each sampling day.

2.2. Chemical and Physical—Chemical Determinations

The chemical composition and NaCl content of the products were determined at T1 by the following methods: moisture content by oven drying at 125 °C for 2 h (method 950.46) [15]; the Kjeldahl method for protein content (method 992.15) [15], using a nitrogen to protein conversion factor of 6.25; ether solvent extraction method for lipid content (method 960.30) [15]; high-performance anion-exchange chromatography with pulsed amperometric detection for carbohydrate content [16]; muffle furnace at 600 °C for the ash content (method 923.03) [15]; total soluble and insoluble dietary fibre (method 991.43) [17]; and the Volhard method for NaCl content (method 935.43) [14].

The thiobarbituric acid-reactive substances (TBARs) value was assayed according to Tarladgis et al. [18] at T1, T30 and T60, respectively, with measurement at 532 nm wavelength using a spectrophotometer (Ultrospec 2100 Pro, Amersham Pharmacia Biotech, Piscataway, NJ, USA), and the data was reported as milligrams of malondialdehyde per kilogram. The pH and water activity (a_w) were determined every 15 days at the core of three sausages per group, using a pH meter equipped with an insertion probe (Crison 25, Crison, Barcelona, Spain) and a hygrometer (AquaLab Series 3 model TE, Decagon Devices, Inc., Pullman, WA, USA), respectively. The same measurements were performed in both batches of products.

2.3. Microbiological Analysis of Sausages

Sausages were aseptically sampled (25 g collected from the inner and outer parts of the products) in triplicate per group and homogenised in a stomacher (Stomacher 400 Circulator, Seward Ltd., Norfolk, UK) with 225 mL of sterile peptone water, followed by enumerating the microorganisms present.

Microbiological analyses of the total viable count (TVC), Enterobacteriaceae count and psychrotrophic microbial count (PMC) were performed according to ISO methods [19–21]; lactic acid bacteria (LAB) count on MRS agar incubated at 37 °C for 24 h under anaerobic conditions; and sulphite-reducing anaerobes on iron sulphite agar (Biolife, Milan, Italy) after anaerobic incubation at 37 °C for 48 h, were performed. Results of microbial analyses were expressed as log CFU/g.

Salmonella spp. were isolated [22], and *Listeria monocytogenes* presence was investigated [23].

2.4. Sensory Analysis

The samples of the three different sausages were tested at the various storage times by a panel of 51 untrained regular sausage consumers (27 female, 24 male), recruited among students and staff of the University of Perugia (Perugia, Italy), with ages ranging from 19 to 65 years [24]. The assessors provided their consent prior the tests, they did not receive any incentives for their participation, and the questionnaires were returned anonymously. No ethical approval was requested. The samples were prepared in a pre-heated oven at 200 °C for a time necessary to reach an internal temperature of 72 °C for 2 min (measured with a temperature probe). The samples were then placed in an isothermal container to maintain the temperature until serving, sliced in pieces of 2 cm length, assigned with a random three digit code and served. For each storage time, each judge, blind to the condition, evaluated the samples of each group (C, MIX5 and MIX10) three times. The overall acceptability of each product, which included the odour, texture and flavour (especially regarding rancid and fermented off-odours and off-flavours) attributes, was scored on a five-point hedonic scale, ranging from *Dislike very much* to *Like very much*. For the same samples, assessors were asked to evaluate acceptance or rejection of the sausages for the different storage periods, by answering the question: *'Would you normally consume this product?'* with a *'Yes'* or *'No'* [25,26].

2.5. Statistical Analysis

Data were analysed by descriptive statistics (mean value and standard error of the mean), and an analysis of variance (ANOVA) model was defined using the GLM procedure in SAS version 2001 (SAS institute inc., Cary, NC, USA) considering the group (C, MIX5 and MIX10) and the time (T1, T2, T3, T4 and T5) as the fixed factor and including replicate (batch) as a random factor nested within the treatment and the time. Tukey's post hoc test was used to compare the means with a significance level of $p < 0.05$. The batch effect was not significant for all the parameters tested ($p > 0.05$) and was not reported in the results.

The effects of the extract mix on the growth of the targeted microorganisms were evaluated through the DMFit function of ComBase online freeware, by fitting the experimental data obtained to the Baranyi–Roberts model, as automatically proposed by the software. For some of the growth curves elaborated by the DMFit, the lag phase was not defined. Therefore, in the absence of specific additional information concerning the actual initial physiological state of microbial populations, and to avoid bias in the growth kinetic evaluation, the lag phase parameter was not considered [27,28]. The results of fitting were analysed by one-way ANOVA (with the sausage group as fixed variable), and Tukey's test ($p < 0.05$).

Survival analysis methodology was used to estimate the shelf-life by analysing the answers of the consumers to the question above using XLStat2015 software (Addinsoft, New York, NY, USA). The cut-offpoint was set by Weibull distribution, considering a 50% rejection probability by the assessors.

3. Results and Discussion

3.1. Chemical Composition and Physical—Chemical Determinations

The results of the chemical composition of the products are reported in Table 1. No difference in the registered values between groups was observed, and the data were consistent with those reported in Ranucci et al. [8] for a similar product.

Table 1. Chemical composition of the sausages with and without pomegranate and citrus mix.

%	C	MIX5	MIX10	SEM
Moisture	51.08	51.21	51.15	0.022
Protein	14.55	14.96	14.56	0.080
Lipid	17.25	16.96	17.04	0.049
Carbohydrate	15.02	14.79	15.11	0.055
Fibre	7.62	7.56	7.64	0.014
Ash	2.10	2.07	2.14	0.012
NaCl	1.51	1.48	1.52	0.008

C = sausage made without *Punica granatum* and *Citrus* spp. extract mix; MIX5 = sausage made with 5‰ *P. granatum* and *Citrus* spp. extract mix; MIX10 = sausage made with 10‰ *P. granatum* and *Citrus* spp. extract mix. SEM = standard error of the mean.

The pH and a_w values are reported in Table 2. The pH of all the products decreased as storage progressed, with lower values registered in group C than MIX10 at 45 days, and both MIX groups at 60 days. The pH decline during storage was registered in a similar product not subjected to post-packaging pasteurisation but at an early stage of 6–12 days [8]. The effect of LAB on the pH drop is reported in other fermented and cooked meat products [29–32], and these microorganisms that exert a favourable technological aspect in dry-cured sausages [33] could be considered spoilage bacteria in cooked meat products, especially if high levels of carbohydrate are present [9]. The increase in LAB concentration during storage is shown in Table 3. No differences were registered for a_w values, which remained almost stable as storage progressed.

Table 2. pH and water activity (a_w) values of the sausages with and without pomegranate and citrus mix.

		n	Storage Time (Days)					SEM	p-Value		
			1	15	30	45	60		T	ST	T × ST
pH	C	10	6.61 w	6.23 x	6.18 x	6.08 ya	5.60 za		<0.001	<0.001	<0.001
	MIX5	10	6.64 w	6.20 x	6.17 x	6.12 yab	6.02 yb	0.030			
	MIX10	10	6.65 w	6.27 x	6.21 x	6.20 xb	6.14 yc				
a_w	C	10	0.966	0.964	0.965	0.965	0.965		0.063	0.072	0.068
	MIX5	10	0.966	0.966	0.966	0.967	0.966	0.001			
	MIX10	10	0.970	0.966	0.967	0.966	0.965				

T = treatment (C, MIX5 and MIX10); ST = storage time; C = sausage made without *Punica granatum* and *Citrus* spp. extract mix; MIX5 = sausage made with 5‰ *P. granatum* and *Citrus* spp. extract mix; MIX10 = sausage made with 10‰ *P. granatum* and *Citrus* spp. extract mix. SEM = standard error of the mean. Values followed by different letters (a, b) in the same column for the same parameter are statistically different ($p < 0.01$); values followed by different letters (v, w, x, y, z) in the same row are statistically different ($p < 0.01$).

Table 3. Microbial counts of the sausages with and without pomegranate and citrus mix. Results are reported as log CFU/g.

		Storage Time (Days)					SEM	*p*-Value		
		1	15	30	45	60		T	ST	T × ST
TVC	C	3.05 v	5.19 wc	5.80 xc	6.15 yb	7.18 zc	0.068	<0.001	<0.001	<0.001
	MIX5	2.99 v	4.24 wb	5.11 xb	5.98 yb	6.11 yb				
	MIX10	2.85 v	3.83 wa	4.17 xa	4.87 ya	5.24 za				
Lactobacillus spp. count	C	2.54 v	4.88 wb	5.12 wb	5.58 xb	6.85 yc	0.089	<0.001	<0.001	<0.001
	MIX5	2.86 v	4.11 wa	4.97 xb	5.82 yb	5.93 yb				
	MIX10	2.58 v	3.90 wa	4.15 wa	4.60 xa	5.04 ya				
PMC	C	2.26 v	4.47 wb	5.45 xc	5.97 yb	6.75 zc	0.101	<0.001	<0.001	<0.001
	MIX5	2.26 v	4.31 wb	5.16 xb	6.04 yb	6.27 yb				
	MIX10	2.20 v	3.89 wa	4.04 wa	4.99 xa	5.33 ya				

TVC = total viable count; PMC = psychrotrophic microbial count; T = treatment (C, MIX5 and MIX10); ST = storage time; C = sausage made without *Punica granatum* and *Citrus* spp. extract mix; MIX5 = sausage made with 5‰ *P. granatum* and *Citrus* spp. extract mix; MIX10 = sausage made with 10‰ *P. granatum* and *Citrus* spp. extract mix. SEM = standard error of the mean. Values followed by different letters (a, b) in the same column for the same parameter are statistically different ($p < 0.01$); values followed by different letters (v, w, x, y, z) in the same row are statistically different ($p < 0.01$). $n = 10$ samples per group.

The TBARs data are reported in Figure 1. The values registered were similar on day 1 (T1), then increased significantly at T3 only in C and MIX5 products, and at T5 for all the groups considered ($p < 0.01$). The differences between samples were registered from T3 onwards, with higher values in group C, followed by MIX5 and MIX10 products.

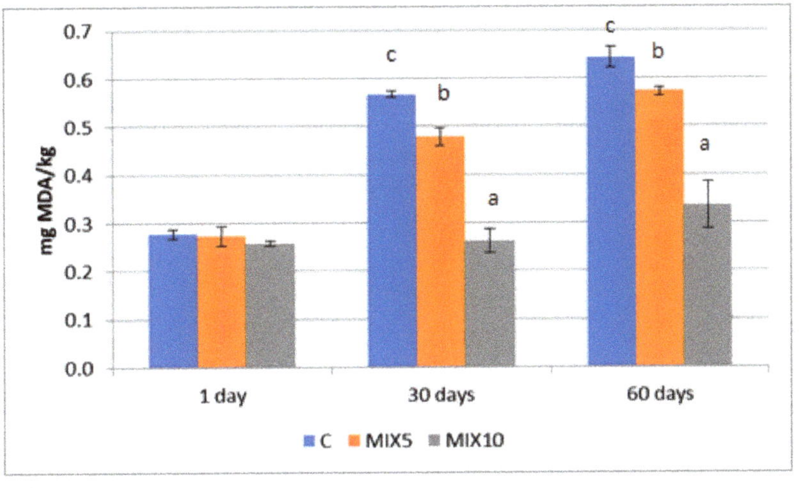

Figure 1. TBARs levels in the three groups of sausages. Different letters (a, b, c) are statistically different ($p < 0.05$); $n = 10$ samples per group of sausages. TBARs = thiobarbituric acid-reactive substances; MDA = malondihaldehyde.

There is a general consensus in the literature that the antioxidant efficacy of pomegranate is effective in preventing food oxidation. The radical scavenging activity of pomegranate extracts is mainly due to the presence of polyphenols [34,35]. A dose-effect association with lipid oxidation is highlighted in cooked meat [36] in which the use of pomegranate extract delays the oxidation process during storage [37,38]. Other phenolic compounds are present in *Citrus* spp. extracts, in both free and bound forms that exert antioxidant activity both in vitro and in food systems [39,40]. Limited antioxidant activity may also be due to almonds and hazelnuts in the products, which contain phenolics and flavonoids, but in limited quantities compared with other nuts, like walnuts and pecans [41].

3.2. Microbiological Growth in the Sausages

Microorganism analyses (Table 3) revealed that all counts increased throughout storage, with higher loads for TVC, LAB and PMC in group C than in the treated groups after 60 days of storage. The presence of non-meat ingredients could be a valuable source of nutrients (carbohydrates) for bacteria to grow and proliferate in the products during storage, even at refrigeration temperatures [8]. These differences are noticed during storage of other meat products added with cereals, but do not always affect the microbial load at the end of the shelf-life [9]. Nonetheless, similar trends are reported in beef frankfurter-type sausages packaged under different modified atmosphere packaging and vacuum conditions without the addition of carbohydrate-containing ingredients [42]. The use of the blended extracts of pomegranate and *Citrus* spp. affected the microbial loads, as evidenced by the lower values in group MIX10 than MIX5 and C products from day 15. These data agree with the findings of Kannatt et al. [38], who noticed that the increased TVC in chicken products during storage was alleviated by incorporation of 1% and 5% pomegranate peel extracts, and with those of Firuzi et al. [43], who used different concentrations of pomegranate juice concentrate and rind powder extracts in frankfurters. This same trend was highlighted in raw meat added with pomegranate extracts [44,45]. The PMC was not affected by the pomegranate extracts in ground beef patties [46], but a reduction in the loads was registered in shrimp stored under refrigeration for 10 days and treated with different concentrations of pomegranate peel extracts [47]. Regarding *Citrus* spp. extracts, Mexis et al. [48] noticed a reduced growth rate in TVC, LAB, and pseudomonads in ground chicken meat. *Mortadella* meat products with incorporated citrus fibre, thyme essential oil and rosemary essential oil lowered the growth rate of both TVC and LAB during storage [48].

No literature data are yet available on the effects of *Citrus* spp. extracts alone on the microbial loads of cooked meat products.

The results from ComBase application to the microbial population are reported in Figure 2.

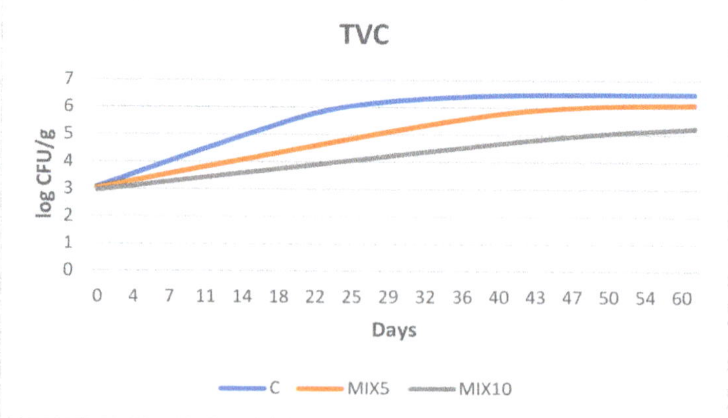

Figure 2. *Cont.*

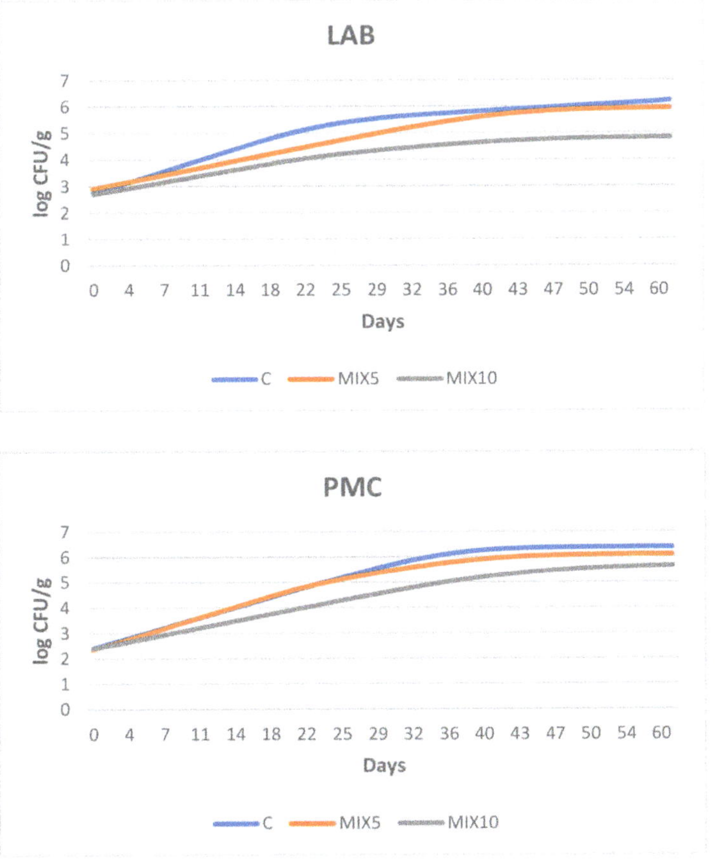

Figure 2. Predictive models of total viable count (TVC), lactic acid bacteria (LAB) and psychrotrophic microbial count (PMC) generated by the Baranyi–Roberts model in sausages with and without pomegranate and citrus mix. C = sausage made without *Punica granatum* and *Citrus* spp. extract mix; MIX5 = sausage made with 5‰ *P. granatum* and *Citrus* spp. extract mix; MIX10 = sausage made with 10‰ *P. granatum* and *Citrus* spp. extract mix.

The Baranyi–Roberts no-lag model highlights differences in the μmax values (maximal growth rate) between groups for the microbial population considered. In particular, the values were 0.00539, 0.00299 and 0.00181 log CFU/h for TVC; 0.00483, 0.00302 and 0.00266 log CFU/h for LAB, and 0.00463, 0.00489 and 0.00313 log CFU/h for PMC in C, MIX5 and MIX10 products, respectively. The effect of the mix on the bacterial growth rate is highlighted by the difference found between the final values registered on day 60 (Table 3). The use of 10 g/1000 g of the mix decreased the maximum growth rate to a greater extent than the lower dose (5 g/1000 g).

Regarding the other microbiological analyses performed, the Enterobacteriaceae counts were always under the detection limit (1 log CFU/g) in the MIX10 group, while in the other groups, the counts increased throughout storage up to 1.33 and 3.35 log CFU/g in group MIX5 and C, respectively. The anti-microbial activity has already been reported for both pomegranate and *Citrus* spp. extracts [49–51], even inducing a decrease in the coliforms count in frozen beef sausages [52]. According to the values detected, a dose-dependent effect of the mix could not be established for the Enterobacteriaceae counts.

No sulphite-reducing anaerobes, such as *Salmonella* spp. and *L. monocytogenes*, were detected in the samples at any time considered. However, *Listeria* spp. were isolated from the products, as the pasteurisation protocol adopted was insufficient to eliminate the bacteria occasionally contaminating products after cooking [53], and both the pH and a_w values of the products favoured *Listeria* spp. growth [54]. *Listeria* spp. were isolated from group C from T2 onwards, and group MIX5 at T5 and T6, but not in group MIX10. As reported by other authors, pomegranate extracts may exert activity against *Listeria* both in vitro [55] and in refrigerated pork meat [56] or delay *Listeria* growth when experimentally inoculated in cooked meat products [57]. The anti-listerial activity is related to the polyphenolic compounds, such as tannins [58], which cause bacterial protein precipitation, including disruption of the cell membrane [38,59] and enzyme inhibition [55]. Inhibition of *Listeria* spp. by *Citrus* spp. extracts are also reported [60], but some authors highlighted a limited in vitro activity for orange and lemon extracts [61].

3.3. Sensory Analyses for Shelf-Life

The results of the sensory study for defining the shelf-life of the products are shown in Figure 3 and Table 4.

Table 4. Estimated sensory shelf-life analyses of the sausages with and without pomegranate and *Citrus* spp. mix by survival analysis.

Percentage	C (Days)	MIX5 (Days)	MIX10 (Days)
1%	7.595	8.951	11.236
5%	15.000	17.390	21.464
10%	20.259	23.318	28.566
1st Quartile 25%	30.815	35.110	42.567
Median 50%	44.487	50.241	60.357
3rd Quartile 75%	59.419	66.637	79.480
90%	73.441	81.942	97.221
95%	81.971	91.217	107.930
99%	98.092	108.684	128.026

C = sausage made without *Punica granatum* and *Citrus* spp. extract mix; MIX5 = sausage made with 5‰ *P. granatum* and *Citrus* spp. extract mix; MIX10 = sausage made with 10‰ *P. granatum* and *Citrus* spp. extract mix.

The higher stability of groups MIX10 and MIX5 than group C was supported by the sensory data. A linear regression analysis indeed was carried out considering assessors' overall acceptability as a dependent variable and storage time as an explanatory variable (Figure 3). The approach used aimed to estimate sensory shelf-life as the time period during which consumers perceived the food item above the "neither like nor dislike" point (cut off 3.5), on a 5-point hedonic scale, as the acceptability limit for the product. The regression of the three products gave a good fit (R2 = 0.9724, 0.9601 and 0.9677, for C, MIX5 and MIX10 respectively), and using this regression, the control (group C) showed a decrease in assessors' acceptability already at 42 days. Howsoever, at 60 days of storage the acceptability of MIX10 remained quite stable, scoring a value of 3.80 followed by MIX5, which was considered stable up to 50 days. During the storage the degradation processes were most evident in the C group, causing defects mainly in the taste as reported in the assessors' comments (data not shown).

Shelf-life decisions based only on acceptability limit might be taken with caution, as they do not always reflect the consumer's decision to accept or reject the product [62]. Indeed, survival analysis of food acceptability has been used to determine a proper shelf-life of the sausage.

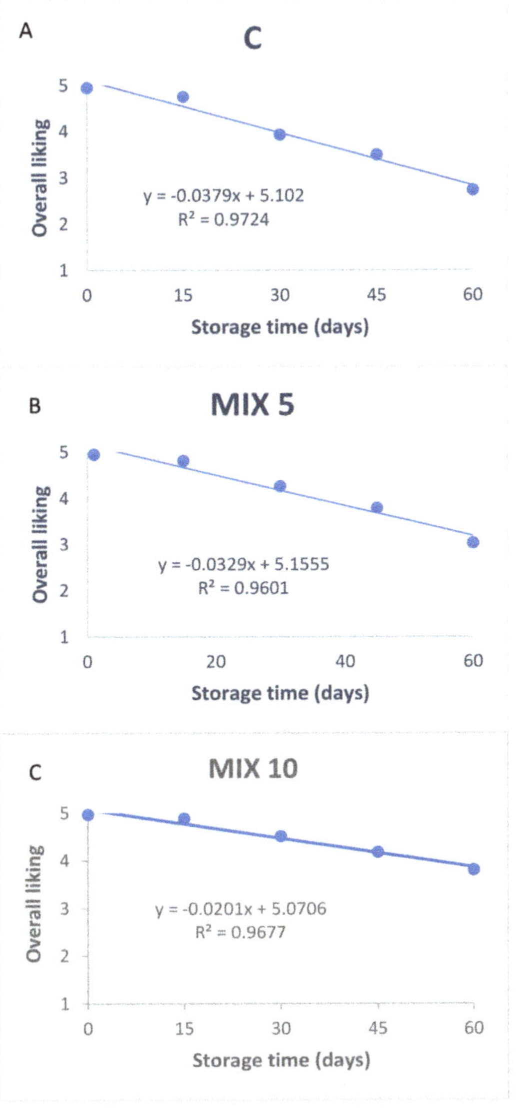

Figure 3. Assessors' average overall liking scores as a function of storage time in sausages with and without pomegranate and *Citrus* spp. mix. (**A**) C = sausage made without *Punica granatum* and *Citrus* spp. extract mix; (**B**) MIX5 = sausage made with 5‰ *P. granatum* and *Citrus* spp. extract mix; (**C**) MIX10 = sausage made with 10‰ *P. granatum* and *Citrus* spp. extract mix.

The estimated shelf life of the products, set by the Weibull distribution of the sensory data (Table 4), revealed that 25% of the assessors reject the products at 30.815, 35.110 and 42.567 days for C, MIX5 and MIX10, respectively. Considering 50% probability of consumer rejection set as cut off point, the estimated shelf life was 44.487, 50.241 and 60.357 for C, MIX5 and MIX10, respectively. Thus, the shelf-life of such a product was improved up to 16 days when the concentration of the mix was increased (MIX10 vs. C). Shelf life estimated using survival analysis for a 50% consumer rejection is in line with the estimate determined by using an acceptability score of 3.5 as the failure criterion. This correspondence implies that during the products' shelf lives, the consumers actually appreciate them.

A comparison of the shelf-life extension with other cooked sausage formulations found in the literature is difficult due to the differences in the composition of the products and the processing technology adopted. Authors referred to sausage with rice bran fibre added as having a shelf-life of fewer than 8 days, but no vacuum-packaging and surface pasteurisation were performed [63]. The shelf-life of sausages can be extended if packaged in modified atmosphere packaging and shrink packaging are used (54 and 45 days, respectively, in chicken frankfurters) [64] or stored under vacuum in refrigerating chambers (more than 8 weeks for cooked blood sausage) [65]. Even a post-packaging pasteurisation process could increase the shelf-life of cooked sausages [66].

4. Conclusions

The use of a mix of pomegranate and *Citrus* spp. extracts in a vacuum-sealed, post-packaged pasteurised cooked sausage made with meat and a high proportion of non-meat ingredients could be a valuable strategy to enhance the shelf-life by controlling both the microbial growth and oxidation during refrigerated storage. Food safety could also be improved, but further studies are needed to define the fate of specific foodborne pathogens in such sausages and to determine whether the single extract could exert specific effects against them and extend the products' shelf-life.

Author Contributions: The authors contrinute to the present article as follow: conceptualization, P.B., D.R. and R.B.; formal analysis, D.R., R.R. and R.B.; writing—original draft preparation, D.R., R.B. and R.R.; writing—review and editing, D.R., R.B., E.A.; supervision, D.R.; project administration, P.B. and D.R.; funding acquisition, P.B.

Funding: This research was funded by the European Community, PSR measure 124 Umbria Region, Italy (grant number 44750254045).

Acknowledgments: The authors express their gratitude to the chef Marino Marini for the fundamental contribution during sausage formulation and the setting of the recipes, and his support in the sampling activity.

Conflicts of Interest: The authors declare no conflict of interest.

References

1. Ayo, J.; Carballo, J.; Solas, M.T.; Jiménez-Colmenero, F. Physicochemical and sensory properties of healthier frankfurters as affected by walnut and fat content. *Food Chem.* **2008**, *107*, 1547–1552. [CrossRef]
2. Jiménez-Colmenero, F.; Sánchez-Muniz, F.J.; Olmedilla-Alonso, B. Design and development of meat-based functional foods with walnut: Technological, nutritional and health impact. *Food Chem.* **2010**, *123*, 959–967. [CrossRef]
3. Mehta, N.; Ahlawat, S.S.; Sharma, D.P.; Dabur, R.S. Novel trends in development of dietary fiber rich meat products—a critical review. *J. Food Sci. Technol.* **2015**, *52*, 633–647. [CrossRef] [PubMed]
4. Weiss, J.; Gibis, M.; Schuh, V.; Salminen, H. Advances in ingredient and processing systems for meat and meat products. *Meat Sci.* **2010**, *86*, 196–213. [CrossRef]
5. Conroy, P.M.; O'Sullivan, M.G.; Hamill, R.M.; Kerry, J.P. Impact on the physical and sensory properties of salt-and fat-reduced traditional Irish breakfast sausages on various age cohorts acceptance. *Meat Sci.* **2018**, *143*, 190–198. [CrossRef]
6. Marcos, C.; Viegas, C.; de Almeida, A.M.; Guerra, M.M. Portuguese traditional sausages: Different types, nutritional composition, and novel trends. *J. Ethnic Foods* **2016**, *3*, 51–60. [CrossRef]
7. Tobin, B.D.; O'Sullivan, M.G.; Hamill, R.M.; Kerry, J.P. The impact of salt and fat level variation on the physiochemical properties and sensory quality of pork breakfast sausages. *Meat Sci.* **2013**, *93*, 145–152. [CrossRef]
8. Ranucci, D.; Miraglia, D.; Branciari, R.; Morganti, G.; Roila, R.; Zhou, K.; Jang, H.; Braconi, P. Frankfurters made with pork meat, emmer wheat (*Triticum dicoccum* Schübler) and almonds nut (*Prunus dulcis* Mill.): Evaluation during storage of a novel food from an ancient recipe. *Meat Sci.* **2018**, *145*, 440–446. [CrossRef]
9. Fernández-López, J.; Lucas-González, R.; Viuda-Martos, M.; Sayas-Barberá, E.; Navarro, C.; Haros, C.M.; Pérez-Álvarez, J.A. Chia (*Salvia hispanica* L.) products as ingredients for reformulating frankfurters: Effects on quality properties and shelf-life. *Meat Sci.* **2019**, *156*, 139–145. [CrossRef]
10. Nayeem, M.; Chauhan, K.; Khan, M.; Siddiqui, M.; Sidduqui, H. Development and shelf life studies of buffalo meat sausages incorporated with foxtail millet (*Setaria italica*). *Int. J. Chem. Stud.* **2017**, *5*, 648–654.

11. Agregán, R.; Barba, F.J.; Gavahian, M.; Franco, D.; Khaneghah, A.M.; Carballo, J.; Ferreira, I.C.F.R.; da Silva Barretto, A.C.; Lorenzo, J.M. Fucus vesiculosus extracts as natural antioxidants for improvement of physicochemical properties and shelf life of pork patties formulated with oleogels. *J. Sci. Food Agric.* **2019**, *99*, 4561–4570. [CrossRef] [PubMed]
12. Roila, R.; Branciari, R.; Staccini, B.; Ranucci, D.; Miraglia, D.; Altissimi, M.S.; Mercuri, M.L.; Haouet, N.M. Contribution of vegetables and cured meat to dietary nitrate and nitrite intake in Italian population: Safe level for cured meat and controversial role of vegetables. *Ital. J. Food Saf.* **2018**, *7*, 7692. [CrossRef] [PubMed]
13. Bakels, C.; Jacomet, S. Access to luxury foods in Central Europe during the Roman period: The archaeobotanical evidence. *World Archaeol.* **2003**, *34*, 542–557. [CrossRef]
14. Pagnoux, C.; Celant, A.; Coubray, S.; Fiorentino, G.; Zech-Matterne, V. The introduction of Citrus to Italy, with reference to the identification problems of seed remains. *Veg. Hist. Archaeobot.* **2013**, *22*, 421–438. [CrossRef]
15. AOAC. *Official Methods of Analysis of AOAC International*, 17th ed.; AOAC International: Gathersberg, MD, USA, 2000.
16. Corradini, C.; Canali, G.; Cogliandro, E.; Nicoletti, I. Separation of alditols of interest in food products by high-performance anion-exchange chromatography with pulsed amperometric detection. *J. Chromatogr. A* **1997**, *791*, 343–349. [CrossRef]
17. AOAC. *Official Methods of Analysis of AOAC International*, 16th ed.; AOAC International: Gathersberg, MD, USA, 1995.
18. Tarladgis, B.G.; Watts, B.M.; Yonathan, M. Distillation method for the determination of malonaldehyde ın rancid foods. *J. Am. Oil Chem. Soc.* **1960**, *37*, 44–48. [CrossRef]
19. ISO 4833-1. *Microbiology of the Food Chain—Horizontal Method for the Enumeration of Microorganisms—Part 1: Colony Count at 30 Degrees C by the Pour Plate Technique*; International Organization for Standardization: Geneva, Switzerland, 2013.
20. ISO 21528-2. *Microbiology of the food chain—Horizontal method for the detection and enumeration of Enterobacteriaceae—Part 2: Colony-count technique*; International Organization for Standardization: Geneva, Switzerland, 2017.
21. ISO 17410. *Microbiology of food and animal feeding stuffs—Horizontal method for the enumeration of psychrotrophic microorganisms*; International Organization for Standardization: Geneva, Switzerland, 2001.
22. ISO 6579-1. *Microbiology of the Food Chain—Horizontal Method for the Detection, Enumeration and Serotyping of Salmonella—Part 1: Detection of Salmonella Spp*; International Organization for Standardization: Geneva, Switzerland, 2017.
23. ISO 11290-1. *Microbiology of the food chain—Horizontal method for the detection and enumeration of Listeria monocytogenes and of Listeria spp.—Part 1: Detection method*; International Organization for Standardization: Geneva, Switzerland, 2017.
24. Branciari, R.; Ranucci, D.; Urbani, E.; Valiani, A.; Trabalza-Marinucci, M.; Dal Bosco, A.; Franceschini, R. Freshwater fish burgers made from four different fish species as a valuable strategy appreciated by consumers for introducing EPA and DHA into a human diet. *J. Aquat. Food Prod. Technol.* **2017**, *26*, 686–694. [CrossRef]
25. Aaslyng, M.D.; Vestergaard, C.; Koch, A.G. The effect of salt reduction on sensory quality and microbial growth in hotdog sausages, bacon, ham and salami. *Meat Sci.* **2014**, *96*, 47–55. [CrossRef]
26. Roila, R.; Valiani, A.; Ranucci, D.; Ortenzi, R.; Servili, M.; Veneziani, G.; Branciari, R. Antimicrobial efficacy of a polyphenolic extract from olive oil by-product against "Fior di latte" cheese spoilage bacteria. *Int. J. Food Microbiol.* **2019**, *295*, 49–53. [CrossRef]
27. Bovill, R.; Bew, J.; Cook, N.; D'agostino, M.; Wilkinson, N.; Baranyi, J. Predictions of growth for Listeria monocytogenes and Salmonella during fluctuating temperature. *Int. J. Food Microbiol.* **2000**, *59*, 157–165. [CrossRef]
28. García, M.R.; Vilas, C.; Herrera, J.R.; Bernárdez, M.; Balsa-Canto, E.; Alonso, A.A. Quality and shelf-life prediction for retail fresh hake (*Merluccius merluccius*). *Int. J. Food Microbiol.* **2015**, *208*, 65–74. [CrossRef] [PubMed]
29. Ranucci, D.; Miraglia, D.; Trabalza-Marinucci, M.; Acuti, G.; Codini, M.; Ceccarini, M.R.; Forte, C.; Branciari, R. Dietary effects of oregano (*Origanum vulgaris* L.) plant or sweet chestnut (*Castanea sativa* Mill.) wood extracts on microbiological, chemico-physical characteristics and lipid oxidation of cooked ham during storage. *Ital. J. Food Saf.* **2015**, *4*, 5497.

30. Simion, A.M.C.; Vizireanu, C.; Alexe, P.; Franco, I.; Carballo, J. Effect of the use of selected starter cultures on some quality, safety and sensorial properties of Dacia sausage, a traditional Romanian dry-sausage variety. *Food Control* **2014**, *35*, 123–131. [CrossRef]
31. Vermeiren, L.; Devlieghere, F.; Debevere, J. Evaluation of meat born lactic acid bacteria as protective cultures for the biopreservation of cooked meat products. *Int. J. Food Microbiol.* **2004**, *96*, 149–164. [CrossRef]
32. Fonseca, S.; Cachaldora, A.; Gómez, M.; Franco, I.; Carballo, J. Effect of different autochthonous starter cultures on the volatile compounds profile and sensory properties of Galician chorizo, a traditional Spanish dry fermented sausage. *Food Control* **2013**, *33*, 6–14. [CrossRef]
33. Ranucci, D.; Loschi, A.R.; Miraglia, D.; Stocchi, R.; Branciari, R.; Rea, S. Effect of selected starter cultures on physical, chemical and microbiological characteristics and biogenic amine content in Protected Geographical Indication Ciauscolo salami. *Ital. J. Food Saf.* **2016**, *5*, 5568.
34. Gil, M.I.; Tomás-Barberán, F.A.; Hess-Pierce, B.; Holcroft, D.M.; Kader, A.A. Antioxidant activity of pomegranate juice and its relationship with phenolic composition and processing. *J. Agric. Food Chem.* **2000**, *48*, 4581–4589. [CrossRef]
35. Madrigal-Carballo, S.; Rodriguez, G.; Krueger, C.G.; Dreher, M.; Reed, J.D. Pomegranate (*Punica granatum*) supplements: Authenticity, antioxidant and polyphenol composition. *J. Funct. Foods* **2009**, *1*, 324–329. [CrossRef]
36. Naveena, B.M.; Sen, A.R.; Vaithiyanathan, S.; Babji, Y.; Kondaiah, N. Comparative efficacy of pomegranate juice, pomegranate rind powder extract and BHT as antioxidants in cooked chicken patties. *Meat Sci.* **2008**, *80*, 1304–1308. [CrossRef]
37. Devatkal, S.K.; Narsaiah, K.; Borah, A. Anti-oxidant effect of extracts of kinnow rind, pomegranate rind and seed powders in cooked goat meat patties. *Meat Sci.* **2010**, *85*, 155–159. [CrossRef]
38. Kanatt, S.R.; Chander, R.; Sharma, A. Antioxidant and antimicrobial activity of pomegranate peel extract improves the shelf life of chicken products. *Int. J. Food Sci. Technol.* **2010**, *45*, 216–222. [CrossRef]
39. Bocco, A.; Cuvelier, M.E.; Richard, H.; Berset, C. Antioxidant activity and phenolic composition of citrus peel and seed extracts. *J. Agric. Food Chem.* **1998**, *46*, 2123–2129. [CrossRef]
40. Fernández-López, J.; Fernández-Ginés, J.M.; Aleson-Carbonell, L.; Sendra, E.; Sayas-Barberá, E.; Pérez-Alvarez, J.A. Application of functional citrus by-products to meat products. *Trends Food Sci. Technol.* **2004**, *15*, 176–185. [CrossRef]
41. Yang, J.; Liu, R.H.; Halim, L. Antioxidant and antiproliferative activities of common edible nut seeds. *LWT-Food Sci. Technol.* **2009**, *42*, 1–8. [CrossRef]
42. Gokoglu, N.; Yerlikaya, P.; Uran, H.; Topuz, O.K. The effect of modified atmosphere packaging on the quality and shelf life of frankfurther type-sausages. *J. Food Qual.* **2010**, *33*, 367–380. [CrossRef]
43. Firuzi, M.R.; Niakousari, M.; Eskandari, M.H.; Keramat, M.; Gahruie, H.H.; Khaneghah, A.M. Incorporation of pomegranate juice concentrate and pomegranate rind powder extract to improve the oxidative stability of frankfurter during refrigerated storage. *LWT-Food Sci. Technol.* **2019**, *102*, 237–245. [CrossRef]
44. Devatkal, S.K.; Thorat, P.; Manjunatha, M. Effect of vacuum packaging and pomegranate peel extract on quality aspects of ground goat meat and nuggets. *J. Food Sci. Technol.* **2014**, *51*, 2685–2691. [CrossRef]
45. Qin, Y.Y.; Zhang, Z.H.; Li, L.; Xiong, W.; Shi, J.Y.; Zhao, T.R.; Fan, J. Antioxidant effect of pomegranate rind powder extract, pomegranate juice, and pomegranate seed powder extract as antioxidants in raw ground pork meat. *Food Sci. Biotechnol.* **2013**, *22*, 1063–1069. [CrossRef]
46. Bouarab-Chibane, L.; Ouled-Bouhedda, B.; Leonard, L.; Gemelas, L.; Bouajila, J.; Ferhout, H.; Cottaz, A.; Joly, C.; Degraeve, P.; Oulahal, N. Preservation of fresh ground beef patties using plant extracts combined with a modified atmosphere packaging. *Eur. Food Res. Technol.* **2017**, *243*, 1997–2009. [CrossRef]
47. Basiri, S.; Shekarforoush, S.S.; Aminlari, M.; Akbari, S. The effect of pomegranate peel extract (PPE) on the polyphenol oxidase (PPO) and quality of Pacific white shrimp (*Litopenaeus vannamei*) during refrigerated storage. *LWT-Food Sci. Technol.* **2015**, *60*, 1025–1033. [CrossRef]
48. Mexis, S.F.; Chouliara, E.; Kontominas, M.G. Shelf life extension of ground chicken meat using an oxygen absorber and a citrus extract. *LWT-Food Sci. Technol.* **2012**, *49*, 21–27. [CrossRef]
49. Viuda-Martos, M.; Ruiz-Navajas, Y.; Fernández-López, J.; Pérez-Álvarez, J.A. Effect of added citrus fibre and spice essential oils on quality characteristics and shelf-life of mortadella. *Meat Sci.* **2010**, *85*, 568–576. [CrossRef] [PubMed]

50. Alsaggaf, M.S.; Moussa, S.H.; Tayel, A.A. Application of fungal chitosan incorporated with pomegranate peel extract as edible coating for microbiological, chemical and sensorial quality enhancement of Nile tilapia fillets. *Int. J. Biol. Macromol.* **2017**, *99*, 499–505. [CrossRef] [PubMed]
51. Tsiraki, M.I.; Yehia, H.M.; Elobeid, T.; Osaili, T.; Sakkas, H.; Savvaidis, I.N. Viability of and Escherichia coli O157: H7 and Listeria monocytogenes in a delicatessen appetizer (yogurt-based) salad as affected by citrus extract (Citrox©) and storage temperature. *Food Microbiol.* **2018**, *69*, 11–17. [CrossRef] [PubMed]
52. Saleh, A.; Morshdy, A.E.M.; Abd-El-Salam, E.H.; Hussein, M.A.; Elewa, E.S.; Mahmoud, A.F.A. Effect of pomegranate peel powder on the hygienic quality of beef sausage. *J. Microbiol. Biotech. Food Sci.* **2017**, *6*, 1300. [CrossRef]
53. Houben, J.H.; Eckenhausen, F. Surface pasteurization of vacuum-sealed precooked ready-to-eat meat products. *J. Food Prot.* **2006**, *69*, 459–468. [CrossRef]
54. Commission Regulation (EC). N. 2073/2005 of 15 November 2005 on Microbiological Criteria for Foodstuffs. Official Journal of the European Union L 338/1. Available online: https://eur-lex.europa.eu/legal-content/EN/ALL/?uri=CELEX%3A32005R2073 (accessed on 10 November 2019).
55. Al-Zoreki, N.S. Antimicrobial activity of pomegranate (*Punica granatum* L.) fruit peels. *Int. J. Food Microbiol.* **2009**, *134*, 244–248. [CrossRef]
56. Shan, B.; Cai, Y.Z.; Brooks, J.D.; Corke, H. Antibacterial and antioxidant effects of five spice and herb extracts as natural preservatives of raw pork. *J. Sci. Food Agric.* **2009**, *89*, 1879–1885. [CrossRef]
57. Hayrapetyan, H.; Hazeleger, W.C.; Beumer, R.R. Inhibition of Listeria monocytogenes by pomegranate (*Punica granatum*) peel extract in meat paté at different temperatures. *Food Control* **2012**, *23*, 66–72. [CrossRef]
58. Li, G.; Xu, Y.; Wang, X.; Zhang, B.; Shi, C.; Zhang, W.; Xia, X. Tannin-rich fraction from pomegranate rind damages membrane of Listeria monocytogenes. *Foodborne Path. Dis.* **2014**, *11*, 313–319. [CrossRef]
59. Kang, J.H.; Song, K.B. Effect of pomegranate (*Punica granatum*) pomace extract as a washing agent on the inactivation of Listeria monocytogenes inoculated on fresh produce. *Int. J. Food Sci. Technol.* **2017**, *52*, 2295–2302. [CrossRef]
60. Casquete, R.; Castro, S.M.; Martín, A.; Ruiz-Moyano, S.; Saraiva, J.A.; Córdoba, M.G.; Teixeira, P. Evaluation of the effect of high pressure on total phenolic content, antioxidant and antimicrobial activity of citrus peels. *Innov. Food Sci. Emerg. Technol.* **2015**, *31*, 37–44. [CrossRef]
61. Fernandez-Lopez, J.; Zhi, N.; Aleson-Carbonell, L.; Pérez-Alvarez, J.A.; Kuri, V. Antioxidant and antibacterial activities of natural extracts: Application in beef meatballs. *Meat Sci.* **2005**, *69*, 371–380. [CrossRef] [PubMed]
62. Giménez, A.; Ares, F.; Ares, G. Sensory shelf-life estimation: A review of current methodological approaches. *Food Res. Int.* **2012**, *49*, 311–325. [CrossRef]
63. Heo, C.; Kim, H.W.; Choi, Y.S.; Kim, C.J.; Paik, H.D. Shelf-life estimation of frankfurter sausage containing dietary fiber from rice bran using predictive modeling. *Korean J. Food Sci. Anim. Resour.* **2009**, *29*, 47–54. [CrossRef]
64. Tovunac, I.; Galić, K.; Prpić, T.; Jurić, S. Effect of packaging conditions on the shelf-life of chicken frankfurters with and without lactate addition. *Food Sci. Technol. Int.* **2011**, *17*, 167–175. [CrossRef]
65. Cachaldora, A.; García, G.; Lorenzo, J.M.; García-Fontán, M.C. Effect of modified atmosphere and vacuum packaging on some quality characteristics and the shelf-life of "morcilla", a typical cooked blood sausage. *Meat Sci.* **2013**, *93*, 220–225. [CrossRef]
66. Balamurugan, S.; Inmanee, P.; Souza, J.D.; Strange, P.; Pirak, T.; Barbut, S. Effects of High-Pressure Processing and Hot Water Pasteurization of Cooked Sausages on Inactivation of Inoculated *Listeria monocytogenes*, Natural Populations of Lactic Acid Bacteria, *Pseudomonas* spp., and Coliforms and Their Recovery during Storage at 4 and 10 °C. *J. Food Prot.* **2018**, *81*, 1245–1251.

foods

MDPI

Review

Safety, Quality and Analytical Authentication of ḥalāl Meat Products, with Particular Emphasis on Salami: A Review

Maria Martuscelli [1,*], Annalisa Serio [1], Oriana Capezio [2] and Dino Mastrocola [1]

[1] Faculty of Bioscience and Technology for Food, Agriculture and Environment, University of Teramo, Via R. Balzarini 1, 64100 Teramo, Italy; aserio@unite.it (A.S.); dmastrocola@unite.it (D.M.)

[2] Department Asian, African and Mediterranean, University of Naples "L'Orientale", Piazza San Domenico Maggiore 12, 80134 Napoli, Italy; ocapezio@unior.it

* Correspondence: mmartuscelli@unite.it

Received: 15 July 2020; Accepted: 7 August 2020; Published: 13 August 2020

check for updates

Abstract: Only some animal species could be transformed into ḥalāl salami and the raw meat must be obtained from ritually slaughtered animals. Several scientific studies have been conducted on ritual slaughtering practices and manufacturing of meat products for Jewish and Muslim religious communities; furthermore, many projects have been funded by the European Community on this topic. The authenticity and traceability of meat is one of the priorities of ḥalāl food certification systems. The pig matrix (meat and/or lard) may be fraudulently present in ḥalāl processed meat, as well as salami, for both economic and technological purposes; in fact, the use of these raw materials reflects the easier availability and their lower cost; furthermore, it allows manufacturers to obtain final products with better quality (sensory properties) and stability (especially with respect to oxidative reactions). The aim of this review is to discuss the qualitative and technological aspects of ḥalāl raw meat for dry fermented sausages (salami); moreover, this study focuses on the most recent studies carried out on the certification system and on the analytical methods performed in order to solve problems such as fraud and adulteration of ḥalāl salami and other halal meat foods.

Keywords: ḥalāl salami; fermented sausages; ḥalāl assurance; authenticity

---

1. Ḥalāl Meat Products and Regulations in Europe

Today Muslim people constitute about 25% of the population in the world and this is expected to increase further; the size of the global ḥalāl market could reach around $2.6 trillion [1,2]. Muslims are projected to increase as a share of Europe's population. The share of Muslims in Europe's population as of 2050 would be depending on three possible future migration scenarios: it would be expected to be somewhere among 7.4% (zero migration) or between 11.2% ("regular" migration) and 14% (high migration) [3].

The ḥalāl food market is a considerable economic opportunity for agro-food enterprises. The migratory flows and the substantial rise of the Muslim population in Europe not only affects the socio-cultural aspects but also from an economic point of view, regarding foodstuffs and other products (drugs, cosmetics, etc.) that must be permissible (ḥalāl) to Muslims consumers, following specific religious laws [4–8]. Malaysia has been the first country in the world to establish ḥalāl -related laws [9]; it has a documented and systematic ḥalāl assurance system; besides conventional legal texts, *fatwa* is a legal opinion issued by Islamic scholars based on interpretation and adaptation of verses from Koran and *Ahadith* [10].

Traditional ḥalāl meat products can be processed in five different ways: salted and/or marinated but not dried, dried not fermented, fermented semidry/dried, smoked, cooked and/or candied [8].

The production and consumption of ḥalāl meat products, as well as ḥalāl salami, obtained through Islamic slaughter rites, have been growing steadily in Europe over recent decades and even the food safety legislation had to consider religious slaughter practices to avoid conflicts [11]. Salami is one of the most representative dry meat products of Mediterranean countries; it involves long ripening periods, with different final quality due, most of all, to their different compositions as the variable ratio of meat and fat, autochthonous or selected microbiota, additives and ingredients, all of them representative of risks to Ḥalāl assurance [12,13]. Figure 1 depicts the most important issues to be considered when ḥalāl salami products are produced (Figure 1).

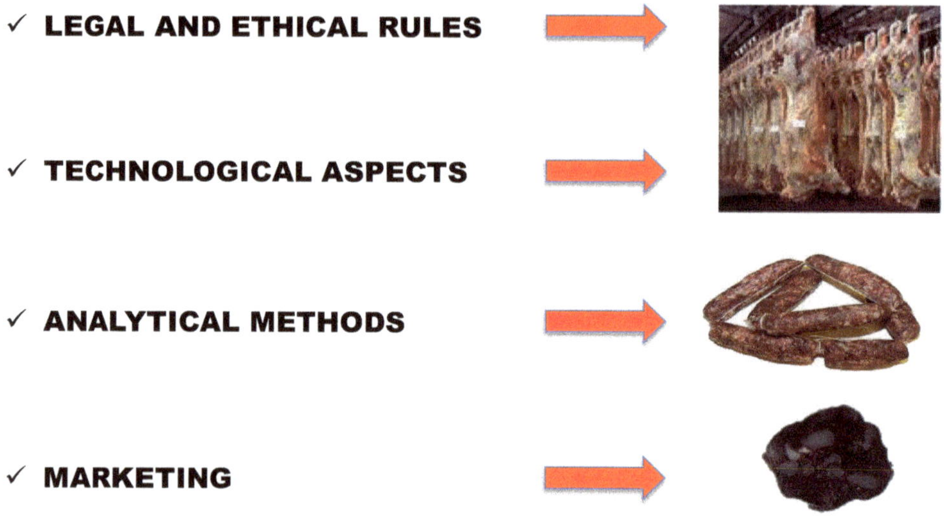

Figure 1. Main issues regarding the quality and authenticity of halal salami.

The European Union law on slaughter imposes stunning of the animal (applied through electrical, mechanical or gaseous means) before slaughtering, with the exception of ritual slaughter (Regulation EC n. 1099/2009) [14]; for this reason it is important to involve Islamic scholars in the research to enable stakeholders in the meat industry to make decisions regarding the aspects of pre-slaughter stunning [15].

Another aspect that needs to be taken into account is the effect of halal bleeding on meat quality and animal welfare. Aghwan et al. [16] reported that an efficient bleeding process after ḥalāl slaughter not only maintains the quality and wholesomeness of meat but also potentially reduces suffering and pain of animals.

According to the most recent data, it is estimated that about 26 million of Muslims and 1.1 million of Jews live in EU [3], therefore, the number of ritual-slaughtered animals is rather high in Member States [17,18], and this fact requires the particular attention of the Institutions. For example, recently, the General Advocate of the European Court of Justice (ECJ) affirmed that stating that ritual slaughter is incompatible with organic farming this would mean "adding a condition not provided for by current legislation", while consumers of kosher or ḥalāl products have the right to benefit from the guarantees provided by products labelled "organic farming" in terms of food quality and safety [19].

In Table 1, the rules providing the regulation of ḥalāl foods in Europe are collected. In the present review, the considered rules are those related to safety, hygiene, and quality aspects of ḥalāl salami and other meat products.

Table 1. Rules providing the regulation of ḥalāl foods in Europe.

Subject	Law	Regulation Issue
Food safety	Regulation (EC) n. 178/2002 [20]	on laying down the general principles and requirements of food law, establishing the European Food Safety Authority and laying down procedures in matters of food safety;
Food hygiene	Regulation (EC) n. 852/2004 [21]	on the hygiene of foodstuffs;
	Regulation (EC) n. 853/2004 [22]	laying down specific hygiene rules for food of animal origin;
Food contact materials	Regulation (EC) n. 1935/2004 [23]	on materials and articles intended to come into contact with food;
Animal slaughter (including ritual one)	Regulation (EC) n. 1099/2009 [14]	on the protection of animals at the time of killing;
Label statement	Regulation (EC) n. 1169/2011 [24]	on the provision of food information to consumers;

2. Ḥalāl Salami Processing

Halāl meat is generally consumed as fresh meat, or as processed products, such as pasties, sausages, luncheon meat, turkey breast, bologna or salami.

Generally, salami means fermented and dried sausages manufactured with raw pork, beef or a mixture of pork and beef meat, although other animal species (goat, sheep, goose, horse, donkey, turkey, wild boar) can be used, depending on the typical products and on the geographic area of production [25,26]. Nevertheless, only some of these species could be transformed to halāl salami, and the raw meat must be obtained from ritually slaughtered animals. The different raw materials, as well as the ingredients which should be in compliance with halāl prescriptions raise issues that need to be addressed.

2.1. *Ḥalāl Raw Material*

Ante-mortem treatments and slaughter management could affect carcass traits and meat quality [17,27]. Although most scientists would accept the fact that the meat quality in stunned animals is comparable to that of animals slaughtered without stunning, it has been recently disproved: in fact, when lambs were slaughtered without stunning, their meat developed lower drip before cooking, and had less cooking loss, compared to meat from electrically or CO_2 stunned lambs [28].

A recent study has demonstrated that three ḥalāl slaughter methods have no substantial effect on lamb meat quality [29], while it has not been tested whether or not ḥalāl slaughter influences the quality of halal dry meat products. The broiled chicken meat sausage was investigated for the effect of ritual slaughter on microbiota: ḥalāl samples were not contaminated with either coliforms, *E. coli* or *Salmonella*, while the non-halal meat sausage contained 1.50×10^6, 2.33×10^5 and 1.50×10^5 CFU g^{-1} of coliforms, *E. coli* and *Salmonella*, respectively [30]. Results of this study highly recommended to follow the Islamic rule in slaughtering poultry and to apply hazard analysis and food hygiene rules to reduce the risk of cross contamination with food-borne pathogens in poultry farms.

Several authors have proposed sheep meat as suitable for ripening processes [31–33]; in addition, other studies have been conducted on the influence of animal nutrition, in order to improve the composition quality of meat and fat [34], with a possible influence also on the sensory acceptability and stability of the transformed products.

The use of lard is preferred in batter making (or even dough processing) both for its ready availability and for its functional properties (in particular higher melting point, able to improve organoleptic properties such as texture and succulence), although some limitations exist. In fact, as pork is forbidden in the diet of many people for religious reasons, in ḥalāl salami both meat and fat have to be replaced in agreement with all the other imposed requirements on the manufacture and on the use of ingredients and additives [12]. In Italy, a recent study has collected data regarding market demand for salami with ḥalāl certification; in particular, interviews were carried out with two groups of consumers, of which 103 of Muslim and 151 non-Muslims faith. Both groups of consumers showed a high interest in purchasing equally goat and sheep salami with ḥalāl certification [35]. Unfortunately, some critical issues are related to the replacement of pork meat in dry fermented sausages, due to the low oxidative stability of other fats and the strong sensory impact of the raw materials [26]. Several studies have been conducted on the quality of Turkish sausages made from sheep in order to improve sensory characteristics and texture [36,37], but only few researchers reported on ḥalāl salami: for example, Indian salami, prepared with meat and fat of buffaloes slaughtered according to ḥalāl rites, have been investigated [38]. Furthermore, in ḥalāl salami the influence of spices (es. pepper, paprika, cumin, garlic) and their essential oils, on the inhibition and/or control of alterative phenomena should be considered. For example, during the ripening and storage of dry fermented mutton sausages formulated with pepper and cumin, a significant increase in level of MUFA and PUFA/SFA ratio was observed in respect to the control [39]. Therefore, autochthonous microbiota of fermented sausages could be related to free fatty acids profile as well as to the production of secondary metabolites with toxic action (biogenic amines) [40,41].

To the best of our knowledge, in the literature only few studies declare the preparation of samples according to the ḥalāl procedures, but it would be correct to assume that published researches on Turkish traditional dry-fermented sausage (sucuk) [42,43] concern ḥalāl salami.

Therefore, bez sucuk is a type of Turkish fermented beef sausage, mainly produced by butchers and small-scale facilities that use traditional technologies without adding starter cultures, in which a few manufacturers use curing agents such as sodium nitrite. Bez sucuk differs from other Turkish-type fermented sausages due to the use of cloth casings sewn to size of 7 × 25 cm from uncolored cloth with 42 threads per cm^2; therefore, the formulation and process conditions (temperature, humidity, and ripening period) show differences among all manufacturers of bez sukuk [44]. Bez sucuk processing has three production steps: mixing the sucuk batter, filling the cloth casings, and ripening for 10–14 days [45].

2.2. Preservatives

In general, the effect of the use of additives (glucose, sodium nitrite, sodium nitrate, sodium ascorbate and sodium citrate) on the safety and quality of dry cured meat products has been studied [37,46]. Many chemical ingredients are added in the ḥalāl food production process to enhance the food characteristics, and also preservatives could be added in salami formulation. A halāl food additives checker system has been optimized to provide consumers with a useful result on the product safety meeting Halalan Toyyiban criteria, where the latter indicate that processed foods or ingredients shall be safe for consumption, non-hazardous and non-intoxicating, thus emphasizing quality aspects [47].

In recent years, microorganisms have been a remarkable option for ḥalāl production. Halāl principles must be followed in the manufacturing of bioproducts, therefore the adding of microbial ingredients must match this specification too. Assuming that in the future the share of ḥalāl microbial products will increase in the biotechnology market, Karahalil et al. [48] evaluated the steps of a fermentation process from an Islamic point of view and determined the control points for ḥalāl requirements.

2.3. Sensory Profile

The sensory characteristics of processed meats could be affected by several factors, such as the kind and the quality of the raw material, the ingredients (other than meat), the eventual addition of starter cultures and the processing. As the raw materials are different from classical sausages, their impact on the final sensory traits of the product has to be considered.

Recently, different percentages of mutton (from adult female sheep, over four years old) and additional autochthonous starter cultures (*Staphylococcus xylosus* LQ3 and *Pedioccoccus pentosaceus* P38) were tested in a study on fermented Turkish sausages [49]. The results of this research showed that the use of indigenous microbial cultures attributed positive and typical characteristics to the fermented sausages, with a high hedonic score for sensory acceptance; furthermore, a positive effect of mutton on the reduction of unsaturated fatty acids and an increase in red tonality were proved.

Moreover, the volatile profile of fermented meat sausages containing 90% of mutton was characterized by a higher abundance of butyric (C4: 0), hexanoic (C6: 0) and octanoic (C8: 0) acid, related to hydrolytic rancidity or to the oxidation of fatty acids; moreover, butanal and a high level of hexanal were detected, too. Generally, low concentrations of short chain saturated fatty acids (up to 10 carbon atoms) are desirable in fermented meat [50]; furthermore, autoxidation of long chain unsaturated fatty acids can generate aldehydes and other aliphatic volatile compounds [51].

With the aim of meeting the growing need for meat in developing countries, several research projects have been carried out on new formulations of ḥalāl salami, such as sucuk reformulated with camel meat and hump fat [52]. The camel meat, especially from young animals, contains less fat and cholesterol and relatively higher PUFA than other meats; therefore, camel-hump fat is used for the production of a cocoa-butter analogue, so its use in dry sausages provides final products of high-quality. In fact, results of this investigation showed a good potentiality of these innovative raw materials, such as sucuk made from camel meat and hump fat showed physical-chemical, fatty-acid and volatile-compounds profiles and sensory qualities similar to sausages made from beef and beef fat (traditional sukuc).

2.4. Biogenic Amines

Biogenic amines have been implicated as the causative agent in several food poisoning outbreaks. Fermented food, such as Turkish style fermented sausages, can also contain biogenic amines; in fact, microorganisms possessing the enzymes and amino acids decarboxylases, which convert amino acids into biogenic amines, are responsible for the formation of these compounds in fermented meats. In addition, in dry fermented mutton sausages, safety and quality have been proved to be difficult to guarantee, particularly because of the presence of biogenic amines, which can accumulate, as a consequence of the presence of producing bacteria [53]. High concentrations of BAs have been found in industrial dry sausages added with starter cultures and not only in artisanal ones, because pure or starter cultures could not be sufficiently competitive in suppressing the growth of wild amine-producing microbiota [41]. Thus, the quality of the raw materials and ingredients and the hygienic processing practices are crucial to control the BAs production in fermented meat products; nevertheless, selected starter cultures could also help in containing the BAs amount. In fact, although the amino acids decarboxylase potential is strain specific, starter species such as *Lactobacillus sakei*, *Lb. plantarum* and *Staphylococcus xylosus*, are generally described as weak or non-aminogenic bacteria. Moreover, different studies have been conducted to evaluate the effect of a combination of negative amine producer starter cultures (*Lactobacillus* spp., *Pediococcus* spp., *Staphylococcus* spp. and *Micrococcus* spp.) in the reduction of the biogenic amines amount during fermented sausages manufacture, with interesting results, proving a BAs reduction from 9% up to about 100%, depending on the specific biogenic amine [54]. These studies underline the importance to test the starter culture strains with the aim of improving the quality and safety of the final product.

As a whole, the sum of vasoactive biogenic amines (tyramine, histamine, tryptamine, 2-phenylethylamine) results not exceeding 200 mg kg^{-1} when dry fermented sausages have been manufactured according to excellent hygienic conditions and good manufacturing practices (GMP) [55].

Ekici and Omer [56] investigated the biogenic amines concentration reached in 120 sukuc samples collected from 10 different brands sold in the local markets of Van (Turkey). Tryptamine (0–129.4 mg/kg), 2-phenylethylamine (0–65.6 mg/kg), putrescine (0–255.6 mg/kg), cadaverine (0–1148.8 mg/kg), histamine (0–469.4 mg/kg), tyramine (0–438.1 mg/kg), spermidine (0–554.4 mg/kg) and spermine (0–614.4 mg/kg) were detected, showing that the occurrence of biogenic amines represent a real risk associated with the fermentation of ḥalāl salami.

Other studies were carried out on bez sucuks produced with different meat:fat ratios (90:10, 80:20 and 70:30, respectively); the results showed that bez sucuks with the highest meat ratio (90:10) had the highest tryptamine, putrescine, and tyramine levels at the end of the processing and storage period [57].

Spices and other plant materials used in fermented meat for their flavoring effect, as well as for the antioxidant and bacteriostatic activity, due to the content in essential oils, phenolic compounds and organic acids, can also reduce the formation of biogenic amines [42,58] (see next section).

2.5. Use of Spices and/or Plant Extracts

Often added to fermented meat products with the aim of enriching the taste and the sensory characteristics, spices and plant extracts also exert interesting bioactivities. In detail, the phenolic constituents of spices and plant extracts are able to interact with the cytoplasmic membrane modifying its fluidity and permeability [59] up to the rupture, with consequent impairment of energy production and leakage of cytoplasmic material [60]. These effects could be useful in contrasting the viability and the metabolic activity of biogenic amines-producing bacteria, with greater effects than nitrites [53]. For example, Jia et al. [53] investigated the inhibitory effect of several spices including clove, cassia, bay leaf, fennel, star anise and nutmeg on the biogenic amines accumulation in dry fermented mutton sausages, revealing that particularly cassia and fennel were very effective in reducing the biogenic amines amount. In detail, reductions up to 27.5% were observed for spermidine, followed by 24.6% for 2-phenylethylamine, 21.8% for tryptamine, 18.7% for tyramine and even 24.4% for histamine, thus proving the importance of spices for the safety of fermented meat products, at least from this point of view.

Nevertheless, spices were demonstrated to improve also the safety profile of pastrami, a dry-cured meat product traditionally produced in Egypt with beef, lamb, water buffalo or camel meat, and very common also in Mediterranean and Middle East countries [61]. In detail, spices contained in a seasoning paste made of salt, sweet and hot pepper, fresh garlic, clove, coriander, rosemary, fenugreek seeds and nutmeg, decreased *Escherichia coli* and aerobic microbial counts and reduced aflatoxins content below the permission limit of 20 ppb [62].

Due to their antioxidant activity, spices such as curry leaves, torch ginger and cinnamon have been proved to maintain the quality of lamb meat also during cooking processes, reducing the formation of heterocyclic aromatic amines, poly aromatic hydrocarbons and trans fatty acids [63]. In 2015, the International Agency for Research on Cancer from the World Health Organization, recognized the above mentioned compounds as responsible for cancerogenicity for consumption of red meat and processed meat. An overview of the effectiveness of spices and natural products in counteracting the development of potential carcinogenic substances in meat products has been recently provided by Lee et al. [64].

Most of all, spices in meat products are essential to contain the oxidative reactions at the expense of lipid and protein fractions, leading to pigment, flavor, and texture deterioration and to the shelf-life reduction. As raw materials often rich in unsaturated fatty acids are used to produce ḥalāl fermented meat products, the role of spices in this kind of product is particularly important to improve the oxidative state of the final product. Mediterranean plants exert antioxidant activity due to the presence of phenolic compounds, terpenes, organosulfur compounds, acids and other molecules, able to contrast proteins and lipids oxidation, decreasing metal ions and scavenging radicals [65]. The same chemical species allow the spices to exert antibacterial and fungicidal activity, against spoilage and pathogenic microorganisms, acting as biopreservatives, improving the safety profile and extending the shelf-life of processed meat products. Therefore, although they are traditionally used in meat and meat products to enrich and enhance the sensory profile of the products, spices and plant extracts have an important impact on many aspects of the products, thus protecting the consumers health, and resulting in a clean label also for ḥalāl meat products [66]. Necessarily, for ḥalāl products, the spices must be ḥalāl-suitable and particular attention has to be paid to spices blends, where non-certified animal-based ingredients have to be avoided, and the risk of cross-contamination should be carefully checked.

2.6. Ḥalāl Casing

Currently, the use of ḥalāl meats increases the request for ḥalāl casings. In fact, while the traditional pork casings are obviously forbidden, those obtained by other animals are allowed, as long as these animals have been slaughtered in compliance with ḥalāl provisions. Moreover, beside non animal casings such as those made of cellulose and other plant materials, innovative solutions are actually under study. For example, the production of a chitosan casing could be well-suited for commercial application in ḥalāl sausages. A study on a novel chitosan-based casing provided an alternative packaging material to collagen to be used as a sausage casing for the meat industry, showing similar mechanical properties as the collagen casing, but lower water solubility, superior transparency, and better UV light barrier [67]. Recently Marcos et al. [68] proposed the co-extruded alginate coating as a feasible alternative to collagen casing: in fact they observed a regular evolution of pH values during the fermentation step (from the initial value about 6.0, the pH decreased just below 5.0) and the control of spoilage microorganisms; no significant difference resulted on the final a_w value (<0.92), but a faster drying kinetic was observed in sausages with alginate coating compared with the ones stuffed into collagen casings; finally, authors reported no significant differences on the sensory properties between different casing types.

Sezer and Bozkurt [43] tested the applications of active packaging on the stability of traditional Turkish type fermented sausage; these authors carried out a study concerning the effect of the incorporation of antimicrobials (chitosan and silver substituted zeolite, AgZeo) into multilayer films as a novel casing. Chitosan has an antimicrobial spectrum against Gram(+)/Gram(−) bacteria, molds, and yeasts [69], whereas Ag-ions exert high antimicrobial activities due to their inactivation effect towards a series of metabolic enzymes [70].

Aerobic plate count and lactic acid bacteria were decreased significantly ($p < 0.05$) by chitosan-incorporated casing; moreover, antimicrobial plastic casings including chitosan and AgZeo decreased ($p < 0.05$) putrescine, histamine, and tyramine formation in sucuks, therefore, these novel casings could be used to improve quality and safety of ḥalāl salami [43].

Finally, a very important aspect to be studied is the evolution of dehydration processes to assess the diffusive phenomena of salt and water, in order to build simple predictive models concerning the safety and quality of ḥalāl salami and other ḥalāl cured meat products [71].

3. Food Safety in Ḥalāl Assurance

Nowadays food safety is a responsibility of government agencies and organizations. European Community (EC) legislation (see Table 1) is primarily geared towards ensuring the production of safe foods for human health but also for ensuring free competition in the food market.

For foodstuffs of animal origin, further specific hygiene requirements are necessary (prescribed by Reg. 853/2004) [22], as these products may present macrobiotic and chemical risks for the human health and therefore, they require the application of specific rules. The rules dictated by Reg. 853/2004 are added to those related to animal welfare [72], without posing particular issues to ritual slaughter operators.

In addition to the above mentioned laws, Regulation (EC) N. 1935/2004 of the European Parliament and of the Council of 27 October 2004 on the regulation of materials and articles intended to come into contact with foodstuffs [23] should also be taken in account. In general, once these materials come into contact with food, they must not cause unacceptable changes in foodstuffs; nor should these provisions pose particular problems to ḥalāl slaughter operators.

Hazard Analysis and Critical Control point (HACCP) system is considered to be effective for enhancing food safety; furthermore, other standards, such as ISO series, Approved Quality Assurance (AQA), Good Agricultural Practices (GAP), Good Manufacturing Practices (GMP) and the Food Safety

Management System (FSMS) could be considered for food quality and safety. These standards could be contemplated by more than one hundred active certifying bodies, governmental or non-governmental organizations, for ḥalāl compliance [73]. In a recent study, many areas for potential research in halāl assurance in the food industry have been identified, and critical issues have been highlighted [74]. Incorporating halāl features into the HACCP system could be a plausible tool for halāl assurance. As the HACCP and Ḥalāl certification processes are similar, the integration of the halāl assurance scheme into the HACCP system could be feasible [75].

Although the knowledge of all factors influencing the ḥalāl assurance systems is a determinant to help companies to identify intervention strategies to improve their performance, limited literature is available on this issue [76]. Recently, Malaysian researchers have applied an interesting study design to explore critical factors affecting the ḥalāl assurance systems: different factors in every country, region or food chain can be found, with differences resulting between food sectors and subsectors and among small or medium (SME) and large-sized enterprises [77]. These studies are particularly important for SMEs that have limited resources.

4. Authentication of ḥalāl Meat for Salami and Other Meat Products

With the increasing population, the demand for ḥalāl food products also increases, putting a responsibility upon government, jurisprudence and companies to certify ḥalāl products [78].

The matching of each product with the label statement is a quality requirement; in European countries it is mandatory that the products are labelled in accordance with Regulation (EC) n. 1169/2011 of the European Parliament and of the Council of 25 October 2011 on the provision of food information to consumers [24]. Furthermore, for ḥalāl foods, the need for clear ḥalāl labels, ensuring that the product (from the ingredients to the processing and handling) meets the appropriate requirements, is a critical issue. Due to the repeated discoveries of non-ḥalāl ingredients in food otherwise labelled as ḥalāl, the status concerning the determination of ḥalāl and non-ḥalāl food products needs to be carefully read.

The analytical authentication of ḥalāl foods has the purpose of solving problems such as fraud and adulteration of ḥalāl products that are highly critical both for importing countries (such as Malaysia, Saudi Arabia, Singapore and Brunei Darussalam) and for the top ḥalāl food exporters (Brazil, Australia, USA and France) [79–81]. Various ḥalāl supervision agencies work closely with food industries to obtain the permission to use their supervision agency's trademark symbol on their products.

Ḥalāl authenticity is an issue of major concern in the food industry, and methods of lard detection have been performed for the investigation in food products such as cakes and chocolate [82,83]. Moreover, specific techniques able to exclude their possible contamination or fraud have been developed [84–86]. A frequent adulteration of meat products is the addition of pork to beef products, which is carried out for economic gain and represents a serious problem in the halāl food industry, in particular for the authenticity of minced and homogenized meat products. Moreover, often companies producing halāl meat products also process other kinds of meat, and thus cross-contaminations are possible. Therefore, many researches have been recently carried out for authenticating the species composition of meat products [87–91] and in particular the ḥalāl authentication studies are focused on the detection of pork derivatives (meat or lard) [92,93].

Different commercial kits which investigate porcine protein and DNA have been developed in many countries (such as USA, UK, France and Belgium) in order to establish the ḥalāl authenticity of food products [94]; kits are useful as they generally allow a rapid determination of the contamination, nevertheless the addition of meat different from pork could remain undetected. To date, different techniques are routinely applied for meat species detection and identification in food: in Tables 2 and 3, protein-based and genetic methods suitable for evaluating the authenticity of halāl meat products are reported, respectively. Due to the characteristic of proteins that tend to denature at high temperatures, these methods have limitations in the detection of animal species from cooked, baked or heat-treated food products; on the contrary, DNA-based methods are more sensitive and reliable, as DNA is

found in a majority of cells, it is species-specific and is stable at higher temperatures [95]. However, meat processing could denature short DNA sequences [96] whilst the primary structure of peptides is relatively stable; for this reason a possible approach for highly processed meat authentication could be the combination of chromatography with mass spectrometry (MS), thus investigating the molecular weight and amino acid sequence of meat proteins [13].

Table 2. Examples of protein-based methods for ḥalāl authenticity analysis.

Methods	Aim	References
Immunoassay (ELISA)	Porcine gelatin determination in processed foods	[97]
Isoelectric focusing (IEF)	Meat authentication in raw and cooked meat products	[98]
Chromatography and mass spectrometry (MS)	Meat authentication Meat species determination	[13,79] [99]
Electric nose (EN)	Pork fat detection Pork meat detection	[100] [66]
Mass spectrometry soft ionization	Identification of muscle proteins of different species Horse and pork meat detection Horse and pork meat detection in highly processed food	[101] [102] [103]
Fourier transform infrared spectroscopy (FTIR); Near-infrared spectroscopy (NIR; FT-NIR)	Lard detection Pork detection in sausages Pork derivatives detection Adulteration of meat	[104] [105,106] [107,108] [109]

Table 3. Examples of genetic methods for ḥalāl authenticity analysis.

Methods	Aim	References
Simple sequence repeat (SSR) and Single nucleotide polymorphism (SNP)	Meat traceability Meat fraud	[81] [110]
Polymerase chain reaction (PCR)	Pork derivatives detection Pork derivatives detection in gelatin Meat species identification Authenticity determination	[11,84] [111] [85,112,113] [94]
PCR- Restriction Fragment length polymorphism (PCR-RFLP)	Pork meat detection in meat products Rabbit, rat and squirrel meat detection in frankfurter	[114] [92]
Real Time PCR	Horse and donkey meat detection Species identification of meat Pork meat detection Pork meat detection and quantification	[115,116]
Next Generation Sequencing	Identification of 46 different meat species in pure samples, in spiked samples and in ground meat samples	[117]
Aptamers	Application in analysis of foods	[118]
Isothermal amplification	Meat species identification Detection of meat of different species Review on Isothermal amplification techniques Rapid on-site detection of meat pork Nucleic acid detection	[119] [120] [121] [122] [123,124]

The PCR amplification of pork mitochondrial genes (12S and 18Sribosomal RNA subunits and cytochrome b) and of the displacement loop region (D-loop) was successfully applied for the detection of pork derivatives and was found to be a suitable technique for routine food analysis and halāl certification [115].

In 2018, for the first time a tetraplex polymerase chain reaction-restriction fragment length polymorphism (PCR-RFLP) assay to identify and discriminate rabbit, rat and squirrel meat in frankfurter formulation was developed and validated. The detection limit of the assay was 0.1% meat in frankfurter formulation. Moreover, results shown in this research assessed that variations in food processing treatments could not affect the stability of the optimized assay [92]. However, it should be considered that it is rather unlikely that rabbit or squirrel meat can be used as a substitute for chicken or beef, because their meat is certainly more expensive.

Nevertheless, although classical PCR and real time-PCR are the most frequently used methods, they usually target a limited pool of species, and usually the most used, such as pork, beef, horse and chicken, whereas the potential adulteration with exotic species often remains uncovered. For this purpose, Cottenet and colleagues [117] have recently developed a Next Generation Sequencing method for the identification of different species, both in pure meat samples, with optimal results, and in mixtures, where the species were correctly identified in spiked samples down to 1% (*w*/*w*). Together with the most common species, also yak, donkey, zebra, hare, fallow deer, reindeer, muskrat, fox, weasel, dog, cat pigeon and rat, and up to 46 different species were detected, amplifying and sequencing a mitochondrial DNA fragment of about 120 bp. Unfortunately, the method was less effective when applied to ground meat, suggesting that work is still required to improve the results.

Moreover, Lavelli has suggested a scheme of traceability implementation for the poultry meat supply chain: the author presented a case study to discuss both the advantages and difficulties of setting up a high-warrant traceability procedure conform to "generic" or "specific" traceability systems, depending on many different factors (technological and economical aspects, specific regulations and internal objectives) [125].

The genetic technologies, such as simple sequence repeat (SSR) and single nucleotide polymorphism (SNP) can be very reliable in the traceability management system of ḥalāl foods [126,127]. Some studies were carried out in order to select SNPs panel useful for traceability of ḥalāl beef [81,110].

A recent study reported about a particular pork peptide (signature sequence LVVITAGAR, from lactate dehydrogenase) that was not recorded in other meats; in this study, liquid based chromatography coupled with mass spectrometry (LC-MS) was used for the detection of the identified porcine-specific peptide as a thermostable marker of highly processed haram (that means proscribed by the Islamic law) meat products [99].

Interestingly, proton nuclear magnetic resonance (^{1}H-NMR) has been applied together with HPLC as a high-performance approach to detect lard used to adulterate butter. In fact, the triacylglycerol (TAG) composition of lard can be used as chemical marker for ḥalāl authentication. More precisely, peaks recorded in the region of 2.60–2.84 ppm highlight specific characteristics present only in lard, and these frequencies can be considered specifically for the analysis [128].

A rapid, accurate, convenient, and eco-friendly analysis for the detection of porcine-based ingredients in food is the Electronic Nose (EN), employed to exclude possible contamination or fraud and also to investigate the oxidative status of meat products [85,86,101,102]. The ability to measure and identify the aroma and characteristics of persistent flavors of a product allows to obtain a variety of information, directly related to the acceptability and to nutritional and quality characteristics in terms of product health and safety. This consideration has led to a growing need and interest in non-destructive monitoring systems with high versatility, sensitivity, accuracy, cost-effectiveness, ease of use and above all rapidity of analytical response, such as electronic nose. This artificial olfactory systems can be used not only in the laboratory but also directly in the production plants for continuous monitoring of odors, from the raw material to the final product, analyzing the volatile compounds released from the matrix related to different aspects (sensory acceptability, deterioration, development of off-flavors, etc.). In our laboratories an electronic nose equipped with non-specific sensors arrays (porphyirins based) has been used to discriminate the lard presence in ḥalāl goat meat batter. First, four experimental batches of salami were manufactured with halal goat meat (70%) and fat (30%); each of the four batches had different levels of pork fat (batch 1, Control, with lard 0% *w/w* of total fat; other batches, with lard 5%, 10% and 20% *w/w* of total fat, respectively); then EN analyses were carried out after an equilibration step (for 30 min, at temperature of 41 °C). Principal Component analyses (PCA) scores (Figure 2) evidenced that porphyirins-based EN was able to discriminate the presence of lard even at the lowest experimental concentration (5%, *w/w*) [129].

Furthermore, a gas sensor array based on peptide modified gold nanoparticles deposited onto 20-MHz quartz crystal microbalances [102] has been also applied to discriminate the lard presence in ḥalāl meat products and investigate the shelf life of ḥalāl dry meat sausages (trials are still going on; unpublished data).

Finally, new approaches regarding ḥalāl authentication, including the latest biotechnological innovations, such as assays and the use of smartphones, are being also developed [130,131].

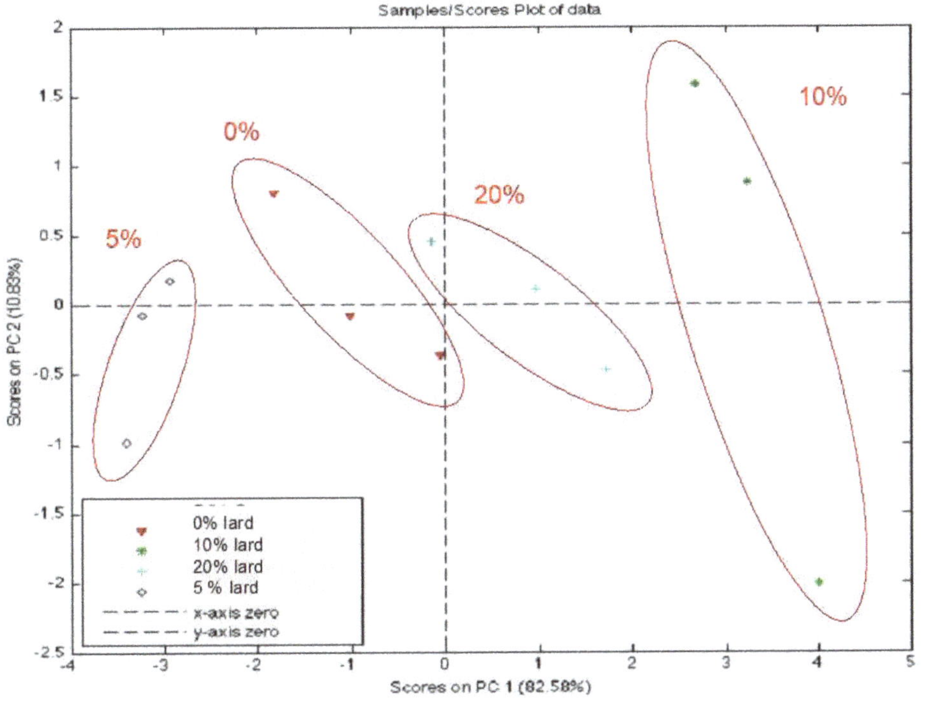

Figure 2. Principal component analyses (PCA) scores on porphyrins-based EN analysis on ḥalāl goat meat dough added with pork fat (0%, 5%, 10% and 20% of lard, *w/w*).

5. Conclusions

The quality assessment and authentication of ḥalāl products are issues raising a growing interest in European Community Countries (France, Sweden, Germany, Greece, Spain, Italy), Switzerland, Russia and other countries in the world (Asia, the UK, South and North America).

Considering that top producers of ḥalāl products (including meat) are countries where Muslims are a minority, future research should take into consideration ḥalāl standards, immigration and integration of qualified Muslim workers, as evidenced by a study recently carried out in Brazil [132]. As regards specifically halāl salami, the origin of the animal raw meat, as well as ingredients and additives are the main concerns for consumers of Islamic faith. The pork matrix may be fraudulently present in processed meat, for both economic and technological purposes. In fact, the use of these raw materials (which are easier to find) has a lower cost; furthermore, it allows manufacturers to obtain final cured meat with better quality (sensory properties) and stability (especially with respect to oxidative reactions).

Many approaches have been proposed in the literature for the evaluation of the authenticity of salami and other meat products with ḥalāl certification. The research of accurate analytical methods for the differentiation of meat species is therefore of great importance for both companies and consumers, and important advances have been made in recent years, while analytical methods to distinguish the type of slaughter applied to obtain the meat are still difficult to optimize. Moreover, the literature is often focused on ḥalāl meat products, while comparisons with non ḥalāl analogue products are still scarce or even missing. Future researchers should carry out further studies on ḥalāl food in order to provide useful information about major factors related to quality and stability, especially for ḥalāl dry cured meat products, such as salami. The market for ḥalāl products is evolving, considering that the non-Islamic consumer would seem to associate the ḥalāl brand with "superior" quality [133]. This would be very important to properly orient the companies that would like to diversify their production and also to guarantee food safety and consumer satisfaction.

Author Contributions: Conceptualization, M.M. and A.S.; methodology, M.M., A.S.; investigation, M.M., A.S.; resources, M.M., A.S.; data curation, M.M., A.S., D.M.; writing—original draft preparation, M.M.; writing—review and editing, M.M., A.S.; visualization, O.C., D.M.; supervision, M.M., D.M.; project administration, M.M.; funding acquisition, M.M. All authors have read and agreed to the published version of the manuscript.

Funding: This research received no external funding.

Acknowledgments: The authors would to thank Daniel Pizzoni for his helpful contribution for processing the data shown in Figure 2. This work is part of the project "Quality and shelf life of dry fermented goat sausage: a new perspective of market for companies of Abruzzo region" supported by funding from Cassa di Risparmio di Teramo (Italy).

Conflicts of Interest: The authors declare no conflict of interest. The funders had no role in the design of the study; in the collection, analyses, or interpretation of data; in the writing of the manuscript, or in the decision to publish the results.

References

1. Pew, R.C. "The Changing Global Religious Landscape", Demographic projections. 2017. Available online: http://assets.pewresearch.org/wp-content/uploads/sites/11/2017/04/07092755/FULL-REPORT-WITH-APPENDIXES-A-AND-B-APRIL-3.pdf (accessed on 15 June 2020).
2. Grand, V.R. Halal Food and Beverage Market Size Report by Product (Meat & Alternatives, Milk & Milk Products, Fruits & Vegetables, Grain Products), by Region, and Segment Forecasts (2018). (Grand View Research). 2017. Available online: https://www.grandviewresearch.com/industry-analysis/halal-food-market2018-2025 (accessed on 15 June 2020).
3. Pew, R.C. Europe's Growing Muslim Population. Available online: https://www.pewforum.org/2017/11/29/europes-growing-muslim-population/ (accessed on 3 August 2020).
4. Aoun, I.; Tournois, L. Building holistic brands: An exploratory study of Ḥuild cosmetics. *J. Islam. Mark.* **2015**, *6*, 109–132. [CrossRef]
5. Mursyidi, A. The role of chemical analysis in the ḥalāl authentication of food and pharmaceutical products. *J. Food Pharm. Sci.* **2013**, *1*, 1–4.
6. Codex Alimentarius. *Joint FAO/WHO Codex Alimentarius Commission: General Guideline for Use of the Term "halal"*; Codex Alimentarius: Rome, Italy, 1997.
7. Jagadeesan, P.; Salem, S. Progress and challenges associated with Halal authentication of consumer packaged goods. *J. Sci. Food Agric.* **2017**, *97*, 4672–4678.
8. Soon, J.M.; Chandia, M.; Regenstein, J.M. Halal integrity in the food supply chain. *Br. Food J.* **2017**, *119*, 39–51. [CrossRef]

9. Riaz, M.N.; Chaudry, M.M. *Ḥalāl Food Production*, 1st ed.; CRC Press: Boca Raton, FL, USA, 2004; pp. 41–58.
10. Ahmad, A.N.; Ungku Zainal Abidin, U.F.; Othman, M.; Abdul Rahman, R. Overview of the halal food control system in Malaysia. *Food Control* **2018**, *90*, 352–363. [CrossRef]
11. Gagaoua, M.; Boudechicha, H.-R. Ethnic meat products of the North African and Mediterranean countries: An overview. *J. Ethn. Foods* **2018**, *5*, 83–98. [CrossRef]
12. Ermis, E. Halal status of enzymes used in food industry. *Trends Food Sci. Technol.* **2007**, *64*, 69–73. [CrossRef]
13. Bohme, K.; Calo-Mata, P.; Barros-Velazquez, J.; Ortea, I. Recent applications of omics-based technologies to main topics in food authentication. *Trends Anal. Chem.* **2019**, *110*, 221–232. [CrossRef]
14. Regulation (EC) n. 1099/2009 of 24 September 2009 on the protection of animals at the time of killing. *Off. J. Eur. Union* **2009**, *L 303*, 1–30.
15. Fuseini, A.; Wotton, S.B.; Hadley, P.J.; Knowles, T.G. The perception and acceptability of pre-slaughter and post-slaughter stunning for Halal production: The views of UK Islamic scholars and Halal consumers. *Meat Sci.* **2017**, *123*, 143–150. [CrossRef]
16. Aghwan, Z.A.; Bello, A.U.; Abubakar, A.A.; Imlan, J.C.; Sazili, A.Q. Efficient halal bleeding, animal handling, and welfare: A holistic approach for meat quality. *Meat Sci.* **2016**, *121*, 420–428. [CrossRef] [PubMed]
17. Velarde, A.; Rodriguez, P.; Dalmau, A.; Fuentes, C.; Llonch, P.; von Holleben, K.V.; Anil, M.H.; Lambooij, J.B.; Pleiter, H.; Yesildere, T.; et al. Religious slaughter: Evaluation of current practices in selected countries. *Meat Sci.* **2014**, *96*, 278–287. [CrossRef] [PubMed]
18. D'Amico, P.; Vitelli, N.; Cenci Goga, B.; Nucera, D.; Pedonese, F.; Guidi, A.; Armani, A. Meat from cattle slaughtered without stunning sold in the conventional market without appropriate labelling: A case study in Italy. *Meat Sci.* **2017**, *34*, 1–6. [CrossRef] [PubMed]
19. Nuthall, K. ECJ Advised non-Stunned Halal and Kosher Meat can be Labelled Organic. Available online: https://www.globalmeatnews.com (accessed on 26 September 2018).
20. Regulation (EC) n. 178/2002 of the European Parliament and of the Council of 28 January 2002 on laying down the general principles and requirements of food law, establishing the European Food Safety Authority and laying down procedures in matters of food safety. *Off. J. Eur. Communities* **2002**, *L 31*, 1–24.
21. Regulation (EC) n. 852/2004 of the European Parliament and of the Council of 29 April 2004 on the hygiene of foodstuffs. *Off. J. Eur. Union* **2004**, *L 139*, 1–54.
22. Regulation (EC) n. 853/2004 of the European Parliament and of the Council of 29 April 2004 on laying down specific hygiene rules for on the hygiene of foodstuffs. *Off. J. Eur. Union* **2004**, *L 139*, 1–54.
23. Regulation (EC) n. 1935/2004 of the European Parliament and of the Council of 27 October 2004 on materials and articles intended to come into contact with food and repealing Directives 80/590/EEC and 89/109/EEC. *Off. J. Eur. Union* **2004**, *L 338*, 4–17.
24. Regulation (EC) n. 1169/2011 of the European Parliament and of the Council of 25 October 2011 on the provision of food information to consumers, amending Regulations (EC) No 1924/2006 and (EC) No 1925/2006 of the European Parliament and of the Council, and repealing Commission Directive 87/250/EEC, Council Directive 90/496/EEC, Commission Directive 1999/10/EC, Directive 2000/13/EC of the European Parliament and of the Council, Commission Directives 2002/67/EC and 2008/5/EC and Commission Regulation (EC) n. 608/2004. *Off. J. Eur. Union* **2011**, *L 304*, 18–63.
25. Demeyer, D.; Raemaekers, M.; Rizzo, A.; Holck, A.; De Smedt, A.; ten Brink, B.; Hagen, B.; Montel, C.; Zanardi, E.; Murbrekk, E.; et al. Control of bioflavour and safety in fermented sausages: First results of a European project. *Food Res. Int.* **2000**, *33*, 171–180. [CrossRef]
26. Tieko Nassu, R.; Guaraldo Goncalves, L.; da Silva, M.A.A.P.; Beserra, F.J. Oxidative stability of fermented goat meat sausages with different levels of natural antioxidant. *Meat Sci.* **2003**, *63*, 43–49. [CrossRef]
27. Gregory, N.G.; von Wenzlawonzlawowicz, M.; Alam, R.M.; Anil, H.M.; Yesildere, T.; Silva-Fletcher, A. False aneurysms in carotid arteries of cattle and water buffalo during shechita and ḥalāl slaughter. *Meat Sci.* **2008**, *79*, 285–288. [CrossRef] [PubMed]
28. Linares, M.B.; Bórnez, R.; Vergara, H. Effect of different stunning systems on meat quality of light lamb. *Meat Sci.* **2007**, *76*, 675–681. [CrossRef]
29. Danso, A.S.; Richardson, R.I.; Khalid, R. Assessment of the meat quality of lamb M. longissimus thoracis et lumborum and M. triceps brachii following three different Ḥalāl slaughter procedures. *Meat Sci.* **2017**, *127*, 6–12. [CrossRef]

30. Ibrahim, S.M.; Abdelgadir, M.A.; Sulieman, A.M.E. Impact of Halal and Non-halal Slaughtering on the Microbiological Characteristics of Broiler Chicken Meat and Sausages. *Food Public Health* **2014**, *4*, 223–228. [CrossRef]
31. Matos, R.A.; Menezes, C.M.; Ramos, E.M.; Ramos, A.L.S.; Gomide, L.A.M. Effects of fermentation types in final quality of mutton cooked fermented sausages. *Bol. Ceppa* **2007**, *25*, 225–234.
32. Paulos, K.; Rodrigues, S.; Oliveira, A.F.; Leite, A.; Pereira, E.; Teixeira, A. Sensory characterization and consumer preference mapping of fresh sausages manufactured with goat and sheep meat. *J. Food Sci.* **2015**, *80*, S1568–S1573. [CrossRef]
33. Cunha de Andrade, J.; Silveira Nalério, E.; Giongo, C.; Dutra de Barcellos, M.; Ares, G.; Deliza, R. Consumer sensory and hedonic perception of sheep meat coppa under blind and informed conditions. *Meat Sci.* **2018**, *137*, 201–210. [CrossRef]
34. Mushi, D.E.; Thomassen, M.S.; Kifaro, G.C.; Eik, L.O. Fatty acid composition of minced meat, longissimus muscle and omental fat from Small East African goats finished on different levels of concentrate supplementation. *Meat Sci.* **2010**, *86*, 337–342. [CrossRef]
35. Martuscelli, M.; Fantini, A.; Bucci, I.; Mastrocola, D. Halal dry fermented goat meat and sheep sausages: Market research and economic prospects. *Ind. Aliment.* **2019**, *58*, 11–19.
36. Aktas, N.; Gencelep, H. Effect of starch type and its modifications on physicochemical properties of bologna-type sausage produced with sheep tail fat. *Meat Sci.* **2006**, *74*, 404–408. [CrossRef]
37. Bakker, W.A.M.; Houben, J.H.; Koolmees, P.A.; Bindrich, U.; Sprehe, L. Effect of initial mild curing, with additives, of hog and sheep sausage casings on their microbial quality and mechanical properties after storage at different temperatures. *Meat Sci.* **1999**, *51*, 163–174. [CrossRef]
38. Ahmad, S.; Srivastava, P.K. Quality and shelf life evaluation of fermented sausages of buffalo meat with different levels of heart and fat. *Meat Sci.* **2007**, *75*, 603–609. [CrossRef] [PubMed]
39. Zhao, L.; Changwei, M.; Song, H.; Li, H.; Wang, Z.; Xiao, S. Physico-chemical characteristics and free fatty acid composition of dry fermented mutton sausages as affected by the use of various combinations of starter cultures and spices. *Meat Sci.* **2011**, *88*, 761–766. [CrossRef] [PubMed]
40. Galgano, F.; Favati, F.; Schirone, M.; Martuscelli, M.; Crudele, M.A. Influence of indigenous starter cultures on the free fatty acids content during ripening in artisanal sausages produced in the Basilicata region. *Food Technol. Biotechnol.* **2003**, *41*, 253–258.
41. Parente, E.; Martuscelli, M.; Gardini, F.; Grieco, S.; Crudele, M.A.; Suzzi, G. Evolution of microbial population and biogenic amine production in dry sausages produced in Southern Italy. *J. Appl. Microbiol.* **2001**, *90*, 882–891. [CrossRef]
42. Kurt, Ş.; Ceylan, H.G. Effects of olive leaf extract on the oxidation stability and microbiological quality of dry fermented sausage (sucuk). *Carpathian J. Food Sci. Technol.* **2017**, *9*, 178–188.
43. Sezer, Y.; Bozkurt, H. Use of novel casing in sucuk production: Antimicrobials incorporated into multilayer plastic film. *Acta Aliment.* **2019**, *48*, 1–8. [CrossRef]
44. Çiçek, Ü.; Köse, T. Physical and biochemical quality parameters of fermented beef sausages: Bez Sucuk. *Acta Aliment. Hung* **2016**, *45*, 363–370. [CrossRef]
45. Çiçek, Ü.; Polat, N. Investigation of physicochemical and sensorial quality of a type of traditional meat product: Bez sucuk. *LWT-Food Sci. Technol.* **2016**, *65*, 145–151. [CrossRef]
46. Lorenzo, J.M.; Martínez, S.; Franco, I.; Carballo, J. Biogenic amine content during the manufacture of dry-cured lac ón, a Spanish traditional meat product: Effect of some additives. *Meat Sci.* **2007**, *77*, 287–293. [CrossRef]
47. Zakaria, M.Z.; Nordin, N.; Malik, A.M.A.; Elias, S.J.; Shahuddin, A.Z. Fuzzy expert systems (fes) for halal food additive. *Indones J. Electr. Eng. Comput. Sci.* **2019**, *13*, 1073–1078. [CrossRef]
48. Karahalil, E. Principles of halal-compliant fermentations: Microbial alternatives for the halal food industry. *Trends Food Sci. Technol.* **2020**, *98*, 1–9. [CrossRef]
49. Eduardo dos Santos, C.; Braun, C.L.K.; Fagundes, M.B.; Gularte, M.A.; Wagner, R.; Padilha da Silva, W.; Fiorentini, Â.M. Development of fermented sausage produced with mutton and native starter cultures. *Lebenson Wiss Technol.* **2018**, *95*, 23–31. [CrossRef]
50. Kurtovic, I.; Marshall, S.N.; Cleaver, H.L.; Miller, M.R. The use of immobilised digestive lipase from Chinook salmon (*Oncorhynchus tshawytscha*) to generate flavour compounds in milk. *Food Chem.* **2016**, *199*, 323–329. [CrossRef]

51. Domínguez, R.; Agregán, R.; Lorenzo, J.M. Role of commercial starter cultures on microbiological, physicochemical characteristics, volatile compounds and sensory properties of dry-cured foal sausage. *Asian Pac. J. Trop. Dis.* **2016**, *6*, 396–403. [CrossRef]
52. Kargozari, M.; Moini, S.; Basti, A.A.; Emam-Djomeh, Z.; Ghasemlou, M.; Revilla, I.; Gandomi, M.H.; Carbonell-Barrachina, A.A.; Szumny, A. Development of Turkish dry-fermented sausages (sucuk) reformulated with camel meat and hump fat and evaluation of physicochemical, textural, fatty acid and volatile compound profiles during ripening. *LWT-Food Sci. Technol.* **2014**, *59*, 849–858. [CrossRef]
53. Jia, W.; Zhang, R.; Shi, L.; Zhang, F.; Chang, J.; Chu, X. Effects of spices on the formation of biogenic amines during the fermentation of dry fermented mutton sausages. *Food Chem.* **2020**, *321*, 126723. [CrossRef]
54. LaTorre-Moratalla, M.L.; Bover-Cid, S.; Veciana-Nogués, M.T.; Vidal-Carou, M.C. Control of biogenic amines in fermented sausageg: Role of starter cultures. *Front. Microbiol.* **2012**, *3*, 169. [CrossRef]
55. Tasic, T.; Ikonic, P.; Jokanovic, M.; Mandic, A.; Tomovic, V.; Sojic, B.; Skaljac, S. Content of Vasoactive Amines in Sremski Kulen and Sremska Kobasica Traditional Dry Fermented Sausages From Vojvodina. *Procedia Food Sci.* **2015**, *5*, 282–284. [CrossRef]
56. Ekici, K.; Omer, A.K. The determination of some biogenic amines in Turkish fermented sausages consumed in *Van. Toxicol. Rep.* **2018**, *5*, 639–643. [CrossRef]
57. Çiçek, Ü.; Tokatli, K. Biogenic Amine Formation in "Bez Sucuk," a Type of Turkish Traditional Fermented Sausage Produced with Different Meat: Fat Ratios. *Korean J. Food Sci. Anim. Resour.* **2018**, *38*, 152–161. [CrossRef] [PubMed]
58. Huang, L.; Ding, B.; Zhang, H.; Kong, B.; Xiong, Y.L. Textural and sensorial quality protection in frozen dumplings through the inhibition of lipid and protein oxidation with clove and rosemary extracts. *J. Sci. Food Agric.* **2019**, *99*, 4739–4747. [CrossRef] [PubMed]
59. Serio, A.; Chiarini, M.; Tettamanti, E.; Paparella, A. Electronic Paramagnetic Resonance investigation of the activity of *Origanum vulgare* L. essential oil on the *Listeria monocytogenes* membrane. *Lett. Appl. Microbiol.* **2010**, *51*, 149–157. [CrossRef] [PubMed]
60. Chan, C.L.; Gan, R.Y.; Shah, N.P.; Corke, H. Polyphenols from selected dietary spices and medicinal herbs differentially affect common food-borne pathogenic bacteria and lactic acid bacteria. *Food Control* **2018**, *92*, 437–443. [CrossRef]
61. Erdemir, E.; Aksu, M.I. Changes in the composition of free amino acid during production of pastirma cured with different levels of sodium nitrite. *J. Food Process. Preserv.* **2017**, *41*. [CrossRef]
62. Abd-Elghany, S.M.; El-Makhzangy, A.M.; El-Shawaf, A.-G.M.; El-Mougy, R.M.; Sallam, K.I. Improving safety and quality of Egyptian pastrami through alteration of its microbial, community. *LWT-Food Sci. Technol.* **2020**, *118*, 108872. [CrossRef]
63. Suleman, R.; Whang, Z.; Aadil, R.M.; Hui, T.; Hopkins, D.L.; Zhang, D. Effect of cooking on the nutritive quality, sensory properties and safety of lamb meat; current challenges and future prospects. *Meat Sci.* **2020**, *167*, 108172. [CrossRef]
64. Lee, S.Y.; Yim, D.G.; Lee, D.Y.; Kim, O.Y.; Kang, H.J.; Kim, H.S.; Jang, A.; Park, T.S.; Jin, S.K.; Hur, S.J. Overview of the effect of natural products on reduction of potential carcinogenic substances in meat products. *Trends Food Sci. Technol.* **2020**, *99*, 568–579. [CrossRef]
65. Alirezalu, K.; Pateiro, M.; Yaghoubi, M.; Alirezalu, A.; Peighambardoust, S.H.; Lorenzo, J.M. Phytochemical constituents, advanced extraction technologies and techno-functional properties of selected Mediterranean plants for use in meat products. A comprehensive review. *Trends Food Sci. Technol.* **2020**, *100*, 292–306. [CrossRef]
66. Granato, D.; Nunes, D.S.; Barba, F.J. An integrated strategy between food chemistry, biology, nutrition, pharmacology, and statistics in the development of functional foods: A proposal. *Trends Food Sci. Technol.* **2017**, *62*, 13–22. [CrossRef]
67. Adzaly, N.Z.; Jackson, A.; Villalobos-Carvajal, R.; Kang, I.; Almenar, E. Development of a novel sausage casing. *J. Food Eng.* **2015**, *152*, 24–31. [CrossRef]
68. Marcos, B.; Gou, P.; Arnau, J.; Guàrdia, M.D.; Comaposada, J. Co-extruded alginate as an alternative to collagen casings in the production of dry-fermented sausages: Impact of coating composition. *Meat Sci.* **2020**, *169*, 108184. [CrossRef] [PubMed]

69. Soysal, Ç.; Bozkurt, H.; Dirican, E.; Güçlü, M.; Bozhüyük, E.D.; Uslu, A.E.; Kaya, S. Effect of antimicrobial packaging on physicochemical and microbial quality of chicken drumsticks. *Food Control* **2015**, *54*, 294–299. [CrossRef]
70. Cerrillo, J.L.; Palomares, A.E.; Rey, F. Silver exchanged zeolites as bactericidal additives in polymeric materials. *Microporous Mesoporous Mater.* **2020**, *305*, 110367. [CrossRef]
71. Martuscelli, M.; Lupieri, L.; Sacchetti, G.; Mastrocola, D.; Pittia, P. Prediction of the salt content from water activity analysis in dry-cured ham. *J. Food Eng.* **2017**, *200*, 29–39. [CrossRef]
72. Grandin, T. Auditing animal welfare and making practical improvements in beef-, pork- and sheep-slaughter plants. *Anim. Welf.* **2012**, *21*, 29–34. [CrossRef]
73. Afifi, M.; Halim, A.; Mahyeddin, M.; Salleh, M. The possibility of uniformity on ḥalāl standards in Organization of Islamic Countries (OIC) country. *World Appl. Sci. J.* **2012**, *17*, 6–10.
74. Abd Rahman, A.; Barau Singhry, H.; Hizam Hanafiah, M.; Abdul, M. Influence of perceived benefits and traceability system on the readiness for Ḥalāl Assurance System implementation among food manufacturers. *Food Control* **2017**, *73*, 1318–1326. [CrossRef]
75. Demirci, N.M.; Soon, J.M.; Wallace, C.A. Positioning food safety in Ḥalāl Assurance. *Food Control* **2016**, *70*, 257–270. [CrossRef]
76. Nasir, K.M.; Pereira, A.A. Defensive dining: Notes on the public dining experiences in Singapore. *Contemp. Islam* **2008**, *2*, 61–73. [CrossRef]
77. Ahmad, A.N.; Abdul Rahman, R.; Othman, M.; Ungku Zainal Abidin, U.F. Critical success factors affecting the implementation of ḥalāl food management systems: Perspective of ḥalāl executives, consultants and auditor. *Food Control* **2017**, *74*, 70–78. [CrossRef]
78. Khan Khattak, J.Z.; Mir, A.; Anwar, Z.; Mustatab Wahedi, H.; Abbas, G.; Khan Kattak, H.Z.; Ismatullah, H. Concept oh Ḥalāl food and biotechnology. *Adv. J. Food Sci. Technol.* **2011**, *3*, 385–389.
79. Ballin, N.Z. Authentication of meat and meat products. *Meat Sci.* **2010**, *86*, 577–587. [CrossRef]
80. Thomson Reuters. *State of the Global Islamic Economy 2014-2015 Report*; Thomson Reuters: Toronto, YYZ, Canada, 2015.
81. Zhao, J.; Chen, A.; You, X.; Xu, Z.; Zhao, Y.; He, W.; Zhao, L.; Yang, S. A panel of SNP markers for meat traceability of Halal beef in the Chinese market. *Food Control* **2018**, *87*, 94–99. [CrossRef]
82. Che Man, Y.B.; Syahariza, Z.A.; Mirghani, M.E.S.; Jinap, S.; Bakar, J. Analysis of potential lard adulteration in chocolate and chocolate products using fourier transform infrared spectroscopy. *Food Chem.* **2005**, *90*, 815–819. [CrossRef]
83. Syahariza, Z.A.; Che Man, Y.B.; Selamat, J.; Bakar, J. Detection of lard adulteration in cake formulation by Fourier Transform Infrared (FTIR) Spectroscopy. *Food Chem.* **2005**, *92*, 365–371. [CrossRef]
84. Che Man, Y.B.; Aida, A.; Raha, A.; Son, R. Identification of pork derivatives in food products by species-specific polymerase chain reaction (PCR) for ḥalāl verification. *Food Control* **2007**, *18*, 885–889. [CrossRef]
85. Murugaiah, C.; Mohd Noor, Z.; Mastakim, M.; Bilung, L.M.; Selamat, J.; Radu, S. Meat species identification and Ḥalāl authentication analysis using mitochondrial DNA. *Meat Sci.* **2009**, *83*, 57–61. [CrossRef]
86. Nurjuliana, M.; Che Man, Y.B.; Mat Hashim, D.; Mohamed, A.K. Rapid identification of pork for ḥalāl authentication using the electric nose and gas chromatography mass spectrometer with headspace analyser. *Meat Sci.* **2011**, *88*, 638–644. [CrossRef]
87. Stachniuk, A.; Sumara, A.; Montowska, M.; Fornal, E. LC-QTOF-MS identification of rabbit-specific peptides for authenticating the species composition of meat products. *Food Chem.* **2020**, *329*, 127185. [CrossRef]
88. Köppel, R.; van Velsen, F.; Ganeshan, A.; Pietsch, K.; Weber, S.; Graf, C.; Murmann, P.; Hochegger, R.; Licina, A. Multiplex real-time PCR for the detection and quantification of DNA from chamois, roe, deer, pork and beef. *Eur. Food Res. Technol.* **2020**, *246*, 1007–1015. [CrossRef]
89. Mei, M.; Chen, R.; Gao, X.; Cao, Y.; Weng, W.; Duan, Y.; Tan, X.; Liu, Z. Establishment and application of a 10-plex liquid bead array for the simultaneous rapid detection of animal species. *J. Sci. Food Agric.* **2020**, *100*, 325–334. [CrossRef] [PubMed]
90. Dobrovolny, S.; Blaschitz, M.; Weinmaier, T.; Pechatschek, J.; Cichna-Markl, M.; Indra, A.; Hufnagl, P.; Hochegger, R. Development of a DNA metabarcoding method for the identification of fifteen mammalian and six poultry species in food. *Food Chem.* **2019**, *272*, 354–361. [CrossRef] [PubMed]

91. Wang, G.-J.; Zhou, G.-Y.; Ren, H.-W.; Xu, Y.; Yang, Y.; Guo, L.-H.; Liu, N. Peptide biomarkers identified by LC–MS in processed meats of five animal species. *J. Food Compos. Anal.* **2018**, *73*, 47–54. [CrossRef]
92. Ali, M.-E.; Ahamad, M.N.U.; Asing, M.A.M.H.; Sultana, S. Multiplex polymerase chain reaction-restriction fragment length polymorphism assay discriminates of rabbit, rat and squirrel meat in frankfurter products. *Food Control* **2018**, *84*, 148–158. [CrossRef]
93. Rohman, A.; Che Man, Y.B. Analysis of pig derivatives for ḥalāl authentication studies. *Food Rev. Int.* **2012**, *28*, 97–112. [CrossRef]
94. Nakyinsige, K.; Man, Y.B.; Sazili, A.Q. Ḥalāl authenticity issues in meat and meat products. *Meat Sci.* **2012**, *91*, 207–214. [CrossRef]
95. Cammà, C.; Domenico, M.D.; Monaco, F. Development and validation of fast Real-Time PCR assays for species identification in raw and cooked meat mixtures. *Food Control* **2012**, *23*, 400–404. [CrossRef]
96. Lo, Y.-T.; Shaw, P.-C. DNA-based techniques for authentication of processed food and food supplements. *Food Chem.* **2018**, *2401*, 767–774. [CrossRef]
97. Doi, H.; Watanabe, E.; Shibata, H.; Tanabe, S. A reliable enzyme-linked immunosorbent assay for the determination of bovine and porcine gelatin in processed foods. *J. Agric. Food Chem.* **2009**, *57*, 1721–1726. [CrossRef]
98. Montowska, M.; Pospiech, E. Species-specific expression of various proteins in meat tissue: Protemic analysis of raw and cooked meat and meat products made from beef, pork ans selected poultry species. *Food Chem.* **2013**, *136*, 1461–1469. [CrossRef] [PubMed]
99. Sarah, S.A.; Faradalila, W.N.; Salwani, M.S.; Amin, I.; Karsani, S.A.; Sazili, A.Q. LC-QTOF-MS identification of porcine-specific peptide in heat treated pork identifies candidate markers for meat species determination. *Food Chem.* **2016**, *199*, 157–164. [CrossRef] [PubMed]
100. Compagnone, D.; Fusella, G.C.; Del Carlo, M.; Pittia, P.; Martinelli, E.; Tortora, L.; Paolesse, R.; Di Natale, C. Gold nanoparticles-peptide based gas sensor arrays for the detection of food aromas. *Biosens. Bioelectron.* **2013**, *42*, 618–625. [CrossRef] [PubMed]
101. Montowska, M.; Rao, W.; Alexander, M.R.; Tucker, G.A.; Barrett, D.A. Tryptic digestion coupled with ambient desorption electrospray ionization and liquid extraction surface analysis mass spectrometry enabling identification of skeletal muscle proteins in mixtures and distinguishing between beef, pork, horse, chicken, and Turkey meat. *Anal. Chem.* **2014**, *86*, 4479–4487.
102. Von Bargen, C.; Dojahn, J.; Waidelich, D.; Humpf, H.U.; Brockmeyer, J. New sensitive high-performance liquid chromatography–tandem mass spectrometry method for the detection of horse and pork in ḥalāl beef. *J. Agric. Food Chem.* **2013**, *61*, 11986–11994. [CrossRef]
103. Von Bargen, C.; Brockmeyer, J.; Humpf, H.U. Meat authentication: A new HPLC–MS/MS based method for the fast and sensitive detection of horse and pork in highly processed food. *J. Agric. Food Chem.* **2014**, *62*, 9428–9435. [CrossRef]
104. Kurniawati, E.; Rohman, A.; Triyana, K. Analysis of lard in meatball broth using Fourier transform infrared spectroscopy and chemometrics. *Meat Sci.* **2014**, *96*, 94–98. [CrossRef]
105. Xu, L.; Cai, C.B.; Cui, H.F.; Ye, Z.H.; Yu, X.P. Rapid discrimination of pork in Ḥalal and non-Ḥalal Chinese ham sausages by Fourier transform infrared (FTIR) spectroscopy and chemometrics. *Meat Sci.* **2012**, *92*, 506–510. [CrossRef]
106. Rohman, A.; Sismindari; Erwanto, Y.; Che Man, Y.B. Analysis of pork adulteration on beef meatball using Fourier transform infrared (FTIR) spectroscopy. *Meat Sci.* **2011**, *88*, 91–95. [CrossRef]
107. Leng, T.; Li, F.; Xiong, L.; Xiong, Q.; Zhu, M.; Chen, Y. Quantitative detection of binary and ternary adulteration of minced beef meat with pork and duck meat by NIR combined with chemometrics. *Food Control* **2020**, *113*, 107203. [CrossRef]
108. Mabood, F.; Boqué, R.; Alkindi, A.Y.; Al-Harrasi, A.; Al Amri, I.S.; Boukra, S.; Jabeen, F.; Hussain, J.; Abbas, G.; Naureen, Z.; et al. Fast detection and quantification of pork meat in other meats by reflectance FT-NIR spectroscopy and multivariate analysis. *Meat Sci.* **2020**, *163*, 108084. [CrossRef] [PubMed]
109. Alamprese, C.; Amigo, J.M.; Casiraghi, E.; Engelsen, S.B. Identification and quantification of turkey meat adulteration in fresh, frozen-thawed and cooked minced beef by FT-NIR spectroscopy and chemometrics. *Meat Sci.* **2016**, *121*, 175–181. [CrossRef] [PubMed]

110. Rogberg-Munoz, A.; Wei, S.; Ripoli, M.V.; Guo, B.L.; Carino, M.H.; Lirón, J.P.; Prando, A.J.; Vaca, R.J.; Peral-García, P.; Wei, Y.M.; et al. Effectiveness of a 95 SNP panel for the screening of breed label fraud in the Chinese meat market. *Meat Sci.* **2016**, *111*, 47–52. [CrossRef] [PubMed]
111. Demirhan, Y.; Ulca, P.; Senyuva, H.Z. Detection of porcine DNA in gelatine and gelatine-containing processed food products-ḥalāl/Kosher authentication. *Meat Sci.* **2012**, *90*, 686–689. [CrossRef]
112. Soares, S.; Amaral, J.S.; Oliviera, M.B.P.P.; Mafra, I. A SYBR Green real-time PCR assay to detect and quantify pork meat in processed poultry meat products. *Meat Sci.* **2013**, *94*, 115–120. [CrossRef]
113. Ulca, P.; Balta, H.; Çağın, I.; Senyuva, H.Z. Meat species identification and Ḥalāl authentication using PCR analysis of raw and cooked traditional Turkish food. *Meat Sci.* **2013**, *94*, 280–284. [CrossRef]
114. Ali, M.E.; Hashim, U.; Mustafa, S.; Che Man, Y.B. Swine-specific PCR-RFLP assay targeting mitochondrial cytochrome B gene for semiquantitative detection of pork in commercial meat products. *Food Anal. Methods* **2012**, *5*, 613–623. [CrossRef]
115. Kim, M.; Yoo, I.; Lee, S.Y.; Hong, Y.; Kim, H.Y. Quantitative detection of pork in commercial meat products by TaqMan ®real-time PCR assay targeting the mitochondrial D-loop region. *Food Chem.* **2016**, *210*, 102–106. [CrossRef]
116. Cai, Y.; He, Y.; Lv, R.; Chen, H.; Wang, Q.; Pan, L. Detection and quantification of beef and pork materials in meat products by duplex droplet digital PCR. *PLoS ONE* **2017**, *12*, e0181949. [CrossRef]
117. Cottenet, G.; Blancpain, C.; Chuah, P.F.; Cavin, C. Evaluation and application of a next generation sequencing approach for meat species identification. *Food Control* **2020**, *110*, 107003. [CrossRef]
118. Song, K.M.; Lee, S.; Ban, C. Aptamers and their biological applications. *Sensors* **2012**, *12*, 612–631. [CrossRef] [PubMed]
119. Roy, S.; Rahman, I.A.; Santos, J.H.; Ahmed, M.U. Meat species identification using DNA-redox electrostatic interactions and non-specific adsorption on graphene biochips. *Food Control* **2016**, *61*, 70–78. [CrossRef]
120. Amaral, J.S.; Santos, C.G.; Melo, V.S.; Oliveira, M.B.P.P.; Mafra, I. Authentication of a traditional game meat sausage (Alheira) by species-specific PCR assays to detect hare, rabbit, red deer, pork and cow meats. *Food Res. Int.* **2014**, *60*, 140–145. [CrossRef]
121. Notomi, T.; Mori, Y.; Tomita, N.; Kanda, H. Loop-mediated isothermal amplification (LAMP): Principle, features, and future prospects. *J. Microbiol.* **2015**, *53*, 1–5. [CrossRef] [PubMed]
122. Lee, S.Y.; Kim, M.J.; Hong, Y.; Kim, H.Y. Development of a rapid on-site detection method for pork in processed meat products using real-time loop-mediated isothermal amplification. *Food Control* **2016**, *66*, 53–61. [CrossRef]
123. Roy, S.; Wei, S.X.; Ying, J.L.Z.; Safavieh, M.; Ahmed, M.U. A novel, sensitive and label-free loop-mediated isothermal amplification detection method for nucleic acids using luminophore dyes. *Biosens. Bioelectron.* **2016**, *86*, 346–352. [CrossRef]
124. Safavieh, M.; Kanakasabapathy, M.K.; Tarlan, F.; Ahmed, M.U.; Zourob, M.U.; Asghar, W.; Shafiee, H. Emerging loop-mediated isothermal amplification-based microchip and microdevice technologies for nucleic acid detection. *ACS Biomater. Sci. Eng.* **2016**, *2*, 278–294. [CrossRef]
125. Lavelli, V. High-warranty traceability system in the poultry meat supply chain: A medium-sized enterprise case study. *Food Control* **2013**, *33*, 148–156. [CrossRef]
126. Arana, A.; Soret, B.; Lasa, I.; Alfonso, L. Meat traceability using DNA markers: Application to the beef industry. *Meat Sci.* **2002**, *61*, 367–373. [CrossRef]
127. Goffaux, F.; China, B.; Dams, L.; Clinquart, A.; Daube, G. Development of a genetic traceability test in pig based on single nucleotide polymorphism detection. *Forensic Sci. Int.* **2005**, *151*, 239–247. [CrossRef]
128. Fazdillah, N.A.; Rohman, A.; Arief Salleh, R.; Amin, I.; Shuhaimi, M.; Farahwahida, M.Y.; Rashidi, O.; Mohammad Aizat, J.; Khatib, A. Authentication of butter from lard adultertationusing high-resolution of nuclear magnetic resonance spectroscopy and high-performance liquid chromatography. *Int. J. Food Prop.* **2017**, *20*, 2147–2156. [CrossRef]
129. Stefano, A. On "Valutazione Dell'autenticità dei Prodotti Carnei Halal: Uso di Metodi Rapidi e Innovativi nei Processi di Trasformazione dei Salumi". Bachelor's Thesis, University of the Study of Teramo, Teramo TE, Italy, 2012.
130. Long, K.D.; Yu, H.; Cunningham, B.T. Smartphone instrument for portable enzyme-linked immunosorbent assay. *Biomed. Opt. Express* **2014**, *5*, 3792–3806. [CrossRef]

131. Orduna, R.A.; Ghosh, D.; Husby, E.; Yang, C.T.; Beaudry, F. Assessment of meat authenticity using bioinformatics, targeted peptide biomarkers and high-resolution mass spectrometry. *Food Addit. Contam. Part A Chem. Anal. Control Expo. Risk Assess.* **2015**, *32*, 1709–1717. [CrossRef] [PubMed]
132. Husseini de Araújo, S. Assembling halal meat and poultry production in Brazil: Agents, practices, power and sites. *Geoforum* **2019**, *100*, 220–228. [CrossRef]
133. Wibowo, M.W.; Ahmad, F.S. Non-Muslim Consumers' Halal Food Product Acceptance Model. *Procedia Econ. Financ.* **2016**, *37*, 276–283. [CrossRef]

MDPI
St. Alban-Anlage 66
4052 Basel
Switzerland
Tel. +41 61 683 77 34
Fax +41 61 302 89 18
www.mdpi.com

Foods Editorial Office
E-mail: foods@mdpi.com
www.mdpi.com/journal/foods

www.ingramcontent.com/pod-product-compliance
Lightning Source LLC
Chambersburg PA
CBHW061449210926
51850CB00008B/1658
* 9 7 8 3 0 3 6 5 1 3 6 4 5 *